SURFACTANTS AND POLYMERS IN AQUEOUS SOLUTION

SURFACTANTS AND POLYMERS IN AQUEOUS SOLUTION

SECOND EDITION

Krister Holmberg

Chalmers University of Technology, S-412 96, Göteborg, Sweden,

Bo Jönsson

Chemical Centre, Lund University, POB 124, S-221 00, Lund, Sweden

Bengt Kronberg

Institute for Surface Chemistry, POB 5607, S-114 87, Stockholm, Sweden

and

Björn Lindman

Chemical Centre, Lund University, POB 124, S-221 00, Lund, Sweden

JOHN WILEY & SONS, LTD

Other Wiley Editorial Offices

John Wiley & Sons Inc., 111 River Street, Hoboken, NJ 07030, USA

Jossey-Bass, 989 Market Street, San Francisco, CA 94103–1741, USA

Wiley-VCH Verlag GmbH, Boschstr. 12, D-69469 Weinheim, Germany

John Wiley & Sons Australia Ltd, 33 Park Road, Milton, Queensland 4064, Australia

John Wiley & Sons (Asia) Pte Ltd, 2 Clementi Loop #02–01, Jin Xing Distripark, Singapore 129809

John Wiley & Sons Canada Ltd, 22 Worcester Road, Etobicoke, Ontario, Canada M9W 1L1

Library of Congress Cataloging-in-Publication Data

Surfactants and polymers in aqueous soltion.–2nd ed./ Krister Homberg . . . [et al.].
    p. cm.
Includes bibliographical references and index.
ISBN 0–471–49883–1 (acid-free paper)
    1. Surface active agents. 2. Polymers. 3. Solution (Chemistry) I. Holmberg, Krister, 1946-
TP994 .S863 2002
668′.1–dc21

2002072621

British Library Cataloguing in Publication Data

A catalogue record for this book is available from the British Library

ISBN 0 471 49883 1 Cloth: 2nd edition
(ISBN 0 471 974422 6 Cloth: 1st edition)
(ISBN 0 471 98698 5 Paper: 1st edition)

Typeset in 10/12pt Times by Kolam Information Services Pvt Ltd, Pondicherry, India.
Printed and bound in Great Britain by Biddles Ltd, Guildford, Surrey.
This book is printed on acid-free paper responsibly manufactured from sustainable forestry
in which at least two trees are planted for each one used for paper production.

CONTENTS

PREFACE TO SECOND EDITION

The basic concept behind 'Surfactants and Polymers in Aqueous Solution', i.e. to combine in one book the physicochemical behaviours of both surfactants and water-soluble polymers, has evidently been attractive. The first edition of this book has sold well and has found a place as a course book at universities and as a reference book for researchers in the area. We, ourselves, use it extensively in our own teaching and research and receive constant feedback from course participants and from research colleagues. The additions and revisions made in this new edition of 'Surfactants and Polymers in Aqueous Solution' are based on suggestions that we have obtained through these years and also from our own ambition to keep the content up-to-date with respect to recent developments in the field.

The interaction between surfactants and polymers is a core topic of the book and constituted one chapter in the previous edition. Surfactant–protein interaction is a related theme of major importance in the life sciences area and one new chapter now deals with this issue. Rheology related to the behaviour of amphiphiles in solution is a subject of practical interest in many areas. This issue was only marginally covered in the first edition but is now the topic of a complete chapter.

Surfactants are widely used as wetting agents and we have received many comments on the fact that the first edition did not cover this aspect. A chapter treating both the wetting of a liquid on another liquid and on a solid, and also discussing the role of the wetting agent, has now been included.

In order to keep up with recent developments in the surfactant area, a contribution on novel surfactants has now been added. This chapter includes polymerizable surfactants, which were also covered in the first edition, but now contains, in addition, new sections on gemini surfactants and cleavable surfactants.

All of the chapters from the first edition that reappear in this second volume have been fully up-dated and revised. In most of these, new material has been added, usually describing the results obtained from recent research. A section on the dermatological aspects of surfactants has been included in the general chapter on surfactants. The chapter dealing with polymers in solution has been extended to include a section which describes different types of water-soluble polymers. In the chapter on interaction of polymers with surfaces the polyelectrolyte adsorption has been restructured. Within the chapter that deals with emulsifiers a general treatment of emulsions has been included, while in the chapter discussing chemical reactions in microheterogeneous media a section has been added on mesoporous materials made via surfactant self-assembly.

Finally, mistakes and indistinct descriptions in the first edition that have been brought to our attention have been taken care of. We believe that this second edition is a more complete and a more coherent book than the first edition. However, we also realize that there is still a long way to go until the book is 'perfect' and therefore encourage comments and suggestions for further improvements.

Göteborg, Lund and Stockholm   Krister Holmberg
April, 2002   Bo Jönsson
                                                                   Bengt Kronberg
                                                                   Björn Lindman

PREFACE TO THE FIRST EDITION

Surfactants are used together with polymers in a wide range of applications. In areas as diverse as detergents, paints, paper coatings, food and pharmacy, formulations usually contain a combination of a low molecular weight surfactant and a polymer which may or may not be highly surface active. Together, the surfactant and the polymer provide the stability, rheology, etc., needed for specific application. The solution behaviour of each component is important, but the performance of the formulated product depends to a large extent on the interplay between the surfactant and the polymer. Hence, knowledge about physicochemical properties of both surfactants and polymers and not least about polymer–surfactant interactions, is essential in order to make formulation work more of a science than an art.

There are books on surfactants and books dealing with water-soluble polymers, but to our knowledge no single work treats both in a comprehensive way. Researchers in the areas involved need to go to different sources to obtain basic information about surfactants and polymers. More serious than the inconvenience of having to consult several books is the considerable variation in the description of physiochemical phenomena from one book to another. Such differences in the treatments can make it difficult to get a good understanding of the solution behaviour of surfactant–polymer combinations. In our opinion there has been a long-standing need for a book covering both surfactants and water-soluble polymers and bringing the two topics together. This book is intended to fill that gap.

This book is practical rather than theoretical in scope. It is written as a reference book for scientists and engineers both in industry and academia. It is also intended as a textbook for courses for employees in industry and for undergraduate courses at universities. It has already been used as such, at the manuscript stage, at the University of Lund.

The book originates from a course on 'Surfactants and Polymers in Aqueous Solution' that we have been giving annually at different places in southern Europe since 1992. The course material started with copies of overhead pictures, grew into extended summaries of the lectures and developed further into a compendium which after several rounds of polishing has become this volume.

We thank the course participants throughout these years for many valuable comments and suggestions.

We would also like to thank Akzo Nobel Surface Chemistry AB, and in particular Dr Lennart Dahlgren, for economic support towards the production of the book. We are grateful to Mr Malek Khan, for his skilful drawing of the figures. We thank many colleagues in Lund and in Stockholm for providing material and for helpful discussions.

Stockholm and Lund   Krister Holmberg
November, 1997   Bo Jönsson
                                                      Bengt Kronberg
                                                      Björn Lindman

1 INTRODUCTION TO SURFACTANTS

Surfactants Adsorb at Interfaces

Surfactant is an abbreviation for surface active agent, which literally means active at a surface. In other words, a surfactant is characterized by its tendency to absorb at surfaces and interfaces. The term interface denotes a boundary between any two immiscible phases; the term surface indicates that one of the phases is a gas, usually air. Altogether five different interfaces exist:

Solid–vapour   *surface*
Solid–liquid
Solid–solid
Liquid–vapour   *surface*
Liquid–liquid

The driving force for a surfactant to adsorb at an interface is to lower the free energy of that phase boundary. The interfacial free energy per unit area represents the amount of work required to expand the interface. The term interfacial tension is often used instead of interfacial free energy per unit area. Thus, the surface tension of water is equivalent to the interfacial free energy per unit area of the boundary between water and the air above it. When that boundary is covered by surfactant molecules, the surface tension (or the amount of work required to expand the interface) is reduced. The denser the surfactant packing at the interface, then the larger the reduction in surface tension.

Surfactants may adsorb at all of the five types of interfaces listed above. Here, the discussion will be restricted to interfaces involving a liquid phase. The liquid is usually, but not always water. Examples of the different interfaces and products in which these interfaces are important are given in Table 1.1.

In many formulated products several types of interfaces are present at the same time. Water-based paints and paper coating colours are examples of familiar but, from a colloidal point of view, very complicated systems containing both solid-liquid (dispersed pigment particles) and liquid–liquid (latex or other binder droplets) interfaces. In addition, foam formation is a common

Table 1.1   Examples of interfaces involving a liquid phase

Interface	Type of system	Product
Solid–liquid	Suspension	Solvent-borne paint
Liquid–liquid	Emulsion	Milk, cream
Liquid–vapour	Foam	Shaving cream

(but unwanted) phenomenon at the application stage. All of the interfaces are stabilized by surfactants. The total interfacial area of such a system is immense: the oil–water and solid–water interfaces of one litre of paint may cover several football fields.

As mentioned above, the tendency to accumulate at interfaces is a fundamental property of a surfactant. In principle, the stronger the tendency, then the better the surfactant. The degree of surfactant concentration at a boundary depends on the surfactant structure and also on the nature of the two phases that meet at the interface. Therefore, there is no universally good surfactant, suitable for all uses. The choice will depend on the application. A good surfactant should have low solubility in the bulk phases. Some surfactants (and several surface active macromolecules) are only soluble at the oil–water interface. Such compounds are difficult to handle but are very efficient in reducing the interfacial tension.

There is, of course, a limit to the surface and interfacial tension lowering effect by the surfactant. In the normal case that limit is reached when micelles start to form in bulk solution. Table 1.2 illustrates what effective surfactants can do in terms of lowering of surface and interfacial tensions. The values given are typical of what is attained by normal light-duty liquid detergents. With special formulations, so-called ultra-low interfacial tensions, i.e. values in the range of 10^{-3} mN/m or below, can be obtained. An example of a system giving ultra-low interfacial tensions is a three-phase system comprising a microemulsion in equilibrium with excess water and oil phases. Such systems are of interest for enhanced oil recovery and are discussed in Chapter 22.

Table 1.2   Typical values of surface and interfacial tensions (mN/m)

Air–water	72–73
Air–10% aqueous NaOH	78
Air–aqueous surfactant solution	40–50
Aliphatic hydrocarbon–water	28–30
Aromatic hydrocarbon–water	20–30
Hydrocarbon–aqueous surfactant solution	1–10

Surfactants Aggregate in Solution

As discussed above, one characteristic feature of surfactants is their tendency to adsorb at interfaces. Another fundamental property of surface active agents is that unimers in solution tend to form aggregates, so-called micelles. (The free or unassociated surfactant is referred to in the literature either as 'monomer' or 'unimer'. In this text we will use 'unimer' and the term 'monomer' will be restricted to the polymer building block.) Micelle formation, or micellization, can be viewed as an alternative mechanism to adsorption at the interfaces for removing hydrophobic groups from contact with water, thereby reducing the free energy of the system. It is an important phenomenon since surfactant molecules behave very differently when present in micelles than as free unimers in solution. Only surfactant unimers contribute to surface and interfacial tension lowering and dynamic phenomena, such as wetting and foaming, are governed by the concentration of free unimers in solution. The micelles may be seen as a reservoir for surfactant unimers. The exchange rate of a surfactant molecule between micelle and bulk solution may vary by many orders of magnitude depending on the size and structure of the surfactant.

Micelles are already generated at very low surfactant concentrations in water. The concentration at which micelles start to form is called the critical micelle concentration, or CMC, and is an important characteristic of a surfactant. A CMC of 1 mM, a reasonable value for an ionic surfactant, means that the unimer concentration will never exceed this value, regardless of the amount of surfactant added to the solution. Surfactant micellization is discussed in detail in Chapter 2.

Surfactants are Amphiphilic

The name amphiphile is sometimes used synonymously with surfactant. The word is derived from the Greek word *amphi*, meaning both, and the term relates to the fact that all surfactant molecules consist of at least two parts, one which is soluble in a specific fluid (the lyophilic part) and one which is insoluble (the lyophobic part). When the fluid is water one usually talks about the hydrophilic and hydrophobic parts, respectively. The hydrophilic part is referred to as the head group and the hydrophobic part as the tail (see Figure 1.1).

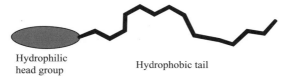

Hydrophilic head group   Hydrophobic tail

Figure 1.1   Schematic illustration of a surfactant

In a micelle the surfactant hydrophobic group is directed towards the interior of the cluster and the polar head group is directed towards the solvent. The micelle, therefore, is a polar aggregate of high water solubility and without much surface activity. When a surfactant adsorbs from aqueous solution at a hydrophobic surface, it normally orients its hydrophobic group towards the surface and exposes its polar group to the water. The surface has become hydrophilic and, as a result, the interfacial tension between the surface and water has been reduced. Adsorption at hydrophilic surfaces often results in more complicated surfactant assemblies. Surfactant adsorption at hydrophilic and hydrophobic surfaces is discussed in Chapter 17.

The hydrophobic part of a surfactant may be branched or linear. The polar head group is usually, but not always, attached at one end of the alkyl chain. The length of the chain is in the range of 8–18 carbon atoms. The degree of chain branching, the position of the polar group and the length of the chain are parameters of importance for the physicochemical properties of the surfactant.

The polar part of the surfactant may be ionic or non-ionic and the choice of polar group determines the properties to a large extent. For non-ionic surfactants the size of the head group can be varied at will; for the ionics, the size is more or less a fixed parameter. As will be discussed many times throughout this book, the relative size of the hydrophobic and polar groups, not the absolute size of either of the two, is decisive in determining the physicochemical behaviour of a surfactant in water.

A surfactant usually contains only one polar group. Recently, there has been considerable research interest in certain dimeric surfactants, containing two hydrophobic tails and two head groups linked together with a short spacer. These species, generally known under the name gemini surfactants, are not yet of commercial importance. They show several interesting physicochemical properties, such as very high efficiency in lowering surface tension and very low CMC. The low CMC values of gemini surfactants can be illustrated by a comparison of the value for the conventional cationic surfactant dodecyltrimethylammonium bromide (16 mM) and that of the corresponding gemini surfactant, having a 2 carbon linkage between the monomers (0.9 mM). The difference in CMC between monomeric and dimeric surfactants could be of considerable practical importance. A typical gemini surfactant is shown in Figure 1.2. Gemini surfactants are discussed further in Chapter 11.

Weakly surface active compounds which accumulate at interfaces but which do not readily form micelles are of interest as additives in many surfactant formulations. They are referred to as hydrotropes and serve the purpose of destroying the ordered packing of ordinary surfactants. Thus, addition of a hydrotrope is a way to prevent the formation of highly viscous liquid crystalline phases which constitutes a well-known problem in surfactant formulations. Xylene sulfonate and cumene sulfonate are typical examples of hydrotropes

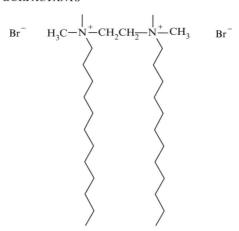

Figure 1.2   A gemini surfactant

used, for instance, in detergent formulations. Short-chain alkyl phosphates have found specific use as hydrotropes for longer-chain alcohol ethoxylates.

Surface Active Compounds are Plentiful in Nature

Nature's own surfactants are usually referred to as polar lipids. These are abundant in all living organisms. In biological systems the surface active agents are used in very much the same way as surfactants are employed in technical systems: to overcome solubility problems, as emulsifiers, as dispersants, to modify surfaces, etc. There are many good examples of this in biological systems: bile salts are extremely efficient solubilizers of hydrophobic components in the blood, while mixtures of phospholipids pack in ordered bilayers of the surfactant liquid crystal type and such structures constitute the membranes of cells. Figure 1.3 gives examples of important polar lipids. The only important example of a surfactant being obtained directly, without chemical conversion, from nature is lecithin. (The term lecithin is not used in a strict way in the surfactant literature. It is sometimes used synonymously with phosphatidylcholine and it sometimes refers to phospholipids in general.) Lecithin is extracted from phospholipid-rich sources such as soybean and egg.

Micro-organisms are sometimes efficient producers of surface active agents. Both high molecular weight compounds, e.g. lipopolysaccharides, and low molecular weight polar lipids can be produced in good yields, particularly when the micro-organism is fermented on a water-insoluble substrate. Surface active polymers of this type are dealt with in Chapter 12. Figure 1.4 gives the structure of a low molecular weight acylated sugar, a trehalose lipid, which has

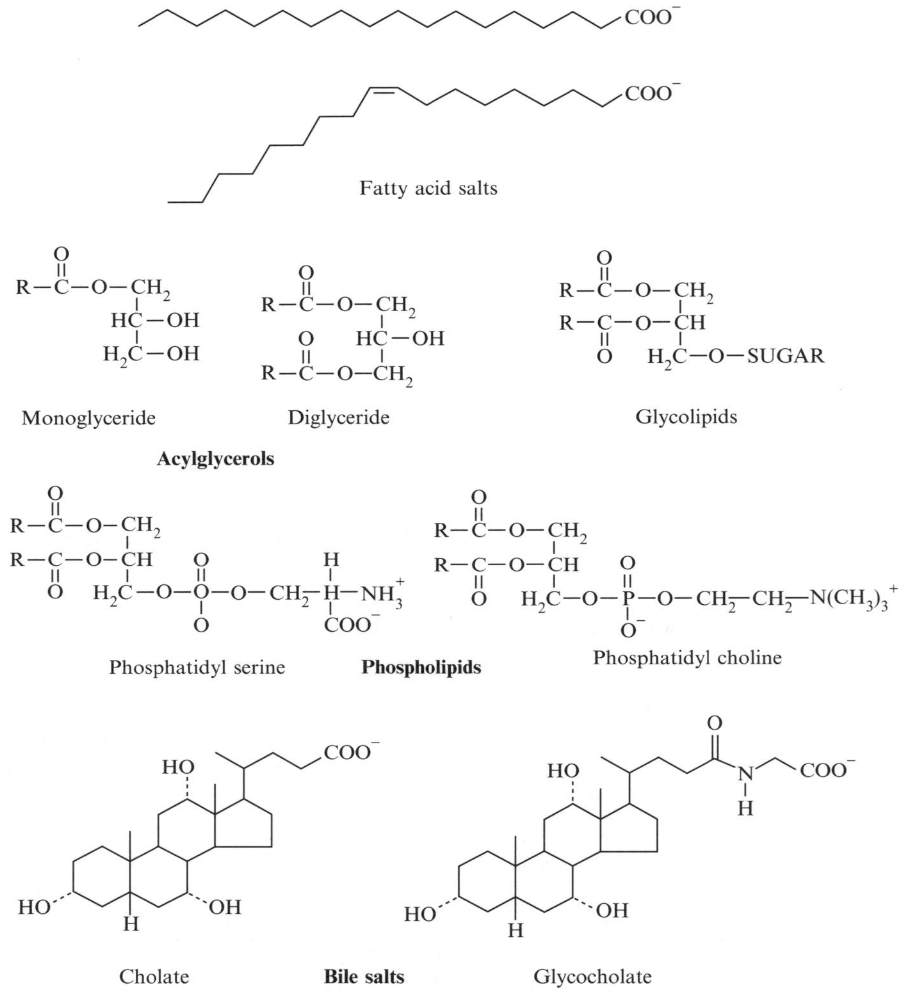

Figure 1.3 Examples of polar lipids

proved to be an effective surfactant. Trehalose lipids and several other surface active agents produced from bacteria and yeasts have attracted considerable interest in recent years and much effort has been directed towards improving the fermentation and, not least, the work-up procedure. Although considerable process improvements have been made, commercial use of these products is still very limited due to their high price.

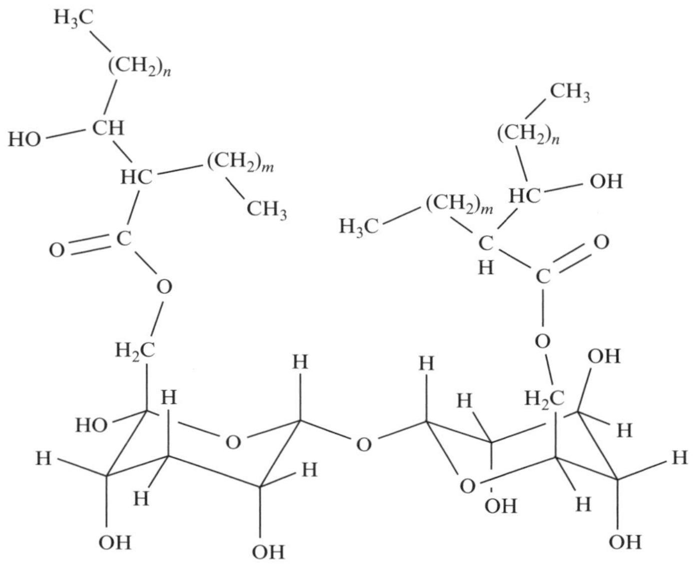

Figure 1.4 A surface active trehalose lipid produced by fermentation

Surfactant Raw Materials May be Based on Petrochemicals or Oleochemicals

For several years there has been a strong trend towards 'green' surfactants, particularly for the household sector. In this context the term 'natural surfactant' is often used to indicate some natural origin of the compound. However, no surfactants used in any substantial quantities today are truly natural. With few exceptions they are all manufactured by organic synthesis, usually involving rather hard conditions which inevitably give by-products. For instance, monoglycerides are certainly available in nature, but the surfactants sold as monoglycerides are prepared by glycerolysis of triglyceride oils at temperatures well above 200°C, yielding di-and triglycerol derivatives as by-products. Alkyl glucosides are abundant in living organisms but the surfactants of this class, often referred to as APGs (alkyl polyglucosides), are made in several steps which by no means are natural.

A more adequate approach to the issue of origin is to divide surfactants into oleochemically based and petrochemically based surfactants. Surfactants based on oleochemicals are made from renewable raw materials, most commonly vegetable oils. Surfactants from petrochemicals are made from small building blocks, such as ethylene, produced by cracking of naptha. Quite commonly, a surfactant may be built up by raw materials from both origins. Fatty acid ethoxylates are one example out of many.

Sometimes the oleochemical and the petrochemical pathways lead to essentially identical products. For instance, linear alcohols in the C10–C14 range, which are commonly used as hydrophobes for both non-ionics (alcohol ethoxylates) and anionics (alkyl sulfates, alkyl phosphates, etc.), are made either by hydrogenation of the corresponding fatty acid methyl esters or via Ziegler–Natta polymerization of ethylene using triethyl aluminium as the catalyst. Both routes yield straight-chain alcohols and the homologue distribution is not very different since it is largely governed by the distillation process. Both pathways are used in very large scale operations.

It is not obvious that the oleochemical route will lead to a less toxic and more environmentally friendly surfactant than the petrochemical route. However, from the carbon dioxide cycle point of view chemical production based on renewable raw materials is always preferred.

Linear long-chain alcohols are often referred to as fatty alcohols, regardless of their source. Branched alcohols are also of importance as surfactant raw material. They are invariably produced by synthetic routes, the most common being the so-called oxo process, in which an olefin is reacted with carbon monoxide and hydrogen to give an aldehyde, which is subsequently reduced to the alcohol by catalytic hydrogenation. A mixture of branched and linear alcohols is obtained and the ratio between the two can be varied to some extent by the choice of catalyst and reaction conditions. The commercial 'oxo alcohols' are mixtures of linear and branched alcohols of specific alkyl chain length ranges. The different routes to higher molecular weight primary alcohols are illustrated in Figure 1.5.

Surfactants are Classified by the Polar Head Group

The primary classification of surfactants is made on the basis of the charge of the polar head group. It is common practice to divide surfactants into the classes anionics, cationics, non-ionics and zwitterionics. Surfactants belonging to the latter class contain both an anionic and a cationic charge under normal conditions. In the literature they are often referred to as amphoteric surfactants but the term 'amphoteric' is not always correct and should not be used as synonymous to zwitterionic. An amphoteric surfactant is one that, depending on pH, can be either cationic, zwitterionic or anionic. Among normal organic substances, simple amino acids are well-known examples of amphoteric compounds. Many so-called zwitterionic surfactants are of this category. However, other zwitterionic surfactants retain one of the charges over the whole pH range. Compounds with a quaternary ammonium as the cationic group are examples of this. Consequently, a surfactant that contains a carboxylate group and a quaternary ammonium group, a not uncommon combination as we shall see later in this chapter, is zwitterionic unless the pH is very low, but is not an amphoteric surfactant.

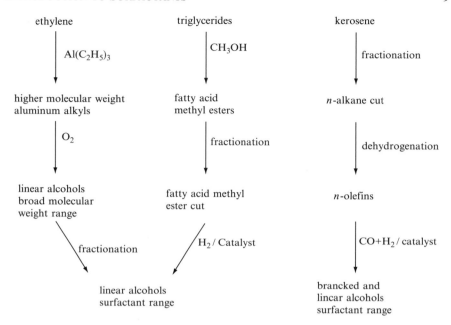

Figure 1.5 Different pathways for preparation of primary alcohols of interest as surfactant raw materials. From left to right: Ziegler–Natta polymerization of ethylene; reduction of fatty acid methyl esters; hydoformylation of higher olefins (the Oxo process)

Most ionic surfactants are monovalent but there are also important examples of divalent anionic amphiphiles. For the ionic surfactants the choice of counterion plays a role in the physicochemical properties. Most anionic surfactants have sodium as counterion but other cations, such as lithium, potassium, calcium and protonated amines, are used as surfactant counterions for speciality purposes. The counterion of cationic surfactants is usually a halide or methyl sulfate.

The hydrophobic group is normally a hydrocarbon (alkyl or alkylaryl) but may also be a polydimethylsiloxane or a fluorocarbon. The two latter types of surfactants are particularly effective in non-aqueous systems.

For a few surfactants there is some ambiguity as to classification. For example, amine oxide surfactants are sometimes referred to as zwitterionics, sometimes as cationics and sometimes as non-ionics. Their charge is pH dependent and in the net neutral state they may either be seen as having distinct anionic and cationic charges or as dipolar non-ionic compounds. Fatty amine ethoxylates which contain both an amino nitrogen atom (cationic polar group) and a polyoxyethylene chain (non-ionic polar group) may be included in either the cationics or the non-ionics class. The non-ionic character dominates when the polyoxyethylene chain is very long, whereas for medium and short chains the physicochemical properties are mainly those of cationic surfactants. Surfactants containing both an anionic group, such as sulfate, phosphate or carboxylate, and a polyoxyethylene chain are also common. These surfactants, known as ether sulfates, etc., invariably contain short polyoxyethylene chains, typically two or three oxyethylene units, and are therefore always categorized as anionics.

Anionics

Carboxylate, sulfate, sulfonate and phosphate are the polar groups found in anionic surfactants. Figure 1.6 shows structures of the more common surfactant types belonging to this class.

Anionics are used in greater volume than any other surfactant class. A rough estimate of the worldwide surfactant production is 10 million tons per year, out of which approximately 60% are anionics. One main reason for their popularity is the ease and low cost of manufacture. Anionics are used in most detergent formulations and the best detergency is obtained by alkyl and alkylarye chains in the C12–C18 range.

The counterions most commonly used are sodium, potassium, ammonium, calcium and various protonated alkyl amines. Sodium and potassium impart water solubility, whereas calcium and magnesium promote oil solubility. Amine/alkanol amine salts give products with both oil and water solubility.

Soap is still the largest single type of surfactant. It is produced by saponification of natural oils and fats. Soap is a generic name representing the alkali

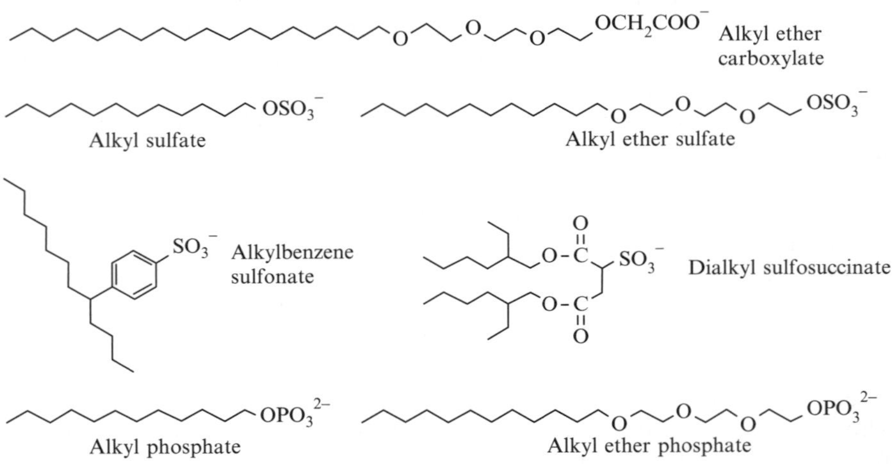

Figure 1.6   Structures of some representative anionic surfactants

metal salt of a carboxylic acid derived from animal fats or vegetable oils. Soap bars are usually based on mixtures of fatty acids obtained from tallow, coconut and palm oil. Under the right conditions soaps are excellent surfactants. Their sensitivity to hard water is a major drawback, however, and has constituted a strong driving force for the development of synthetic surfactants. A very specific use of the lithium salt of a fatty acid, i.e. lithium 12-hydroxystearic acid, is as the major constituent of greases.

Alkylbenzene sulfonates have traditionally been the work-horse among synthetic surfactants. They are widely used in household detergents as well as in a variety of industrial applications. They are made by sulfonation of alkylbenzenes. In large-scale synthesis, sulfur trioxide is the sulfonating agent of choice but other reagents, such as sulfuric acid, oleum ($H_2SO_4 n SO_3$), chlorosulfonic acid ($ClSO_3H$) or amidosulfonic acid (sulfamic acid, H_2NSO_3H), may also be used and may be preferred for specific purposes. Industrial synthesis is usually carried out in a continuous process, using a falling film reactor. The first step of the synthesis results in the formation of pyrosulfonic acid, which slowly and spontaneously reacts further to give the sulfonic acid.

$$R-\!\!\bigcirc + 2\ SO_3 \xrightarrow[\text{fast}]{} R-\!\!\bigcirc\!\!-SO_2OSO_3H \xrightarrow[\text{slow}]{R-\bigcirc} R-\!\!\bigcirc\!\!-SO_3H$$

The sulfonic acid is subsequently neutralized, usually by caustic soda, to give the surface active alkylbenzene sulfonate salt. Due to the bulkiness of the alkyl substituent, the process gives almost exclusively *p*-sulfonation. R in the scheme above is typically an alkyl group of 12 carbon atoms. Originally, alkylbenzenes as surfactant intermediates were based on branched alkyls, but these have now almost entirely been replaced by their linear counterparts, thus giving the name linear alkylbenzene sulfonate (LABS or LAS). Faster biodegradation has been the main driving force for the transition to chains without branching. Alkylbenzenes are made by alkylation of benzene with an *n*-alkene or with alkyl chloride using HF or $AlCl_3$ as catalyst. The reaction yields a mixture of isomers with the phenyl group attached to one of the non-terminal positions of the alkyl chain.

Other sulfonate surfactants that have found use in detergent formulations are paraffin sulfonates and α-olefin sulfonates, with the latter often referred to as AOSs. Both are complex mixtures of compounds with varying physicochemical properties. Paraffin sulfonates, or secondary *n*-alkane sulfonates, are mainly produced in Europe. They are usually prepared by sulfoxidation of paraffin hydrocarbons with sulfur dioxide and oxygen under UV (ultraviolet) irradiation. In an older process, which is still in use, paraffin sulfonates are made by sulfochlorination. Both processes are free radical reactions and since secondary carbons give much more stable radicals than primary, the sulfonate group will be introduced more or less randomly on all non-terminal carbon atoms along the alkane chain. A C14–C17 hydrocarbon cut, sometimes called the 'Euro cut', is normally used as hydrophobe raw material. Thus, the product obtained will be a very complex mixture of both isomers and homologues.

α-Olefin sulfonates are prepared by reacting linear α-olefins with sulfur trioxide, typically yielding a mixture of alkene sulfonate (60–70%), 3- and 4-hydroxyalkane sulfonates (around 30%) and some disulfonate and other species. The two main α-olefin fractions used as starting material are C12–C16 and C16–C18. The ratio of alkene sulfonate to hydroxyalkane sulfonate is to some degree governed by the ratio of SO_3 to olefin: the higher the ratio, then the more alkene sulfonic acid will be formed. Formation of hydroxyalkane sulfonic acid proceeds via a cyclic sultone, which is subsequently cleaved by alkali. The sultone is toxic and it is important that its concentration in the end-product is very low. The route of preparation can be written as follows:

$$R-CH_2CH_2CH=CH_2 + SO_3 \longrightarrow R'-CH=CH(CH_2)_nSO_3H + R-CH_2CHCH_2CH_2$$
$$\underset{O \underline{\quad\quad} SO_2}{}$$

$$+ R-\underset{\underset{O \underline{\quad\quad\quad} SO_2}{|}}{CH}CH_2CH_2CH_2 \xrightarrow{\text{NaOH}} R'-CH=CH(CH_2)_nSO_3^-Na^+$$

$$+ R-\underset{\underset{OH \quad\quad SO_3^-Na^+}{|}}{CH_2}CHCH_2CH_2 \quad + R-\underset{\underset{OH \quad\quad\quad SO_3^-Na^+}{|}}{CH}CH_2CH_2CH_2$$

An alkyl sulfonate surfactant widely used in surface chemistry research is sodium di(2-ethylhexyl)sulfosuccinate, often referred to by its American Cyanamid trade name Aerosol OT, or AOT. This surfactant, with its bulky hydrophobe structure (see Figure 1.6), is particularly useful for preparation of water/oil (w/o) microemulsions, as is discussed in Chapter 6.

Isethionate surfactants, with the general formula $R - COOCH_2CH_2 SO_3^-Na^+$, are fatty acid esters of the isethionic acid salt. They are among the mildest sulfonate surfactants and are used in cosmetics formulations.

Very crude sulfonate surfactants are obtained by sulfonation of lignin, petroleum fractions, alkylnaphthalenes or other low-cost hydrocarbon fractions. Such surfactants are used in a variety of industrial applications as dispersants, emulsifiers, demulsifiers, defoamers, wetting agents, etc.

Sulfated alcohols and alcohol ethoxylates constitute another important group of anionics, widely used in detergent formulations. These are monoesters of sulfuric acid and the ester bond is a labile linkage which splits with particular ease at low pH where hydrolysis is autocatalytic. Both linear or branched alcohols, typically with eight to sixteen carbon atoms, are used as raw materials. The linear 12-carbon alcohol leads to the dodecylmonoester of sulfuric acid and, after neutralization with caustic soda, to sodium dodecyl sulfate (SDS), which is by far the most important surfactant within this category. The alcohol ethoxylates used as intermediates are usually fatty alcohols with two or three oxyethylene units. The process is similar to the sulfonation discussed above. Sulfur trioxide is the reagent used for large-scale production and, in analogy to sulfonation, the reaction proceeds via an intermediate pyrosulfate:

$$R-OH + 2\ SO_3 \xrightarrow{\text{fast}} R-O-SO_2OSO_3H \xrightarrow[\text{slow}]{R-OH} R-O-SO_3H$$

Synthesis of sulfate esters of ethoxylated alcohols proceeds similarly. In this reaction 1,4-dioxane is usually formed in non-negligible amounts. Since dioxane is toxic, its removal by evaporation is essential. These surfactants are usually referred to as ether sulfates. Such surfactants are good at producing foams and have a low toxicity to the skin and eye. They are popular in hand dishwashing and shampoo formulations.

Ethoxylated alcohols may also be transformed into carboxylates, i.e. to give ether carboxylates. These have traditionally been made from sodium monochloroacetate by using the Williamson ether synthesis:

$$R-(OCH_2CH_2)_n - OH + ClCH_2COO^-Na^+ \longrightarrow$$
$$R-(OCH_2CH_2)_n - O - CH_2COOH + NaCl$$

The Williamson synthesis usually does not proceed quantitatively. A more recent synthetic procedure involves oxygen or peroxide oxidation of the alcohol ethoxylate in alkaline solution using palladium or platinum catalyst. This reaction gives conversion of the ethoxylate in very high yield, but may also lead to oxidative degradation of the polyoxyethylene chain. Ether carboxylates have found use in personal care products and are also used as a consurfactant in various liquid detergent formulations. Like ether sulfates, ether carboxylates are very tolerant to high water hardness. Both surfactant types also exhibit good lime soap dispersing power, which is an important property for a surfactant in personal care formulations. Lime soap dispersing power is usually defined as the number of grams of surfactant required to disperse the lime soap formed from 100 g of sodium oleate in water with a hardness equivalent of 333 ppm of $CaCO_3$.

Phosphate-containing anionic surfactants, both alkyl phosphates and alkyl ether phosphates, are made by treating the fatty alcohol or alcohol ethoxylate with a phosphorylating agent, usually phosphorus pentoxide, P_4O_{10}. The reaction yields a mixture of mono-and diesters of phosphoric acid, and the ratio between the esters is governed by the ratio of the reactants and the amount of water in the reaction mixture:

$$6 \ R—OH + P_4O_{10} \longrightarrow 2 \ R—O-\overset{\overset{\displaystyle O}{\|}}{\underset{\underset{\displaystyle OH}{|}}{P}}-OH \ + \ 2 \ R—O-\overset{\overset{\displaystyle O}{\|}}{\underset{\underset{\displaystyle OH}{|}}{P}}-O-R$$

All commercial phosphate surfactants contain both mono-and diesters of phosphoric acid, but the relative amounts vary from one producer to another. Since the physico-chemical properties of the alkyl phosphate surfactants depend on the ratio of the esters, alkyl phosphates from different suppliers are less interchangeable than other surfactants. Phosphorus oxychloride, $POCl_3$, can also be used as a phosphorylating agent to produce alkyl phosphate surfactants. Also with $POCl_3$ a mixture of mono-and diesters of phosphoric acid is obtained.

Phosphate surfactants are used in the metal working industry where advantage is taken of their anticorrosive properties. They are also used as emulsifiers in plant protection formulations. Some important facts about anionic surfactants are given in Table 1.3.

Non-Ionics

Non-ionic surfactants have either a polyether or a polyhydroxyl unit as the polar group. In the vast majority of non-ionics, the polar group is a polyether consisting of oxyethylene units, made by the polymerization of ethylene oxide. Strictly speaking, the prefix 'poly' is a misnomer. The typical number of oxyethylene units in the polar chain is five to ten, although some surfactants, e.g. dispersants, often have much longer oxyethylene chains. Ethoxylation is usually carried out under alkaline conditions. Any material containing an active hydrogen can be ethoxylated. The most commonly used starting materials are fatty alcohols, alkylphenols, fatty acids and fatty amines. Esters, e.g. triglyceride oils, may be ethoxylated in a process that, in a one-pot reaction, involves alkaline ester hydrolysis, followed by ethoxylation of the acid and

Table 1.3 Important facts about anionic surfactants

---

1. They are by far the largest surfactant class.
2. They are generally not compatible with cationics (although there are important exceptions).
3. They are generally sensitive to hard water. Sensitivity decreases in the order carboxylate > phosphate > sulfate $\simeq$ sulfonate.
4. A short polyoxyethylene chain between the anionic group and the hydrocarbon improves salt tolerance considerably.
5. A short polyoxypropylene chain between the anionic group and the hydrocarbon improves solubility in organic solvents (but may reduce the rate of biodegradation).
6. Sulfates are rapidly hydrolysed by acids in an autocatalytic process. The other types are stable unless extreme conditions are used.

---

alcohol formed and subsequent partial condensation of the ethoxylated species. Castor oil ethoxylates, used for animal feed applications, constitute an interesting example of triglyceride-based surfactants.

Examples of polyhydroxyl (polyol)-based surfactants are sucrose esters, sorbitan esters, alkyl glucosides and polyglycerol esters, the latter type actually being a combination of polyol and polyether surfactant. Polyol surfactants may also be ethoxylated. A common example is fatty acid esters of sorbitan (known under the Atlas trade name of Span) and the corresponding ethoxylated products (known as Tween). The five-membered ring structure of sorbitan is formed by dehydration of sorbitol during manufacture. The sorbitan ester surfactants are edible and, hence, useful for food and drug applications. Acetylenic glycols, surfactants containing a centrally located acetylenic bond and hydroxyl groups at the adjacent carbon atoms, constitute a special type of hydroxyl-based surfactant, which have found use as antifoam agents, particularly in coatings applications.

Figure 1.7 gives structures of the more common non-ionic surfactants. As mentioned below, a commercial oxyethylene-based surfactant consists of a very broad spectrum of compounds, broader than for most other surfactant types. Fatty acid ethoxylates constitute particularly complex mixtures with high amounts of poly(ethylene glycol) and fatty acid as by-products. The single most important type of non-ionic surfactant is fatty alcohol ethoxylates. They are used in liquid and powder detergents as well as in a variety of industrial applications. They are particularly useful to stabilize oil-in-water emulsions and their use as an emulsifier is discussed in some detail in Chapter 21. Fatty alcohol ethoxylates can be regarded as hydrolytically stable in the pH range 3–11. They undergo a slow oxidation in air, however, and some oxidation products, e.g. aldehydes and hydroperoxides, are more irritating to the skin than the intact surfactant. Throughout this text fatty alcohol ethoxylates are referred to as $C_m E_n$, with m being the number of carbon atoms in the alkyl chain and n being the number of oxyethylene units. Some important facts about non-ionic surfactants are given in Table 1.4.

Table 1.4 Important facts about non-ionic surfactants

1. They are the second largest surfactant class.
2. They are normally compatible with all other types of surfactants.
3. They are not sensitive to hard water.
4. Contrary to ionic surfactants, their physicochemical properties are not markedly affected by electrolytes.
5. The physicochemical properties of ethoxylated compounds are very temperature dependent. Contrary to ionic compounds they become less water soluble—more hydrophobic—at higher temperatures. Sugar-based non-ionics exhibit the normal temperature dependence, i.e. their solubility in water increases with temperature.

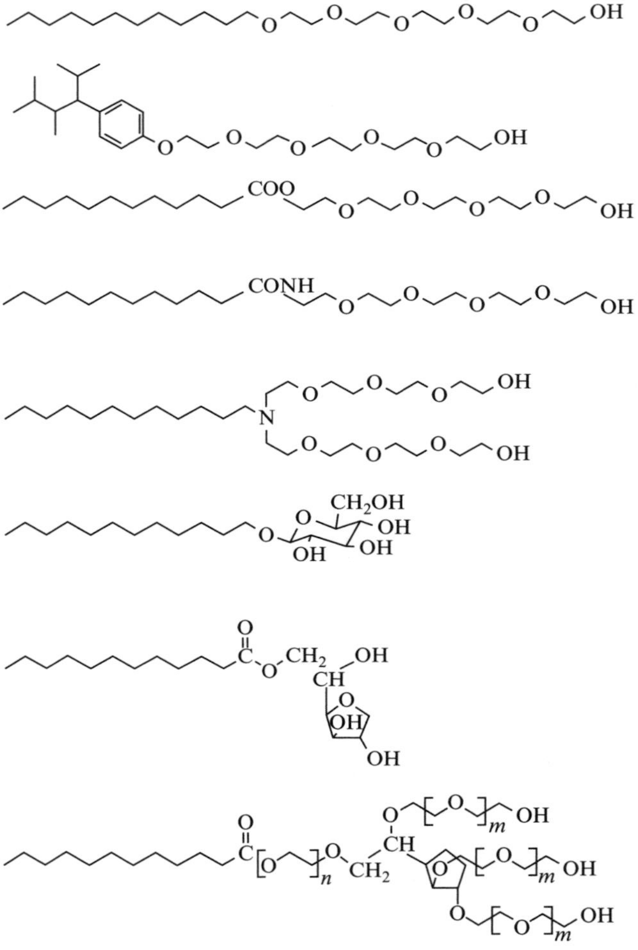

Fatty alcohol ethoxylate

Alkylphenol ethoxylate

Fatty acid ethoxylate

Fatty amide ethoxylate

Fatty amine ethoxylate

Alkyl glucoside

Sorbitan alkanoate

Ethoxylated sorbitan alkanoate

Figure 1.7 Structures of some representative non-ionic surfactants

Ethoxylated surfactants can be tailor-made with high precision with regard to the average number of oxyethylene units added to a specific hydrophobe, e.g. a fatty alcohol. However, the ethoxylation invariably gives a broad distribution of chain lengths. If all hydroxyl groups, i.e. those of the starting alcohol and the glycol ethers formed, had the same reactivity, a Poisson distribution of oligomers would be obtained. Since the starting alcohol is slightly less acidic than the glycol ethers, its deprotonation is disfavoured, leading to a lower probability for reaction with ethylene oxide. The reaction scheme is given in Figure 1.8.

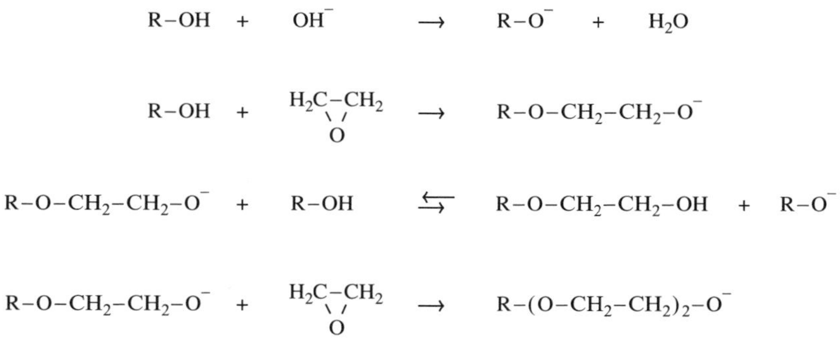

R–OH + OH⁻ → R–O⁻ + H2O
R–OH + H2C–CH2 (O) → R–O–CH2–CH2–O⁻
R–O–CH2–CH2–O⁻ + R–OH ⇆ R–O–CH2–CH2–OH + R–O⁻
R–O–CH2–CH2–O⁻ + H2C–CH2 (O) → R–(O–CH2–CH2)2–O⁻

Figure 1.8 Base catalysed ethoxylation of a fatty alcohol, R—OH

Hence, a considerable amount of unethoxylated alcohol will remain in the reaction mixture, also with relatively long ethoxylates. This is sometimes a problem and considerable efforts have been made to obtain a more narrow homologue distribution. The distribution can be affected by the choice of ethoxylation catalyst and it has been found that alkaline earth hydroxides, such as $Ba(OH)_2$ and $Sr(OH)_2$, give a much more narrow distribution than KOH, probably due to some coordination mechanism. Also Lewis acids, e.g. $SnCl_4$ and BF_3, give narrow distributions. Acid catalysed ethoxylation suffers from the drawback of 1,4-dioxane being formed in considerable quantities as by-product. Therefore, this process can only be used to prepare short ethoxylates. In Figure 1.9, the homologue distribution of a conventional alcohol ethoxylate, using KOH as catalyst, is compared with ethoxylates prepared using a Lewis acid and an alkaline earth hydroxide as catalyst.

SnCl4
Sr(OH)2
KOH
Mole (%)
0
5
10
15
20
25
0
2
4
6
8
10
Mole EO/mole fatty alcohol

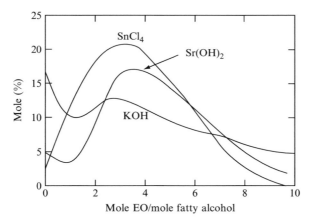

Figure 1.9 Typical homologue distribution of a fatty alcohol reacted with 4 moles of ethylene oxide (EO) using different ethoxylation catalysts

So-called peaked ethoxylates have a growing share of the market. Typical advantages of ethoxylates with peaked distribution are that:

(1) The low content of free alcohol reduces smell.

(2) The low content of free alcohol reduces 'pluming' during spray-drying.

(3) The low content of low oxyethylene homologues increases solubility.

(4) The low content of high oxyethylene homologues reduces viscosity.

(5) In alkyl ether sulfates, the low content of alkyl sulfate reduces skin irritation.

As mentioned in Table 1.4, non-ionic surfactants containing polyoxyethylene chains exhibit reverse solubility versus temperature behaviour in water. On raising the temperature two phases eventually appear. The temperature at which this occurs is referred to as the cloud point, alluding to the fact that the solution becomes turbid. The cloud point depends on both the hydrophobe chain length and the number of oxyethylene units, and it can be determined with great accuracy. In the manufacture of polyoxethylene-based surfactants, cloud point determination is used as a way to monitor the degree of ethoxylation. The onset of turbidity varies somewhat with surfactant concentration and in the official test method the cloud point is determined by heating a 1% aqueous solution to above clouding and then monitoring the transition from turbid to clear solution on slow cooling of the sample. For surfactants with long polyox-yethylene chains the cloud point may exceed 100°C. For such surfactants deter-minations are often made in electrolyte solutions since most salts lower the cloud point. Clouding of non-ionic surfactants is discussed in detail in Chapter 4.

Ethoxylated triglycerides, e.g. castor oil ethoxylates, have an established position on the market and are often regarded as 'semi-natural' surfactants. In recent years there has been a growing interest in fatty acid methyl ester eth-oxylates, made from the methyl ester by ethoxylation using a special type of catalyst, e.g. hydrotalcite, a magnesium-aluminium hydroxycarbonate. The methyl ester ethoxylate has the advantage over alcohol ethoxylate in being much more soluble in aqueous solution. Surfactants which combine high water solu-bility with proper surface activity, are needed in various types of surfactant concentrates.

$$R-COOCH_3 + CH_2\overset{O}{-}CH_2 \longrightarrow R-\overset{O}{\underset{\|}{C}}-O(CH_2CH_2O)_n-CH_3$$

Alcohol ethoxylates with the terminal hydroxyl group replaced by a methyl or ethyl ether group constitute a category of niche products. Such 'end-capped' non-ionics are made by O-alkylation of the ethoxylate with alkyl chloride or dialkyl sulfate or by hydrogenation of the corresponding acetal. Compared with normal alcohol ethoxylates, the end-capped products are more stable against strong alkali and against oxidation. They are also characterized by unusually low foaming.

Cationics

The vast majority of cationic surfactants are based on the nitrogen atom carrying the cationic charge. Both amine and quaternary ammonium-based products are common. The amines only function as a surfactant in the protonated state; therefore, they cannot be used as high pH. Quaternary ammonium compounds, 'quats', on the other hand, are not pH sensitive. Non-quaternary cationics are also much more sensitive to polyvalent anions. As discussed previously, ethoxylated amines (see Figure 1.7) possess properties characteristic of both cationics and non-ionics. The longer the polyoxyethylene chain, than the more non-ionic the character of this surfactant type.

Figure 1.10 shows the structures of some typical cationic surfactants. The ester 'quat' represents a new, environmentally friendly type which to a large extent has replaced dialkyl 'quats' as textile softening agents.

The main synthesis procedure for non-ester quaternary ammonium surfactants is the nitrile route. A fatty acid is reacted with ammonia at high temperature

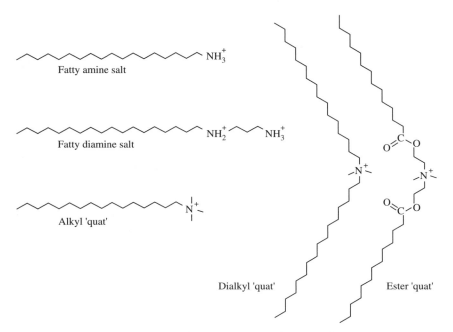

Figure 1.10 Structures of some representative cationic surfactants

to yield the corresponding nitrile, a reaction that proceeds via an intermediate amide. The nitrile is subsequently hydrogenated to primary amine using a cobalt or nickel catalyst:

$$R-COOH + NH_3 \xrightarrow{-H_2O} R-C\equiv N \xrightarrow{H_2} R-CH_2NH_2$$

Secondary amines can be produced either directly from the nitrile or in a two-stage reaction from the primary amine. In the one-stage route, which is believed to proceed via an intermediate imine, ammonia is continuously removed from the reaction in order to promote secondary amine formation:

$$R-C\equiv N + R-CH_2NH_2 + 2H_2 \longrightarrow (R-CH_2)_2NH + NH_3$$

Primary amines can be converted to long-chain 1,3-diamines by cyanoethylation:

$$R-CH_2NH_2 + CH_2=CHCN \longrightarrow R-CH_2NH(CH_2)_2CN \xrightarrow{H_2}$$

$$R-CH_2NH(CH_2)_3NH_2$$

Primary or secondary long-chain alkyl amines can be methylated to tertiary amines, e.g. by reaction with formaldehyde under reducing conditions:

$$(R-CH_2)_2NH + HCHO + H_2 \longrightarrow (R-CH_2)_2NCH_3$$

Ethylene oxide can also be used as an alkylating agent to convert primary or secondary amines to tertiary amines with the general structures $R-CH_2N(CH_2CH_2OH)_2$ and $(R-CH_2)_2NCH_2CH_2OH$.

Quaternary ammonium compounds are usually prepared from the tertiary amine by reaction with a suitable alkylating agent, such as methyl chloride, methyl bromide or dimethyl sulfate, the choice of reagent determining the surfactant counterion:

$$(R-CH_2)_2NCH_3 + CH_3Cl \longrightarrow (R-CH_2)_2N+(CH_3)_2 \; Cl^-$$

Ester-containing quaternary ammonium surfactants, 'ester quats', are prepared by esterifying a fatty acid (or a fatty acid derivative) with an amino alcohol followed by N-alkylation as above. The process is illustrated for triethanol amine as the amine alcohol and dimethyl sulfate as the methylating agent:

$$2\,R-COOH + N(CH_2CH_2OH)_3 \xrightarrow{-H_2O} (R-COOCH_2CH_2)_2NCH_2CH_2OH$$

$$\xrightarrow{(CH_3)_2SO_4} (R-COOCH_2CH_2)_2\overset{+}{\underset{\underset{CH_3}{|}}{N}}CH_2CH_2OH \quad CH_3SO_4^-$$

Nitrogen-based compounds constitute the vast majority of cationic surfactants. However, phosphonium, sulfonium and sulfoxonium surfactants also exist. The first two are made by treatment of trialkyl phosphine or dialkyl sulfide, respectively, with alkyl chloride, as is shown for phosphonium surfactant synthesis:

$$R_3P + R'X \longrightarrow R_3P^+-R' \ X^-$$

Sulfoxonium surfactants are prepared by hydrogen peroxide oxidation of the sulfonium salt. The industrial use of non-nitrogen cationic surfactants is small since only rarely do they give performance advantages over their less expensive nitrogen counterparts. Surface active phosphonium surfactants carrying one long-chain alkyl and three methyl group have found use as biocides.

The majority of surfaces, metals, minerals, plastics, fibres, cell membranes, etc., are negatively charged. The prime uses of cationics relate to their tendency to adsorb at these surfaces. In doing so they impart special characteristics to the surface. Some examples are given in Table 1.5, while some important facts about cationic surfactants are given in Table 1.6.

Table 1.5 Applications of cationic surfactants related to their adsorption at surfaces

Surface	Application
Steel	Anticorrosion agent
Mineral ores	Flotation collector
Inorganic pigments	Dispersant
Plastics	Antistatic agent
Fibres	Antistatic agent, fabric softener
Hair	Conditioner
Fertilizers	Anticaking agent
Bacterial cell walls	Bactericide

Table 1.6 Important facts about cationic surfactants

1. They are the third largest surfactant class.
2. They are generally not compatible with anionics (although there are important exceptions).
3. Hydrolytically stable cationics show higher aquatic toxicity than most other classes of surfactants.
4. They adsorb strongly to most surfaces and their main uses are related to *in situ* surface modification.

Zwitterionics

Zwitterionic surfactants contain two charged groups of different sign. Whereas the positive charge is almost invariably ammonium, the source of negative charge may vary, although carboxylate is by far the most common. Zwitterionics are often referred to as 'amphoterics', but as was pointed out on p. 0, the terms are not identical. An amphoteric surfactant is one that changes from net cationic via zwitterionics to net anionic on going from low to high pH. Neither the acid nor the base site is permanently charged, i.e. the compound is only zwitterionic over a certain pH range.

The change in charge with pH of the truly amphoteric surfactants naturally affects properties such as foaming, wetting, detergency, etc. These will all depend strongly on solution pH. At the isoelectric point the physicochemical behaviour often resembles that of non-ionic surfactants. Below and above the isoelectric point there is a gradual shift towards the cationic and anionic character, respectively. Surfactants based on sulfate or sulfonate to give a negative charge remain zwitterionic down to very low pH values due to the very low pK_a values of monoalkyl sulfuric acid and alkyl sulfonic acid, respectively.

Common types of zwitterionic surfactants are *N*-alkyl derivatives of simple amino acids, such as glycine (NH_2CH_2COOH), betaine ((CH_2)$_2$NCH$_2$COOH) and amino propionic acid ($NH_2CH_2CH_2COOH$). They are usually not prepared from the amino acid, however, but by reacting a long-chain amine with sodium chloroacetate or a derivative of acrylic acid, giving structures with one and two carbons, respectively, between the nitrogen and the carboxylate group. As an example, a typical betaine surfactant is prepared by reacting an alkyldimethyl amine with sodium monochloroacetate:

$$
\begin{array}{ccc}
& CH_3 & CH_3 \\
& | & | \\
R-N & + ClCH_2COO^- Na^+ \longrightarrow R-N^{\pm}CH_2COO^- + NaCl \\
& | & | \\
& CH_3 & CH_3 \\
\end{array}
$$

Amidobetaines are synthesized analogously from an amidoamine:

$$
\begin{array}{ccc}
& CH_3 & CH_3 \\
& | & | \\
RCONH(CH_2)_3-N & + ClCH_2COO^- Na^+ \longrightarrow RCONH(CH_2)_3-N^{\pm}CH_2COO^- + NaCl \\
& | & | \\
& CH_3 & CH_3 \\
\end{array}
$$

Another common type of zwitterionic surfactant, usually referred to as an imidazoline, is synthesized by reaction of a fatty acid with aminoethylethanolamine followed by treatment with chloroacetate. The nomenclature for this surfactant type is a bit confused; it was believed that the products contained an imidazoline ring, but later investigations have shown that the five-membered ring is cleaved during the second synthesis step. A typical reaction sequence is

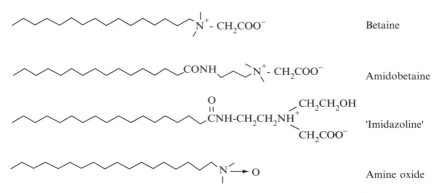

Zwitterionics as a group are characterized by having excellent dermatological properties. They also exhibit low eye irritation and are frequently used in shampoos and other cosmetic products. Since they possess no net charge, zwitterionics, similar to non-ionics, function well in high electrolyte formulations. A traditional use of the products has been in alkaline cleaners. Figure 1.11 shows examples of typical zwitterionics and Table 1.7 summarizes some general information about this surfactant class. As mentioned previously, amine oxide surfactants, or more correctly N-oxides of tertiary amines, are sometimes categorized as zwitterionics, sometimes as non-ionics and sometimes as cationics. They have a formal charge separation on the nitrogen and oxygen atoms. They basically behave as non-electrolytes, but at low pH or in the presence of anionic surfactants they will pick up a proton to form the cationic conjugate acid. A 1:1 salt will form between the anionic surfactant and the protonated amine oxide and this salt is very surface active. Amine oxides are prepared by hydrogen peroxide oxidation of the corresponding tertiary amine.

Figure 1.11   Structures of some representative zwitterionic surfactants

Table 1.7 Important facts about zwitterionic surfactants

---

1. They are the smallest surfactant class (partly due to high price).
2. They are compatible with all other classes of surfactants.
3. They are not sensitive to hard water.
4. They are generally stable in acids and bases. In particular, the betaines retain their surfactant properties in strong alkali.
5. Most types show very low eye and skin irritation. They are therefore well suited for use in shampoos and other personal care products.

Dermatological Aspects of Surfactants are Vital Issues

The dermatological effects of surfactants are important issues which are subject to much current concern. A large fraction of dermatological problems in normal working life can be related to exposure of unprotected skin to surfactant solutions. These solutions are often cleaning formulations of different kinds but may also be cutting fluids, rolling oil emulsions, etc. Skin irritation of various degrees of seriousness are common, and in rare cases allergic reactions may also appear. Whereas skin irritation is normally induced by the surfactant itself, the sensitization causing the allergic reaction is usually caused by a by-product. A well-known example of a severe allergic reaction is the so-called 'Margarine disease' which struck The Netherlands in the 1960s and which was subsequently traced to a by-product of a new surfactant which had been added to a margarine brand as an anti-spattering agent, i.e. a surfactant that keeps the water droplets finely dispersed during frying. Sensitizing agents are electrophiles which react with the nucleophilic groups of proteins, creating unnatural protein derivatives which the body then detects as foreign. The margarine surfactant contained an appreciable amount of an unreacted intermediate, a maleic anhydride derivative, which could be ring-opened in the body by protein amino or thiol groups.

The physiological effects of surfactants on the skin are investigated by various dermatological and biophysical methods, starting with the surface of the skin and progressing via the horny layer and its barrier function to the deeper layer of the basal cells. At the same time, subjective sensations, such as the 'feeling' of the skin, are recorded by verbalization of tactile sense and experience. Surfactant classes that are generally known to be mild to the skin are polyol surfactants, e.g. alkyl glucosides, zwitterionic surfactants, e.g. betaines and amidobetaines, and isethionates. Such surfactants are frequently used in cosmetics formulations.

For homologous series of surfactants there is usually a maximum in skin irritation at a specific chain length of the hydrophobic tail. For instance, in a comparative study of alkyl glucosides in which the C8, C10, C12, C14 and C16 derivatives were evaluated, a maximum irritation effect was obtained with the C12 derivative. Such a maximum is normally also obtained when it comes to the biocidal effects of surfactants. This probably reflects the fact that the biological effect, caused by surfactant action on the mucous membrane or the bacterial surface, respectively, is favoured by high surface activity and high unimer concentration. Since an increasing chain length of the hydrocarbon tail of the surfactant leads to an increased surface activity and to a reduced CMC, i.e. a reduced unimer concentration, there will somewhere be an optimum in hydrocarbon chain length (assuming the same polar head group for all surfactants).

Alcohol ethoxylates are relatively mild surfactants but not as mild as polyol-based non-ionics, such as, for instance, alkyl glucosides. Recent investigations have revealed that the dermatological effect seen with alcohol ethoxylates may not be caused by the intact surfactant but by oxidation products formed during storage. All ethoxylated products have been found to undergo autoxidation to give hydroperoxides on the methylene groups adjacent to the ether oxygen. Hydroperoxides formed at methylene groups in the polyoxyethylene chain seem to be too unstable to be easily isolated. The hydroperoxide with the OOH group at carbon number 2 of the hydrophobic tail is relatively stable, however, and has been isolated in an amount of around 1% after storage of a normal fatty alcohol ethoxylate for one year. This hydroperoxide exhibited considerable skin irritation. Another oxidation product of some dermatological concern that has been isolated is the surfactant aldehyde shown below. This aldehyde is not very stable, however, and further oxidation leads to a break-down of the polyoxyethylene chain with formaldehyde formed as one of many degradation products. Both the surfactant aldehyde and formaldehyde are irritating to the skin and eye:

$$C_nH_{2n+1}(OCH_2CH_2)_{m+1} \longrightarrow C_nH_{2n+1}(OCH_2CH_2)_m - CH_2CHO$$

$$\longrightarrow HCHO + \text{ other degradation products}$$

An indirect way to monitor the autoxidation of alcohol ethoxylates is to measure the change in cloud point with time. Figure 1.12 shows an example of such a storage test. As can be seen from this figure, both of the surfactants investigated show a considerable drop in cloud point on storage at 40°C.

Anionic surfactants are, in general, more skin irritating than non-ionics. For instance, sodium dodecyl sulfate (SDS), although used in some personal care products such as tooth pastes, has relatively high skin toxicity. Sodium alkyl ether sulfates are much milder than sodium alkyl sulfates which is one of the reasons why the ether sulfates are the most commonly used anionic surfactant in hand dishwashing formulations. (Their good foaming ability is another reason for their widespread use in such products.) The better dermatological characteristics of alkyl ether sulfates when compared to alkyl sulfates is one of the main reasons for the interest in ethoxylates with narrow homologue

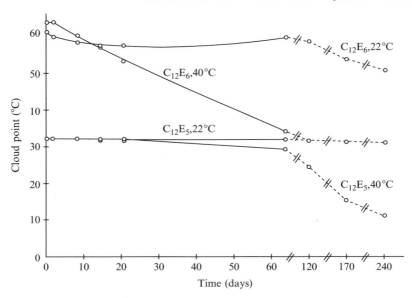

Figure 1.12 Cloud point versus time curves for the storage of 1% solutions of the surfactants $C_{12}E_5$ and $C_{12}E_6$, both 'homologue pure', measured at two different temperatures

distributions, i.e. 'peaked ethoxylates', as discussed above. With a peaked ethoxylate as the intermediate to be sulfated, the content of the more aggressive alkyl sulfate will be considerably lower than when a standard ethoxylate with a broad homologue distribution is used.

The effect of surfactants on the skin is often evaluated in the so-called modified Duhring chamber test. Figure 1.13 shows some typical results obtained for a sodium alkyl ether sulfate (sodium laureth sulfate), an alkyl glucoside (decyl glucoside) and combinations of the two. As can be seen, the skin irritation decreases relatively linearly with increasing amount of polyol surfactant. In other cases, already small additions of a mild surfactant may cause very large improvements in the dermatological properties. Such a 'synergistic effect' is probably related to a strong reduction in the CMC of the formulation due to mixed micelle formation. Some amphoteric surfactants, such as betaines, are known to give a very pronounced reduction of the skin irritation of anionics such as alkyl ether sulfates. This is probably due to a protonation of the carboxyl group of the betaine surfactant, transforming it into a cationic surfactant, and subsequent packing into mixed micelles together with the anionic surfactant. Thus, the energy gain in the mixed micelle formation (see Chapter 5) induces protonation of the carboxylate group of the betaine surfactant already at a pH level far above the pK_a value.

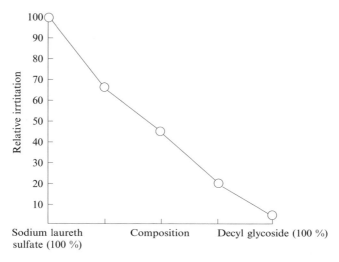

Figure 1.13   Modified Duhring chamber test with relative irritation scores for erythema formation. Reproduced by permission of Wiley-VCH from Hill, K., von Rybinski, W. and Stoll, G., *Alkyl Polyglycosides: Technology, Properties and Applications*, Wiley-VCH, Weinheim, 1997

The Ecological Impact of Surfactants is of Growing Importance

Although concern about the ecological impact of surfactants has been en-shrined in legislation for more than 20 years, it is only recently that the issue has become a major factor in all formulation work involving surfactants. Most of the surfactants used in households and in industry go into the sewers. The rate of biodegradation in the sewage plant will determine the volume of surfactant that reaches the environment. The rate of biodegradation, in com-bination with the degree of aquatic toxicity, will determine the environmental impact. The OECD has issued guidelines and directives regarding:

Aquatic toxicity
Biodegradability
Bioaccumulation

Aquatic Toxicity

Aquatic toxicity may be measured on fish, daphnia or algae. Toxicity is given as LC_{50} (for fish) or EC_{50} (for daphnia or algae), where LC and EC stand for lethal and effective concentrations, respectively. Values below 1 mg/l after 96 h testing on fish and algae and 48 h on daphnia are considered toxic. Environ-mently benign surfactants should preferably be above 10 mg/l.

Biodegradability

Biodegradation is a process carried out by bacteria in nature. By a series of enzymatic reactions, a surfactant molecule is ultimately converted into carbon dioxide, water and oxides of the other elements. If a product does not undergo natural biodegradation then it is stable and persists in the environment. For surfactants the rate of biodegradation varies from 1–2 h for fatty acids, via 1–2 days for linear alkylbenzene sulfonates, to months for branched alkylbenzene sulfonates.

In all testing of biodegradation it is important to realize that the rate of degradation depends on factors such as concentration, pH and temperature. The temperature effect is particularly important. The rate at which chemicals are broken down in sewage plants may vary by as much as a factor of five between summer and winter in Northern Europe.

Two criteria are of importance with respect to testing for biodegradation: primary degradation and ultimate degradation. Primary degradation of surface active agents relates to the loss of surfactants properties. For instance, an ester surfactant may rapidly break down into alcohol and acid, neither of which is very surface active. This type of test is of interest for specific purposes, e.g. to be able to predict whether or not a product persists in giving foams on rivers.

More important from an ecological point of view is the test for ultimate biodegradation. A plethora of test methods exists for so-called ready biodegradability. In most of these, such as the popular modified Sturm test (OECD Test 301 B for ultimate biodegradation), conversion into carbon dioxide is measured as a function of time. The test is performed in closed bottles to which samples of sludge from a sewage plant have been added. The surfactant is added to one set of bottles, while another set of bottles without added surfactant is used as the reference. The gas produced is monitored with time and the difference obtained between the two sets of bottles is related to the degradation of the surfactant. With most surfactants there is an induction period followed by a steep rise of the curve and then an abrupt levelling off. A typical test result, together with the criteria that must be fulfilled to pass the test, are given in Figure 1.14.

Bioaccumulation

Hydrophobic organic compounds are persistent in nature since all biodegradation requires some kind of aqueous environment. Bioaccumulation can be measured directly in fish but is more often calculated from a model experiment. Partitioning of the compound between two phases, octanol and water, is measured and the logarithm of the value obtained, $\log P$, is used. (This is a common procedure for assigning hydrophobicity value to organic substances and tables of $\log P$ values are available in the literature.) A surfactant is considered to bioaccumulate if:

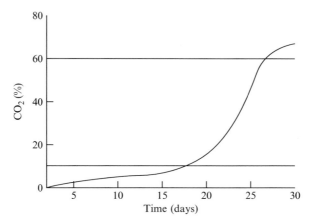

Figure 1.14 Criteria needed to pass the test for ready biodegradability and a typical example of a degradation curve. Note the long induction period before the degradation 'takes off'

$$\log P_{oct/w} > 3$$

The vast majority of surfactants have log P values below 3. Bioaccumulation, therefore, is not considered to be a crucial issue.

Log P values for many surfactants have been collected. It has been found that these values can be used as an indication of the surfactant hydrophilicity; the lower the log P value, then the higher the hydrophilicity. Values of surfactant hydrophilicity are often helpful in formulation work. A more well-known standard for the same purpose is the hydrophilic – lipophilic balance (HLB) concept which is widely used in the selection of emulsifiers. The HLBs of surfactants are discussed in some detail in Chapter 21. There is a reverse relationship between log P and the HLB, i.e. the higher the HLB value, then the lower the log P value. The critical packing parameter (CPP) concept (see Chapter 2) is yet another way to assign a number to the surfactant hydrophilicity.

Labelling

In the OECD guidelines for the labelling of surfactants, values of aquatic toxicity and ready biodegradation are taken into account. Figure 1.15 illustrates the procedure. Many of the commonly used surfactants today are borderline cases. There is a clear trend to replace these by compounds positioned 'upwards and right' in the diagram.

Tests for aquatic toxicity, ready biodegradation and bioaccumulation by no means give the complete picture of the environmental impact of a surfactant. A number of other relevant parameters are listed in Table 1.8. In addition, the

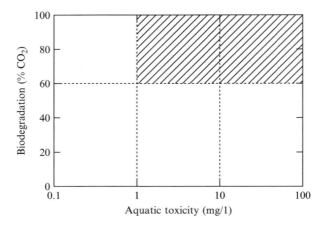

Figure 1.15 Environmental ranking of surfactants is based on the values of ready biodegradation and aquatic toxicity. The shaded areas of the diagram are 'approved areas'

Table 1.8 Relevant factors besides aquatic toxicity, ready biodegradability and bioaccumulation for assessment of surfactant ecotoxicity

---

1. Anaerobic biodegradation (alkylbenzene sulfonates, alkylphenol ethoxylates and EO–PO block copolymers are examples of surfactants with very slow anaerobic biodegradation).
2. Chronic toxicity.
3. Effect on function of sewage plants.
4. Toxicity in the production of surfactants.
5. Life cycle aspects.
6. Technical effect, i.e. the more effective the surfactant, then the smaller the amount needed; therefore, biotoxicity per effect unit is the relevant measure (this is particularly relevant for the formulated end-product).

---

formulated end-product often contains mixtures of surfactants or a blend of surfactant and polymer. Since it is well-known that the physicochemical behaviour of a surfactant may be very different in such mixtures than alone in solution, it is reasonable to believe that the biological effects may also be different. In order to obtain the full picture, ecological tests should be carried out on each individual end-product.

The Rate of Biodegradation Depends on Surfactant Structure

Several parameters are of importance for the rate of biodegradation of a surfactant. First of all, the surfactant needs to have a reasonable solubility in water. Very lipophilic amphiphiles such as fluorinated surfactants accumulate in lipid compartments in organisms and break down very slowly. As discussed above, most surfactants have a high enough water solubility that bioaccumulation of the intact molecule is not a problem. However, the initial degradation may lead to intermediates with very limited water solubility. A well-known example of this is the class of alkylphenol ethoxylates which degrade by oxidative cleavage from the hydroxyl end of the polyoxyethylene chain. This means that alkylphenol ethoxylates with polar head groups of just a few oxyethylene units will form. These are very lipophilic and biodegrade at a very slow rate. Analysis of fish in waters exposed to nonylphenol ethoxylates have shown high levels of nonylphenol with two and three oxyethylene units in lipid tissues. Such findings were one of the reasons for the strong concern about the environmental effects of this surfactant class. Fatty alcohol ethoxylates probably break down via many different pathways (oxidation at the terminal end of the polyoxyethylene chain, oxidation of the terminal end of the alkyl chain, and central cleavage to give fatty alcohol and poly(ethylene glycol)), with the result being that lipophilic metabolites are not formed in large amounts.

Besides water solubility, it is essential that the surfactant contains bonds that can be easily broken down by enzymatic catalysis. Most, if not all, chemical bonds will ultimately break down in nature but it is important that the rate is high enough so that unacceptable concentrations of a surfactant or its metabolites are not generated in the environment. In order to speed up the rate of biodegradation it has now become common practice to build a weak bond into the surfactant structure. Such easily cleavable bonds may be placed randomly in a surfactant but for synthesis reasons they are usually inserted between the hydrophobic tail and the polar head group. Typical examples of such bonds are esters and amides, the breakdown of which are catalysed by esterases/lipases and peptidases/acylases, respectively. The concept of surfactants with easily cleavable bonds is discussed in some detail in Chapter 11. One may have anticipated the ether bond of non-ionic surfactants to be a problem since ether-splitting enzymes are not common in nature. This is evidently not the case, however. Under aerobic conditions, hydroperoxides are generated in the α-position to the ether bond and the breakdown proceeds from there via aldehydes and acids.

A third factor to take into account in addition to water solubility and the presence of cleavable bonds is the degree of branching of the non-polar part of the surfactant molecule. Extensive branching of the hydrocarbon tail often leads to a reduced rate of biodegradation. This is probably due to the side chains causing steric hindrance in the approach of the surfactant into the active site of the enzyme. However, the picture is far from clear. Some branching patterns seem to be more troublesome than others and the effect is probably specific to the enzyme in question. It seems that methyl branching is less of a problem than branching involving longer alkyl chains, but if many methyl branchings appear in a row, such as in poly(propylene glycol) derivatives, they are still problematic. A very good example of the importance of linearity of alkyl chains is the difference in the rates of biodegradation between alkylbenzene sulfonates with linear or branched alkyl chains. As mentioned briefly above, branched alkylbenzene sulfonates, which are based on tetra-1,2-propylene as the alkyl chain, were once the bulk surfactant in household detergents— cheap, efficient and chemically stable—but too stable in the environment. When environmental aspects became an issue in the 1960s and 1970s, these surfactants were rapidly replaced by their counterparts with linear alkyl chains. The linear alkylbenzene sulfonates break down satisfactorily under aerobic conditions. Their rate of unaerobic biodegradation is relatively slow, however, a fact which is currently subject to some concern. In addition, the position of the branching along a hydrocarbon chain seems to be critical. Branching on a carbon two atoms away from a cleavable bond (such as in 2-ethylhexyl ethers, carboxylic esters, acetals and sulfates) is less harmful than branching on the carbon one atom away. This is of importance because the oxo alcohols, which are widely used as surfactant raw materials, contain a large proportion of 2-alkyl branching. It has also been found that the length of the 2-alkyl side chain has a negligible effect on the rate of biodegradation. However, considerably more work is needed before one is able to predict the biodegradation characteristics from inspection of the chemical formula of a surfactant alone.

Environmental Concern is a Strong Driving Force for Surfactant Development

All major types of surfactants, i.e. alkylbenzene sulfonates, alkyl sulfates, alcohol ethoxylates, etc., have been around for decades. Their routes of preparation have been carefully optimized and their physicochemical behaviour is relatively well understood. Besides the constant challenge of finding ways to minimize the manufacturing cost for existing surfactants, the market pull for 'greener' products has been the overriding driving force for surfactant development in later years. Clear trends in current developments are (i) to synthesize the surfactant from natural building blocks, and (ii) to make cleavable surfactants. Cleavable surfactants are covered in Chapter 11. Below is a brief account of current developments in the area of surfactants produced from natural building blocks.

The Polar Head Group

Two main types of natural products have been investigated as surfactant polar head groups, namely sugars and amino acids. Surfactants can be made from both by either organic synthesis or enzymatic synthesis, or by a combination of the two. The most interest has been directed towards sugar-based surfactants and this area will be presented below.

In recent years there has been a focus on three classes of surfactants with sugar or a polyol derived from sugar as the polar head group: alkyl polyglucosides (APGs), alkyl glucamides and sugar esters. Representative structures of the three surfactant types are shown in Figure 1.16.

There is currently a very strong interest in exploring alkyl polyglucosides (APGs) as surfactants for several types of applications. APGs are synthesized by the direct reaction of glucose with fatty alcohol, using a large excess of alcohol in order to minimize sugar oligomerization. Alternatively, they are made by transacetalization of a short-chain alkyl glucoside, such as ethyl or butyl glucoside, with a long-chain alcohol. An acid catalyst is used in both processes. Either glucose or a degraded starch fraction is used as the starting material. Figure 1.17 illustrates the synthesis. Alkyl glucosides can also be made by enzymatic synthesis, using β-glucosidase as the catalyst, which yields only the β-anomer (and gives low yield). The corresponding α-anomer can more readily be obtained by β-glucosidase catalysed hydrolysis of the racemate. There are considerable differences between the α, β mixture obtained by organic synthesis and the pure enantiomers obtained by the bio-organic route.

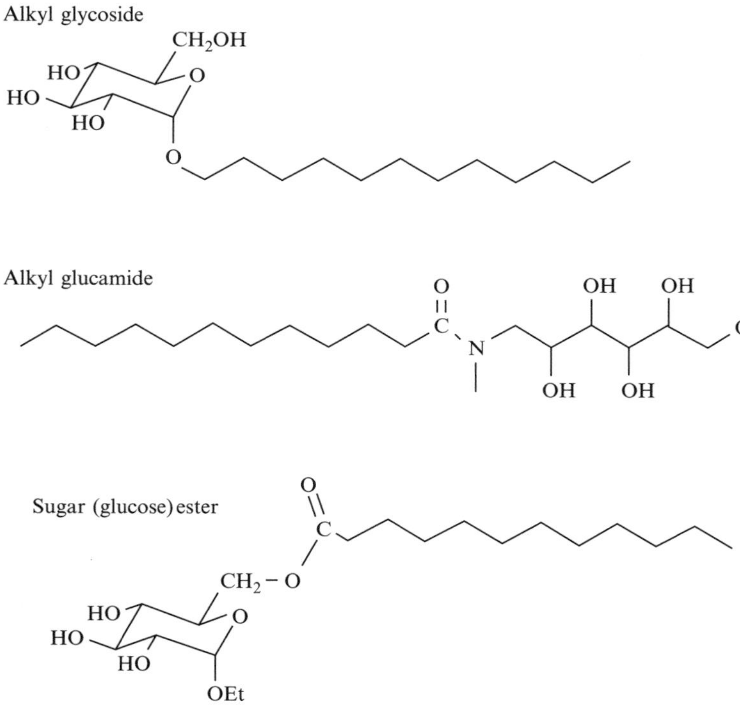

Figure 1.16   Structures of some representative polyol surfactants

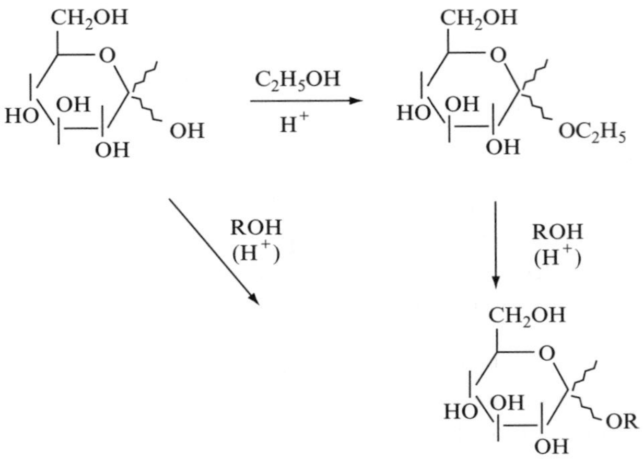

Figure 1.17   Routes of preparation of alkyl glucoside surfactants

The β-anomer of *n*-octyl glucoside has found use as a surfactant in biochemical work.

Alkyl glucosides are stable at high pH and sensitive to low pH where they hydrolyse to sugar and fatty alcohol. A sugar unit is more water-soluble and less soluble in hydrocarbons than the corresponding polyoxyethylene unit; hence, APGs and other polyol-based surfactants are more lipophobic than their polyoxyethylene-based surfactant counterparts. This makes the physico-chemical behaviour of APG surfactants in oil–water systems distinctly different from that of conventional non-ionics. Furthermore, APGs do not show the pronounced inverse solubility vs. temperature relationship that normal non-ionics do. This makes an important difference in solution behaviour bet-ween APGs and polyoxyethylene-based surfactants. The main attractiveness of APGs lies in their favourable environmental profile: the rate of biodegrad-ation is usually high and the aquatic toxicity is low. In addition, APGs exhibit favourable dermatological properties, being very mild to the skin and eye. This mildness makes this surfactant class attractive for personal care products, although APGs have also found a wide range of technical applications.

Alkyl glucamides (see Figure 1.16), more strictly named as *N*-alkanoyl-*N*-methylglucamines, are commercially important products. The product sold in large quantities for the detergent sector is *N*-dodecanoyl-*N*-methylglucamine, i.e. the C12-derivative. This product is prepared from glucose, methyl amine, hydrogen and methyl laurate by a two-step reaction. The physicochemical properties and many other characteristics of this surfactant class are similar to those of APGs. However, whereas alkyl glucosides are very stable to alkali and labile to acids, alkyl glucamides are stable to alkali and also relatively stable to acids.

Esters of glucose (see Figure 1.16) can be made either by enzymatic synthesis, using a lipase catalyst, or by an organic chemical route. Using the right enzyme the bio-organic route can give esterification almost exclusively at the 6-position of the sugar moiety. The organic synthesis requires extensive use of protecting groups to obtain high selectivity. The selective enzymatic synthesis of sugar esters other than glucose has been difficult to achieve without the use of protective groups. Starting from sugar acetals and fatty acids, monoesters of sugars can be obtained in good yield, after a deprotection step, from several starting materials, both mono-and disaccharides. All sugar esters are very labile on the alkaline side and fairly stable to acid. Their degradation products, i.e. sugar and fatty acids, are both very natural products; thus, sugar esters are ideal candidates as food surfactants. Sugar esters seem to undergo rapid biodegradation regardless of the sugar head group size or the acyl chain length. Sulfonated sugar esters have also been prepared. These anionic surfactants were found to undergo less rapid biodegradation.

Polyol surfactants have many attractive propeties: they are mild to the skin, they exhibit a low aquatic toxicity and a high rate of biodegradation, and they are easy to work with in that they show good tolerance to high electrolyte concentrations. Some characteristic properties of polyol surfactants in general are summarized in Table 1.9.

The Hydrocarbon Tail

Fatty acids are the first choice of a natural hydrophobe for surfactants. Fatty acids have been used for a long time, e.g. in fatty acid ethoxylates and in sorbitan esters of fatty acids. In recent years there has been increased interest in fatty acid monoethanolamide ethoxylates, or fatty amide ethoxylates. These surfactants are easily prepared by ethoxylation of the fatty acid monoethanolamide. The ethanolamide, in turn, is prepared by aminolysis of the fatty acid methyl ester by ethanolamine. The fatty amide ethoxylates are of interest as an alternative to fatty alcohol ethoxylates for several reasons: (i) they biodegrade readily to fatty acids and amino-terminated poly(ethylene glycols), (ii) the

Table 1.9   Characteristics of polyol surfactants

1. Aerobic and anaerobic biodegradation is fast.
2. Aquatic toxicity is low.
3. Hydroxyl groups are strongly lipophobic. At the same time, surfactants with long enough hydrocarbon chains are strongly hydrophobic. Polyol surfactants therefore have a high tendency to remain at the oil–water interface.
4. The effect of temperature on solution behaviour is small and opposite to that of ethoxylates. Mixtures of polyol surfactants (larger fraction) and ethoxylates (smaller fraction) can be formulated so that a non-ionic surfactant with phase behaviour unaffected by temperature is obtained.

amide bond in the structure may improve surfactant packing due to hydrogen bond formation, and (iii) double bonds in the fatty acid chain can be preserved in the product (although during the ethoxylation step 1,3-*cis*, *cis*-double bonds undergo migration to conjugated *trans*, *cis* structures).

Double bonds in the hydrocarbon tail have been found to increase the CMC. This may partly be due to the surfactants becoming more hydrophilic and partly to the increased bulkiness of the chains, thus rendering packing into closed aggregates and formation of hydrogen bonds between the amide groups more difficult. The presence of one double bond, *cis* (oleyl) or *trans* (elaidyl), does not greatly affect the molecular cross-sectional area at the surface. Two double bonds give much larger areas, however. Fatty amide ethoxylates with double bonds in the hydrocarbon tail are of interest as polymerizable surfactants, as is discussed further in Chapter 11.

Sterols are another class of natural hydrophobes of potential interest as surfactant building blocks. The special feature of sterol-based surfactants is the large hydrophobic group of fully natural origin which, due to its rather planar four-ring structure, may induce good packing at interfaces. Phytosterol is the common name for sterols of plant origin and they are already today used as surfactant raw materials. Their structure is similar to that of cholesterol, the prime example of sterols from animal sources. The sterols contain a secondary hydroxyl group that can be ethoxylated. The alcohol is sterically hindered, however, and the reaction with ethylene oxide is not straightforward. The best procedure seems to be to start the ethoxylation with a Lewis acid catalyst, cease the reaction after 3–5 moles of ethylene oxide have been added, and continue with KOH as initiator.

Sterol ethoxylates are large, stiff molecules. They need a long time to reach an equilibrium position at an interface. The decay of surface tension can take hours. Most likely, the long equilibrium time is due to exchange reactions of sterol ethoxylates with varying polyoxyethylene chain lengths at the air–water interface and to slow conformational changes of individual molecules at the surface. The multi-ring structure of sterols is rigid and the time required to adapt a favourable conformation at an interface may be long. Sterol surfactants are of interest as solubilizers and emulsifiers in drug and cosmetics applications.

Bibliography

Ash, M. and I. Ash, *Handbook of Industrial Surfactants*, Gower, Aldershot, UK, 1993.

Holmberg, K. (Ed.), *Novel Surfactants. Synthesis, Applications and Biodegradability*, Surfactant Science Series, 74, Marcel Dekker, New York, 1998.

Holmberg, K. (Ed.), *Handbook of Applied Surface and Colloid Chemistry*, Wiley, Chichester, UK, 2001.

Karsa, D. R. and M. R. Porter (Eds), *Biodegradability of Surfactants*, Blackie & Sons, London, 1995.

Kosswig, H. and H. Stache, *Die Tenside*, Carl Hanser Verlag, Munich, Germany, 1993.

Lomax, E. G. (Ed), *Amphoteric Surfactants*, Surfactant Science Series 59, Marcel Dekker, New York, 1996.

Myers, D. *Surfactant Science and Technology*, 2nd Edn, Wiley-VCH, New York, 1999.

Porter, M. R., *Handbook of Surfactants*, Blackie & Sons, London, 1991.

Richmond, J. M. (Ed.), *Cationic Surfactants. Organic Chemistry*, Surfactant Science Series 34, Marcel Dekker, New York, 1990.

Rosen, M. J., *Surfactants and Interfacial Phenomena*, 2nd Edn, Wiley, New York, 1989.

Stache, H. W., *Anionic Surfactants. Organic Chemistry*, Surfactant Science Series 56, Marcel Dekker, New York, 1996.

van Os, N. M., J. R. Haak and L. A. M. Rupert, *Physico-Chemical Properties of Selected Anionic, Cationic and Nonionic Surfactants*, Elsevier, Amsterdam, 1993.

2 SURFACTANT MICELLIZATION

Different Amphiphile Systems

Surfactant self-assembly leads to a range of different structures, of which a few are shown in Figure 2.1. Systems containing amphiphile are best classified into homogeneous (or single-phase) systems and heterogeneous systems of two or more phases. The single-phase systems can in turn be divided into isotropic solutions, solid phases and liquid crystalline phases. The solid crystalline phases have, as do crystals in general, both long-range and short-range order, but the degree of short-range order varies between different phases. Isotropic solution phases are characterized by disorder over both short and long distances, while liquid crystalline phases or mesophases have a short-range disorder but some distinct order over larger distances. In both isotropic solutions and liquid crystals, the state of the amphiphile alkyl chains can be denoted as 'liquid-like'. In crystals, formed below the 'chain melting temperature', the state is more or less 'solid-like'. The more important amphiphile systems can be arranged as shown in Table 2.1.

Surfactants Start to Form Micelles at the CMC

In measuring the different physicochemical properties of an aqueous solution of a surfactant or lipid that is polar enough to be water soluble up to relatively high concentrations, we will encounter many peculiarities, as exemplified in Figure 2.2 for an ionic surfactant. At low concentrations, most properties are similar to those of a simple electrolyte. One notable exception is the surface tension which decreases rapidly with surfactant concentration. At some higher concentration, which is different for different surfactants, unusual changes are recorded. For example, the surface tension, as well as the osmotic pressure, takes on an approximately constant value, while light scattering starts to increase and self-diffusion starts to decrease. All of the observations suggest and are consistent with a change-over from a solution containing single surfactant molecules or ions, i.e. unimers, to a situation where the surfactant occurs more and more in a self-assembled or self-associated state. We will examine in

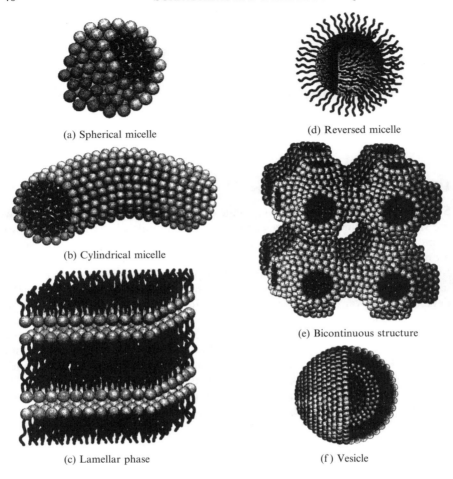

(a) Spherical micelle

(d) Reversed micelle

(b) Cylindrical micelle

(e) Bicontinuous structure

(c) Lamellar phase

(f) Vesicle

Figure 2.1 Surfactant self-assembly leads to a range of different structures for which a few are shown:

(a) Spherical micelles with an interior composed of the hydrocarbon chains and a surface of the polar head groups (pictured as spheres) facing water. Spherical micelles are characterized by a low surfactant number (critical packing parameter) and a strongly positive spontaneous curvature. The hydrocarbon core has a radius close to the length of the extended alkyl chain.

(b) Cylindrical micelles with an interior composed of the hydrocarbon chains and a surface of the polar head groups facing water. The cross-section of the hydrocarbon core is similar to that of spherical micelles. The micellar length is highly variable so these micelles are polydisperse.

(c) Surfactant bilayers which build up lamellar liquid crystals have for surfactant–water systems a hydrocarbon core with a thickness of *ca*. 80% of the length of two extended alkyl chains.

Table 2.1   Different amphiphile systems

Homogeneous systems
  Solid phases
    Many different structures
  Liquid crystalline phases
    Lamellar
    Hexagonal
    Reversed hexagonal
    Cubic: several structures known, which can be grouped into water-continuous,
      hydrophobe-continuous and bicontinuous
    'Intermediate' and 'deformed' phases, including 'nematic lyotropic'
  Isotropic solution phases
    Dilute and concentrated micellar solutions
    Reversed micellar solutions
    Microemulsions
    Vesicle solutions
Heterogeneous systems
  Emulsions
  Suspensions
  Vesicles, liposomes
  Foams
  Adsorbed surfactant layers and other surfactant films
  Gels

---

detail the structures formed, as well as the underlying mechanisms, and will here only note two general features. The concentration for the onset of self-assembly is quite well defined and becomes more so the longer the alkyl chain of the surfactant. The first-formed aggregates are generally approximately spherical in shape. We call such aggregates micelles, and the concentration where they start to form is known as the critical micelle concentration, abbreviated to CMC. An illustration of a micelle's structure is given in Figure 2.3.

The CMC is the single most important characteristic of a surfactant, useful *inter alia* in consideration of the practical uses of surfactants. We will now

---

(d) Reversed or inverted micelles have a water core surrounded by the surfactant polar head groups. The alkyl chains together with a non-polar solvent make up the continuous medium. Like 'normal' micelles, they can grow into cylinders.

(e) A bicontinuous structure with the surfactant molecules aggregated into connected films characterized by two curvatures of opposite sign. The mean curvature is small (zero for a minimal surface structure).

(f) Vesicles are built from bilayers similar to those of the lamellar phase and are characterized by two distinct water compartments, with one forming the core and one the external medium. Vesicles may have different shapes and there are also reversed-type vesicles. Reproduced by permission of Wiley-VCH from D. F. Evans and H. Wennerström, *The Colloidal Domain. Where Physics, Chemistry, Biology, and Technology Meet*, VCH, New York, 1994, pp. 14–15

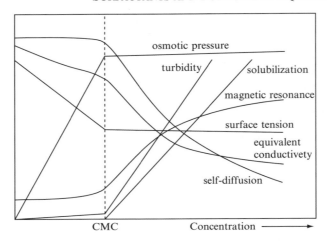

Figure 2.2   Schematic representation of the concentration dependence of some physical properties for solutions of a micelle-forming surfactant. From B. Lindman and H. Wennerström, *Topics in Current Chemistry*, Vol. 87, Springer-Verlag, Berlin, 1980, p. 6

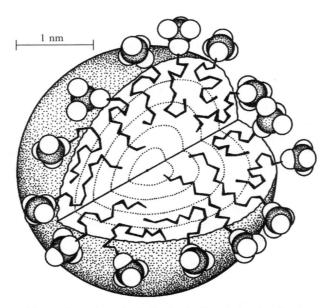

Figure 2.3   An illustration of a spherical micelle (for dodecyl sulfate) emphasizing the liquid-like character with a disordered hydrocarbon core and a rough surface. From J. Israelachvili, *Intermolecular and Surface Forces, with Applications to Colloidal and Biological Systems*, Academic Press, London, 1985, p. 251

consider how different factors influence the CMC, but let us first make a note on how to measure it. The two most common and generally applicable techniques are surface tension and solubilization, i.e. the solubility of an otherwise insoluble compound. For an ionic amphiphile, the conductivity offers a convenient approach to obtain the CMC. However, as a very large number of physicochemical properties are sensitive to surfactant micellization, there are numerous other possibilities, such as self-diffusion measurements and NMR (nuclear magnetic resonance) and fluorescence spectroscopy. As we will see, CMC is not an exactly defined quantity, which causes difficulties in its determination. For long-chain amphiphiles, an accurate determination is straightforward and different techniques give the same results. However, for short-chain, weakly associating, amphiphiles this is not the case and great care must be taken not only in the measurements but also in evaluating the CMC from experimental data.

CMC Depends on Chemical Structure

A list of the CMC values of some selected surfactants at 25°C is given in Table 2.2, while in Table 2.3 a similar list for some non-ionic surfactants is presented.

Table 2.2 List of CMC values of selected surfactants

Surfactant	CMC^a
Dodecylammonium chloride	1.47×10^{-2} M
Dodecyltrimethylammonium chloride	2.03×10^{-2} M
Decyltrimethylammonium bromide	6.5×10^{-2} M
Dodecyltrimethylammonium bromide	1.56×10^{-2} M
Hexadecyltrimethylammonium bromide	9.2×10^{-4} M
Dodecylpyridinium chloride	1.47×10^{-2} M
Sodium tetradecyl sulfate	2.1×10^{-3} M
Sodium dodecyl sulfate	8.3×10^{-3} M
Sodium decyl sulfate	3.3×10^{-2} M
Sodium octyl sulfate	1.33×10^{-1} M
Sodium octanoate	4×10^{-1} M
Sodium nonanoate	2.1×10^{-1} M
Sodium decanoate	1.09×10^{-1} M
Sodium undecanaote	5.6×10^{-2} M
Sodium dodecanoate	2.78×10^{-2} M
Sodium p-octylbenzene sulfonate	1.47×10^{-2} M
Sodium p-dodecylbenzene sulfonate	1.20×10^{-3} M
Dimethyldodecylamineoxide	2.1×10^{-3} M
$CH_3(CH_2)_9(OCH_2CH)_6OH$	9×10^{-4} M
$CH_3(CH_2)_9(OCH_2CH)_9OH$	1.3×10^{-3} M
$CH_3(CH_2)_{11}(OCH_2CH)_6OH$	8.7×10^{-5} M
$CH_3(CH_2)_7C_6H_4(CH_2CH_2O)_6$	2.05×10^{-4} M
Potassium perfluorooctanoate	2.88×10^{-2} M

^{a} In mol/dm^3 (M) or mol/kg H$_2$O (m)

Table 2.3   List of CMC values for non-ionic surfactants

Surfactant	CMC (μM)
C_6E_3	10×10^4
C_8E_4	8.5×10^3
C_8E_5	9.2×10^3
C_8E_6	9.9×10^3
$C_{10}E_5$	9.0×10^2
$C_{10}E_6$	9.5×10^2
$C_{10}E_8$	10×10^2
$C_{12}E_5$	6.5×10
$C_{12}E_6$	6.8×10
$C_{12}E_7$	6.9×10
$C_{12}E_8$	7.1×10
$C_{14}E_8$	9.0×10
$C_{16}E_9$	2.1×10
$C_{16}E_{12}$	2.3×10
$C_{16}E_{21}$	3.9×10
$C_8\phi E_9$	3.4×10^2
$C_8\phi E_{10}$	3.4×10^2
$C_{12}NO$	2.2×10^3
β-D$-C_8$ glucoside	2.5×10^4
β-D$-C_{10}$ glucoside	2.2×10^3
β-D$-C_{12}$ glucoside	1.9×10^2

From these and other data, several general remarks about the variation of the CMC with the surfactant chemical structure can be made:

1. The CMC decreases strongly with increasing alkyl chain length of the surfactant (Figures 2.4 and 2.5). As a general rule, the CMC decreases by a factor of *ca.* 2 for ionics (without added salt) and by a factor of *ca.* 3 for non-ionics on adding one methylene group to the alkyl chain (Table 2.4). Comparisons between different classes of surfactants are best made at a fixed number of carbons in the alkyl chain.

2. The CMCs of non-ionics are much lower than those of ionics. The relationship depends on the alkyl chain length, but two orders of magnitude is a rough starting point.

3. Besides the major difference between ionics and non-ionics, the effects of the head group are moderate. Cationics typically have slightly higher CMCs than anionics. For non-ionics of the oxyethylene variety, there is a moderate increase in the CMC as the polar head becomes larger.

4. The valency of the counterion is significant. While simple monovalent inorganic counterions give roughly the same CMC, increasing the valency to 2 gives a reduction of the CMC by roughly a factor of 4. Organic counterions

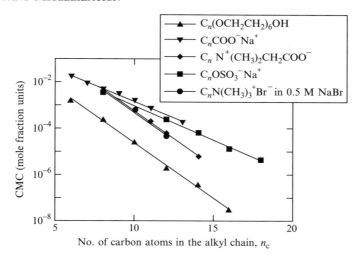

Figure 2.4 The logarithm of the CMC varies linearly with the number of carbon atoms in the alkyl chain of a surfactant. The slope is larger for a non-ionic surfactant or an ionic with added salt than for an ionic surfactant without added electrolyte. From B. Lindman and H. Wennerström, *Topics in Current Chemistry*, Vol. 87, Springer-Verlag, Germany, 1980, p. 8

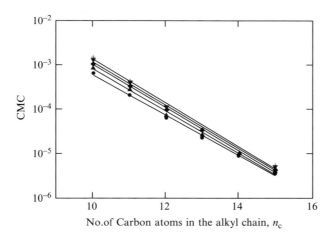

Figure 2.5 The logarithm of the CMC (molar concentration) versus the number of carbons in the alkyl chain for octa(ethylene glycol)monoalkylethers at different temperatures. From top to bottom, the temperatures are 15.0, 20.0, 25.0, 30.0 and 40.0°C. From K. Meguro, M. Ueno and K. Esumi, in *Nonionic Surfactants. Physical Chemistry*, M. J. Schick (Ed.), Marcel Dekker, New York, 1987, p. 134

Table 2.4 The CMC decreases strongly with the alkyl chain length. The decrease follows, to a good approximation, the relationship $\log \text{CMC} = A - Bn_c$, where n_c is the number of carbons in the alkyl chain

Surfactant series	Temperature (°C)	A	B
Na carboxylates (soaps)	20	1.8_5	0.30
K carboxylates (soaps)	25	1.9_2	0.29
Na (K) n-alkyl 1-sulfates or -sulfonates	25	1.5_1	0.30
Na n-alkane-1-sulfonates	40	1.5_9	0.29
Na n-alkane-1-sulfonates	55	1.1_5	0.26
Na n-alkane-1-sulfonates	60	1.4_2	0.28
Na n-alkyl-1-sulfates	45	1.4_2	0.30
Na n-alkyl-1-sulfates	60	1.3_5	0.28
Na n-alkyl-2-sulfates	55	1.2_8	0.27
Na p-n-alkylbenzene sulfonates	55	1.6_8	0.29
Na p-n-alkylbenzene sulfonates	70	1.3_3	0.27
n-Alkylammonium chlorides	25	1.2_5	0.27
n-Alkylammonium chlorides	45	1.7_9	0.30
n-Alkyltrimethylammonium bromides	25	1.7_7	0.30
n-Alkyltrimethylammonium chlorides (in 0.1 M NaCl)	25	1.2_3	0.33
n-Alkyltrimethylammonium bromides	60	1.7_7	0.29
n-Alkylpyridinium bromides	30	1.7_2	0.31
n-$C_nH_{2n+1}(OC_2H_4)_6OH$	25	1.8_2	0.49

reduce the CMC compared to inorganic ones, the more so the larger the non-polar part.

5. While alkyl chain branching and double bonds, aromatic groups or some other polar character in the hydrophobic part produce sizeable changes in the CMC, a dramatic lowering of the CMC (one or two orders of magnitude) results from perfluorination of the alkyl chain. Partial fluorination interestingly may increase the CMC, e.g. fluorination of the terminal methyl group roughly doubles the CMC value. The anomalous behaviour of partially fluorinated surfactants is due to unfavourable interactions between hydrocarbon and fluorocarbon groups.

Temperature and Cosolutes Affect the CMC

It is a characteristic feature of micellization that the CMC is, to a first approximation, independent of temperature. The temperature dependence of the CMC of sodium dodecyl sulfate (SDS) displayed in Figure 2.6 is a good illustration of this. The CMC varies in a non-monotonic way by *ca.* 10–20% over a wide range. The shallow minimum around 25°C can be compared with a similar minimum in the solubility of hydrocarbons in water. Non-ionic surfactants of

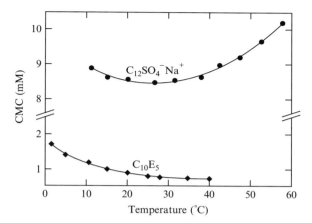

Figure 2.6 Temperature dependence of the CMC of sodium dodecyl sulfate (top) and penta(ethylene glycol)monodecyl ether. From P. H. Elworthy, A. T. Florence and C. B. Macfarlane, *Solubilisation by Surface-Active Agents*, Chapman & Hall, London, 1968

the polyoxyethylene type deviate from this behaviour and show typically a monotonic, and much more pronounced, decrease in CMC with increasing temperature. As will be discussed in some detail in Chapter 4, this class of non-ionics behaves differently from other surfactants with respect to temperature effects. Pressure has little influence on the CMC, even up to high values.

Turning next to the effect of cosolutes on the CMC, this is an important and broad issue that we will come back to repeatedly. A most important matter is the effect of added electrolyte on the CMC of ionics. This is illustrated in Figure 2.7 for the simplest and generally most important case of adding a 1:1 inert electrolyte to a solution of a monovalent surfactant. The following features are noted:

(1) Salt addition gives a dramatic lowering of the CMC, which may amount to an order of magnitude.

(2) The effect is moderate for short-chain surfactants but is much larger for long-chain ones.

(3) As a consequence, at high salt concentrations the variation of CMC with the number of carbons in the alkyl chain is much stronger than without added salt. The rate of change at high salt concentrations becomes similar to that of non-ionics.

(4) The salt effects (as many other aspects of ionic surfactant self-assembly) can be quantitatively reproduced from a simple model of electrostatic interactions, i.e. the Poisson-Boltzmann equation. We will present an account of this in Chapter 8.

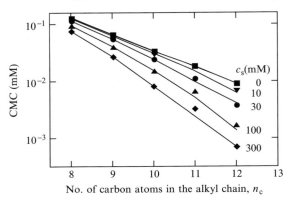

Figure 2.7 Effect of sodium chloride addition on the CMC of different sodium alkyl sulfates. The solid lines are predictions of electrostatic theory (Poisson–Boltzmann equation); c_s gives the salt concentration. Reprinted with permission from G. Gunnarsson, B. Jönsson and H. Wennerström, *J. Phys. Chem.*, **84** (1980) 3114. Copyright (1980) American Chemical Society

Let us further add that:

(5) The effect of added salt depends strongly on the valency of the ions and, in line with what was noted above, it is most sensitive to the valency of added counterions.

(6) For non-ionics, simple salts produce only small variations in the CMC, with both increases and decreases possible.

Other lower molecular weight cosolutes produce changes in the CMC to a very different extent depending on the cosolute polarity. Both increases and decreases in the CMC are possible. Small or moderate increases may result from the addition of highly water-soluble compounds, the reason being that water is the most effective solvent for surfactant self-assembly. More common and more interesting are the decreases in CMC observed for most uncharged cosolutes. The effect will depend on cosolute polarity and on any amphiphilic character it may have and is perhaps best illustrated by the addition of simple alcohols. As seen in Figure 2.8, alcohols lower the CMC but to very different extents. The alcohols are less polar than water and are distributed between the bulk solution and the micelles. The more preference they have for the micelles, than the more they stabilize them. A longer alkyl chain leads to a less favourable location in water and a more favourable location in the micelles. Here it will act basically as any added non-ionic amphiphile, like a non-ionic surfactant, in lowering the CMC.

The Solubility of Surfactants may be Strongly Temperature Dependent

There are many important and intriguing temperature effects in surfactant self-assembly. One, which is of great practical significance, is the dramatic temperature-dependent solubility displayed notably by many ionic surfactants. The solubility may be very low at low temperatures and then increase by orders of magnitude in a relatively narrow temperature range. The phenomenon is generally denoted as the Krafft phenomenon with the temperature for the onset of the strongly increasing solubility being known as the Krafft point or Krafft temperature. The temperature dependence of surfactant solubility in the region of the Krafft point is illustrated in Figure 2.9. The Krafft point may vary dramatically with subtle changes in the surfactant chemical structure, but some general remarks can be made for alkyl chain surfactants:

(a)

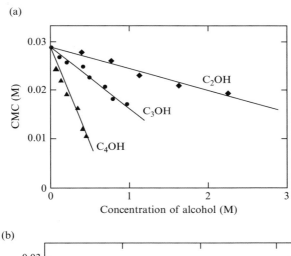

(b)

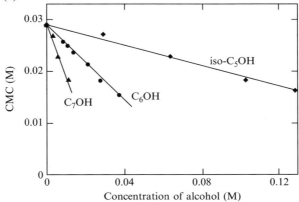

(c)

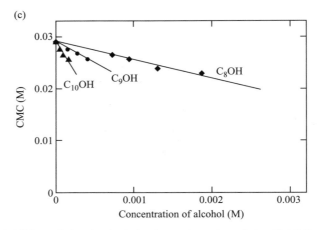

Figure 2.8 Addition of simple alcohols gives a lowering of the CMC (in this case for potassium dodecanoate) which is approximately linear with the cosolute concentration. The slope increases rapidly when the alcohol becomes less polar. From K. Shinoda, T. Nakagawa, B.-I. Tamamushi and T. Isemura, *Colloidal Surfactants, Some Physicochemical Properties*, Academic Press, London 1963

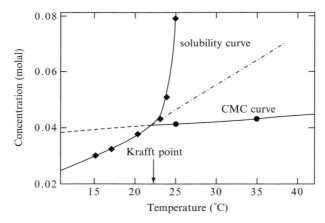

Figure 2.9 Temperature dependence of surfactant solubility in the region of the Krafft point. From K. Shinoda, T. Nakagawa, B.-I. Tamamushi and T. Isemura, *Colloidal Surfactants, Some Physicochemical Properties*, Academic Press, London, 1963

(1) The Krafft point increases strongly as the alkyl chain length increases. The increase is not regular but displays an odd–even effect.

(2) The Krafft point is strongly dependent on the head group and counterion. Salt addition typically raises the Krafft point, while many other cosolutes decrease it. There are no general trends for the counterion dependence.

Thus, for example, for alkali alkanoates the Krafft point increases as the atomic number of the counterion decreases, while the opposite trend is observed for alkali sulfates or sulfonates. For cationics, the Krafft point is typically higher for bromide than for chloride and is still higher for iodide. With divalent counterions, the Krafft point is typically often much higher.

The Krafft phenomenon is best discussed from the interplay between the temperature dependence of the surfactant unimer solubility and the temperature dependence of the CMC. As we have learnt above, the latter temperature dependence is very weak and we can consider here that CMC is, to a good approximation, independent of temperature. On the other hand, we expect the dissolution of the surfactant into the constituent solvated ions to increase markedly with temperature, as observed for simple salts. If this solubility is below the CMC, no micelles can form and the total solubility is limited by the (low) unimer solubility. If, on the other hand, the unimer solubility reaches the CMC, micelles may form. It is a characteristic feature of micellization, as we will see later, that as the micelle concentration increases there is virtually no change in the free unimer activity (or concentration). This, together with a very high micelle solubility, explains why a quite small increase in unimer solubility (resulting here from a temperature increase) leads to a dramatic increase in overall surfactant solubility.

The Krafft point is determined by the energy relationships between the solid crystalline state and the micellar solutions. It appears that the micellar solutions vary only weakly between different cases, like different counterions, while due to packing effects the solid crystalline state may change dramatically. Looking for an understanding of the Krafft phenomenon, we have therefore to look into the packing effects and ionic interactions in the solid state.

If the solubility of a surfactant is very low it will clearly not be operative in various applications. Since a longer-chain surfactant is generally more efficient, there is commonly a delicate compromise in the design of surfactants. Attempts to lower the Krafft point should mainly be directed towards the conditions in the solid state. Besides changing counterion, which is not always possible, we should look into the packing conditions of the hydrophobic chains. The development of surfactants with a lower Krafft point is generally based on making the packing conditions in the solid state less favourable in one of the following ways:

(1) Introduction of a methyl group, or some other chain branching, in the alkyl chain.

(2) Introduction of a double bond in the alkyl chain.

(3) Introduction of a polar segment, usually an oxyethylene group, between the alkyl chain and the ionic group. (Other aspects of insertion of oxyethylene groups are discussed earlier on page 14).

These are also common approaches to manufacturing surfactants compatible with hard water. Control of chain melting is also very important in biological systems, notably biological membranes, and is achieved by controlling chain unsaturation.

Driving Forces of Micelle Formation and Thermodynamic Models

Hydrophobic Interactions

We have returned several times to the important characteristic of surfactants and polar lipids—the amphiphilicity. Water does not interact favourably with the hydrophobic groups and there is a driving force for expelling them from the aqueous environment. This may be achieved by a macroscopic phase separation or by 'hiding' the non-polar groups in some other way. There are numerous other examples of hydrophobic effects and hydrophobic interactions, as illustrated in Figure 2.10. For a hydrocarbon in water there is a strong driving force for transfer to a hydrocarbon phase or some other non-polar environment. When a polar group is attached to the hydrocarbon an opposing force is created, which counteracts phase separation. If the opposing force is weak, phase separation will still result. If it is very strong compared to the hydrophobic effect, on the other hand, the amphiphile will occur as single molecules or as small aggregates, like dimers. It is the common intermediate situation with a balance between hydrophobic and hydrophilic interactions that we are concerned with in surfactant self-assemblies.

The hydrophobic interaction increases with increasing alkyl chain length of an alkane or the hydrophobic group of a surfactant. Indeed, the decrease in solubility of an alkane with the number of carbons very much parallels the change in CMC that was discussed above.

We have noted that micellization (and surfactant self-assembly in general) is somewhere intermediate between phase separation and simple complex formation and this is illustrated in the ways that micellization has been modelled in thermodynamic analyses. Micelle formation is generally discussed in terms of one of the following models.

Phase Separation Model

In this model, micelle formation is considered as akin to a phase separation, with the micelles being the separated (pseudo-)phase and the CMC the saturation concentration of surfactant in the unimeric state. Surfactant addition above the CMC consequently only affects the micelle concentration, but not the unimer concentration. In many physicochemical investigations we observe a number average over the different states that a surfactant molecule can occupy. The phase separation model is particularly simple for interpretation of experimental

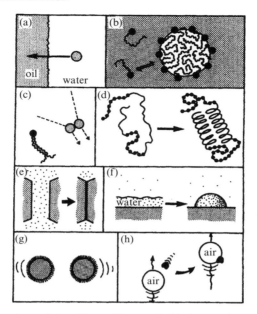

Figure 2.10 Illustrations of the effects of hydrophobic interactions, i.e. the tendency to eliminate contacts between water and non-polar molecules or surfaces:
(a) Water and oil are immiscible, with a strong driving force to expel hydrocarbon molecules from water.
(b) Self-assembly of surfactant molecules.
(c) Other types of association of hydrocarbon chains.
(d) Folding of proteins.
(e) Strong adhesion between non-polar surfaces in water.
(f) Non-wetting of water on hydrophobic surfaces.
(g) Rapid coagulation of hydrophobic particles in water.
(h) Attachment of hydrophobic particles to air bubbles (mechanism of froth flotation).
From J. Israelachvili, *Intermolecular and Surface Forces*, 2nd Edn, Academic Press, London, 1991

observations. Below the CMC we have only unimers and the average of a quantity Q is simply:

$$< Q > = Q_{aq}$$

For a concentration above the CMC we get, since $C_{mic} = C_{tot} - CMC$ and $C_{aq} = CMC$:

$$< Q > = p_{mic}Q_{mic} + p_{aq}Q_{aq} = (1 - CMC/C_{tot})Q_{mic} + (CMC/C_{tot})Q_{aq}$$

For concentrations sufficiently above the CMC, $< Q >$ approaches Q_{mic}.

The phase separation model is simple to apply, illustrative and sufficient for many considerations. As we may expect, it becomes a better approximation the higher the aggregation number.

Mass Action Law Model

Here we assume a single micellar complex in equilibrium with the unimeric surfactant:

$$nA_1 \longleftrightarrow A_n$$

$$(A_n)/(A_1)^n = K$$

In this model, the aggregation number may be obtained from the variation in Q_{obs} around the CMC. The more gradual the change, then the lower is the aggregation number and the more appropriate is the mass action law model with respect to the phase separation model. The fraction of added surfactant that goes to the micelles is plotted in Figure 2.11 as a function of the total surfactant concentration for different aggregation numbers. For very high aggregation numbers, N, there is a close to stepwise change and the variation in the limit is the same as that predicted by the phase separation model.

In reality, micelles are not monodisperse but there is a distribution of aggregation numbers, and micelles are formed in a stepwise process. This is taken into account in the following model.

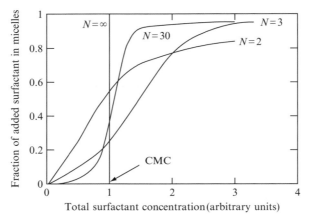

Figure 2.11 The fraction of added surfactant that goes to the micelles as a function of the total surfactant concentration for different aggregation numbers, N. From R. J. Hunter, *Foundations of Colloid Science*, Oxford University Press, Oxford, 1989, p. 576

Multiple Equilibrium Model

$$2A_1 \longleftrightarrow A_2$$
$$A_2 + A_1 \longleftrightarrow A_3$$
$$A_3 + A_1 \longleftrightarrow A_4$$
$$\vdots$$
$$A_{n-1} + A_1 \longleftrightarrow A_n$$
$$(A_n)/(A_{n-1}A_1) = K_n$$

As written, all of these treatments consider non-ionic surfactants. To account for the association of the counterions to the micelles, we can add the relevant (stepwise, if needed) equilibria for the counterions. This is normally not a reasonable approach, however, since the treatment in terms of equilibrium constants assumes short-range interactions and the formation of well-defined complexes. The distribution of counterions is governed by electrostatic interactions which are long range. It is, therefore, not possible to assign definite characteristics of 'micellar-bound' counterions. This does not mean that counterion binding or association may not be a useful concept. However, we should be aware of the limitations and analyse the findings in terms of appropriate models.

The Association Process and Counterion Binding can be Monitored by NMR Spectroscopy

A complete characterization of the self-association of a surfactant would include giving the concentration of all the different species as a function of the total surfactant concentration. We cannot easily measure the concentration of micelles of all different aggregation numbers, but we can obtain some suitable averages. Let us consider the unimer concentration, the concentration of micellized surfactant, the concentration of bound and free counterions, and the hydration of micelles. Using counterion- and surfactant-specific electrodes, we can obtain counterion and surfactant activities that provide information on the distribution between the micellar and unimeric states. Even more complete information is obtained if we measure the self-diffusion coefficients of surfactant molecules, micelles, counterions and water molecules (see p. 147). These can nowadays be obtained in a single fast experiment using NMR methodology. A representative example is given in Figure 2.12 for the case of decylammonium dichloroacetate. The CMC is 26 mM. Self-diffusion coefficients constitute one example where the observed quantity is a weighted average over the different environments (micellar and unimeric) so that:

$$D_{obs} = p_{mic} D_{mic} + p_{free} D_{free}$$

or:

$$D_{obs} = (C_{mic} D_{mic} + C_{free} D_{free})/C_{tot}$$

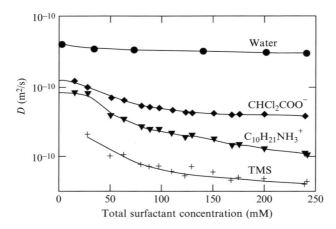

Figure 2.12 Dependence on the surfactant concentration of the self-diffusion coefficients of surfactant ions (decylammonium), counterions (dichloroacetate), water molecules and micelles (approximated by the diffusion of added tetramethylsilane (TMS)). Reprinted with permission from P. Stilbs and B. Lindman, *J. Phys. Chem.*, **85** (1981) 2587. Copyright (1981) American Chemical Society

where D, p and C denote, respectively, the self-diffusion coefficient, the fraction of molecules in a given environment and the concentration; D_{free} is obtained from data below the CMC and D_{mic}, for example, from measurements on probe molecules confined to the micelles. In the study of Figure 2.12, a low (to avoid perturbation of the micellization) concentration of sparingly soluble tetramethylsilane (TMS) is used. Since micelles are large entities, their D-values are very much lower than those of single unimeric molecules.

As can be seen from Figure 2.12, the rates of self-diffusion are very different for the various molecules at concentrations well above the CMC. We can directly read from this that water molecules are least associated with the (slowly moving) micelles while TMS molecules are located in micelles to the largest extent. A quantitative analysis of the surfactant and counterion diffusion data gives the results presented in Figure 2.13. There are some important general features, generally applicable for ionic surfactants, that we should note, as follows:

(1) To a good approximation, all surfactant molecules are in the unimeric state below the CMC.

(2) Above the CMC, the surfactant unimer concentration decreases and may reach values well below the CMC.

(3) The free counterion concentration increases at a lower rate above than below the CMC.

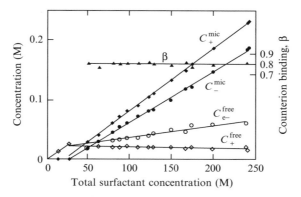

Figure 2.13 Concentrations of micellar and free unimeric surfactant ions (+) and counterions (−) as a function of the total surfactant (decylammonium dichloroacetate) concentration; β is the degree of counterion binding. Reprinted with permission from P. Stilbs and B. Lindman, *J. Phys. Chem.*, **85** (1981) 2587. Copyright (1981) American Chemical Society

(4) If we normalize the micellar bound counterion concentration (C_{mic}^c) to the micellar surfactant concentration (C_{mic}^s), we obtain the degree of counterion binding, β, as follows:

$$\beta = C_{mic}^c / C_{mic}^s$$

It is an important result that β remains constant when we vary the micelle concentration over orders of magnitude. The latter observation is often referred to as a counterion condensation, meaning that counterion association is on a level that gives a certain critical effective charge density. The parameter β then remains approximately constant, even with large variations of the conditions, not only for micelle concentration but also for added salt and temperature. The counterion condensation phenomenon is common to all systems of high charge densities, including also polyelectrolyes and charged surfaces. It is very well understood from electrostatic theory (see Chapter 8).

Hydrophobic Compounds can be Solubilized in Micelles

TMS (see Figure 2.12) is an example of a solubilized compound, a solubilizate, i.e. a compound that becomes soluble due to the presence of micelles. Typically, the solubility stays very low until the CMC is reached, while above the CMC it increases rapidly and almost linearly with the surfactant concentration (Figure 2.14). Solubilization is one of the most important phenomena for surfactant solutions, with a direct bearing on *inter alia* detergency and the formulation of

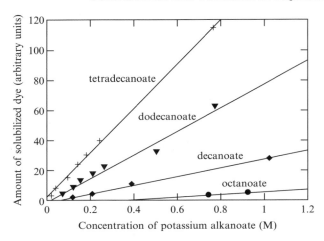

Figure 2.14 Amount of a dye solubilized in solutions of potassium alkanoates. From K. Shinoda, T. Nakagawa, B.-I. Tamamushi and T. Isemura, *Colloidal Surfactants, Some Physicochemical Properties*, Academic Press, London, 1963, p. 26

pharmaceuticals. We will come back to it repeatedly, for example when we discuss phase diagrams, which give quantitative information on the capacity of micelles to incorporate different molecules. In order to study the solubilization equilibria and thus obtain information on the thermodynamics of solubilization, the mentioned self-diffusion technique is the most general and useful approach.

Surface tension measurements on impure surfactants provide an illustration of solubilization. If the surface tension is measured for sodium dodecyl sulfate solutions, one frequently obtains a minimum. On purification, the minimum is eliminated. The explanation is that SDS often contains some dodecanol (due to hydrolysis). The latter is more surface active than SDS and becomes concentrated at the air–water interface. However, as soon as micelles start to form there is another, more favourable, location for dodecanol: in the micelles. Dodecanol is removed from the surface by solubilization and the surface tension increases. We return to these aspects in Chapter 16 when we discuss surfactant absorption.

Micelle Size and Structure may Vary

With a good approximation, micelles can, over a wide concentration range above the CMC, be viewed as microscopic liquid hydrocarbon droplets covered with the polar head groups, which are in interaction with water. It appears that the radius of the micelle core constituted of the alkyl chains is close to the extended length of the alkyl chain, i.e. in the range of 1.5–3.0 nm. Why is this so?

The driving force of micelle formation is the elimination of the contact between the alkyl chains and water. The larger a spherical micelle, then the more efficient this is, since the volume-to-area ratio increases. Decreasing the micelle size always leads to an increased hydrocarbon–water contact. However, if the spherical micelle was made so large that no surfactant molecule could reach from the micelle surface to the centre, one would either have to create a void or some surfactant molecules would lose the contact with the surface, introducing polar groups in the center. Both alternatives are unsatisfactory.

We should note that the fact that the micelle radius equals the length of an extended surfactant molecule does not mean that the surfactant molecules are all extended. Only one molecule needs to be extended (in an all-*trans* state) to fulfil the requirements mentioned, and the majority of the surfactant molecules are in a disordered state with many *gauche* conformations. Spectroscopic studies have been used to characterize the state of the alkyl chains in micelles in detail. This state indeed is very close to that of the corresponding alkane in a neat liquid oil. The liquid-like state is clearly expressed in molecular dynamics. Thus chain isomerism occurs on a time-scale of a few tens of a picosecond, only slightly slower than for liquid alkanes. Due to the constraint the attachment to the micelle surface involves, the motion is slightly anisotropic.

We emphasize that a micelle may for many purposes be considered as a microscopic droplet of oil. This explains the large solubilization capacity towards a broad range of non-polar and weakly polar substances. We note, however, that the locus of solubilization will be very different for different solubilizates. While a saturated hydrocarbon will be rather uniformly distributed over the micelle core, an aromatic compound, being slightly surface active, such as a long-chain alcohol, tends to prefer the surface region and to orient in the same way as the surfactant itself.

At the surface of the micelle we have the associated counterions, which in number amount to 50–80% of the surfactant ions; as noted above, a number quite invariant to the conditions. Simple inorganic counterions are very loosely associated with the micelle. The counterions are very mobile and there is no specific complex formed with a definite counterion–head group distance. Rather, the counterions are associated by long-range electrostatic interactions to the micelle as a whole. They remain hydrated to a great extent; cations especially tend to keep their hydration shell.

Some water of hydration is thus accounted for by the associated counterions and, furthermore, the polar head groups are extensively hydrated. On the other hand, water molecules are effectively excluded from the micelle core. There is, due to packing limitations, some inevitable exposure of the hydrocarbon chains at the micelle surface, but even a short step inwards the probability of finding water molecules becomes extremely low.

The micelle size, as expressed by the radius of a spherical aggregate, may be obtained *inter alia* from various scattering experiments and from micelle self-diffusion. A related and equally important characteristic of a micelle is the micelle aggregation number, i.e. the number of surfactant molecules in one micelle. This is best determined in fluorescence quenching experiments. To take an example, the aggregation number of SDS micelles at 25°C is 60–70. The aggregation number is quite well defined with only a narrow distribution. Micelles of all aggregation numbers exist in equilibrium, but for aggregation numbers deviating markedly from the average the probability is very small. For this reason, the mass action law model offers a good description. In fact, as mentioned above, by analysing experimental observations around the CMC using this model some information on the micelle aggregation numbers is obtained.

Based on the mass action law model and the general relationship between the free energy change and the equilibrium constant,

$$\Delta G^0 = -RT\ln K$$

one can derive the approximate relation for micelle formation:

$$\Delta G^0 = RT\ln \text{ CMC}$$

Here ΔG^0 represents the free energy difference between a unimer in the micelle and some suitably chosen reference state. This is a convenient starting point for thermodynamic considerations.

A Geometrical Consideration of Chain Packing is Useful

We came above to a simple characterization of the micelle core as a hydrocarbon droplet with a radius equalling the length of the extended alkyl chain of the surfactant. We also noted that, since the cross-section area per chain decreases radially towards the centre, only one chain can be fully extended while the others are more or less folded. The aggregation number, N, can be expressed as the ratio between the micellar core volume, V_{mic}, and the volume, v, of one chain:

$$N = V_{\text{mic}}/v = \frac{4}{3}\pi R_{\text{mic}}^3/v$$

We can alternatively express the aggregation number as the ratio between the micellar area, A_{mic}, and the cross-sectional area, a, of one surfactant molecule, as follows:

$$N = A_{\text{mic}}/a = 4\pi R_{\text{mic}}^2/a$$

Putting these aggregation numbers as equal, we obtain:

$$v/(R_{mic}a) = \frac{1}{3}$$

Since R_{mic} cannot exceed the extended length of the surfactant alkyl chain:

$$l_{max} = 1.5 + 1.265n_c$$

we find that:

$$v/(l_{max}a) \le \frac{1}{3} \text{ for a spherical micelle}$$

The ratio $v/(l_{max}a)$, which gives a geometric characterization of a surfactant molecule, will be seen to be very useful when discussing the type of structure formed by a given amphiphile. This is denoted as the critical packing parameter (CPP) or the surfactant number. The concept is discussed further in Chapter 3.

Kinetics of Micelle Formation

We have already noted that micelles are formed in a stepwise process, so the elementary step is the equilibrium between a unimer and a micellar aggregate:

$$A_n + A_1 \underset{k_-}{\overset{k_+}{\longleftrightarrow}} A_{n+1}$$

The 'on' rate constant, k_+, is diffusion controlled and depends little on surfactant and micelle size (cf. Table 2.5). The 'off' rate constant, k_-, on the other hand, is strongly dependent on alkyl chain length, micelle size, etc. Because of the co-operativity in micelle formation there is a very deep minimum in the size distribution curve. This leads to a two-step approach to equilibrium after a perturbation. In a fast step, a quasi-equilibrium is reached under the constraint of a constant total number of micelles. The redistribution of unimers between abundant micelles is a fast process. To reach a true equilibrium, the number of micelles must change. Because of the stepwise process, this also involves the very rare intermediate micelles. Therefore, this process is slow.

From fast kinetic measurements, two relaxation times are determined that characterize molecular processes in micellar solutions: τ_1 measures the rate at which surfactant molecules exchange between micelles, while τ_2 measures the rate at which micelles form and disintegrate. Examples of relaxation times are given in Table 2.6. As can be noted, τ_1 is of the order of 10 μs for SDS, while τ_2 is longer than a millisecond. Both relaxation processes become much slower as the surfactant alkyl chain length increases.

Table 2.5 Kinetic parameters of association and dissociation of alkyl sulfates from their micelles

Surfactant	N	CMC (M)	k^- (s^{-1})	k^+ ($M^{-1}s^{-1}$)
NaC_6SO_4	17	0.42	1.32×10^9	3.2×10^9
NaC_7SO_4	22	0.22	7.3×10^8	3.3×10^9
NaC_8SO_4	27	0.13	1.0×10^8	7.7×10^9
NaC_9SO_4	33	6×10^{-2}	1.4×10^8	2.3×10^9
$NaC_{11}SO_4$	52	1.6×10^{-2}	4×10^7	2.6×10^9
$NaC_{12}SO_4$	64	8.2×10^{-3}	1×10^7	1.2×10^9
$NaC_{14}SO_4$	80	2.05×10^{-3}	9.6×10^5	4.7×10^8

Table 2.6 Relaxation times τ_1 and τ_2 for some sodium alkyl sulfates

Surfactant	Temperature (°C)	Concentration (M)	$\tau_1(\mu s)$	τ_2 (ms)
$NaC_{16}SO_4$	30	1×10^{-3}	760	350
$NaC_{14}SO_4$	25	2.1×10^{-3}	320	41
	30	2.1×10^{-3}	245	19
	35	2.1×10^{-3}	155	7
	25	3×10^{-3}	125	34
$NaC_{12}SO_4$	20	1×10^{-2}	15	1.8
	20	5×10^{-2}	—	50

Since the slow relaxation process is critically dependent on the micelle size distribution, kinetic measurements can be used to determine the standard deviation of the distribution. This distribution is relatively narrow for surfactants with longer alkyl chains.

Surfactants may Form Aggregates in Solvents other than Water

Polar Solvents

In strongly polar solvents, such as formamide and ethylene glycol, micelles are formed with qualitatively the same features as in water. As exemplified by the surface tension studies displayed in Figure 2.15, the CMC is much higher in formamide than in water, in this case 100 mM compared to 1 mM in water. It is a general feature, also exemplified by smaller micelle radii and aggregation numbers, that self-assembly is much less co-operative in alternative polar solvents. As a consequence the degree of counterion binding is also lower.

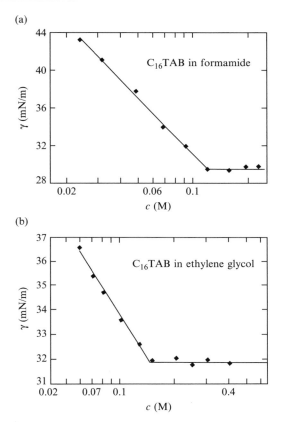

Figure 2.15 Surface tension, γ, of hexadecyltrimethylammonium bromide (C_{16} TABr) in (a) formamide and (b) ethylene glycol as a function of the logarithm of the surfactant concentration (M) at 60°C. From M. Sjöberg, *Surfactant Aggregation in Nonaqueous Polar Solvents*, Ph.D. Thesis, Department of Physical Chemistry, The Royal Institute of Technology, Stockholm, Sweden, 1992

Non-Polar Solvents

For simple amphiphilic compounds, the association is of low co-operativity in non-polar solvents and leads typically only to smaller and polydisperse aggregates. An illustration is given in Figure 2.16. However, introduction of even quite small amounts of water can induce a co-operative self-assembly, leading to reverse micelles. We will come back to this point later.

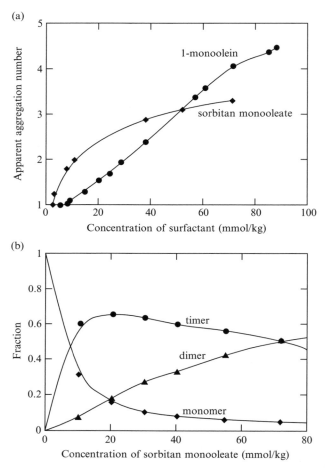

Figure 2.16   The aggregation numbers of amphiphiles in non-polar solvents are generally low. The upper diagram (a) shows the average aggregaton numbers of 1-monoolein and sorbitan monooleate in cyclohexane. The lower diagram (b) shows the fractions of surfactant as monomer, dimer and trimer of sorbitan monooleate in benzene as a function of surfactant concentration. From K. Kon-no and A. Kitahara, in *Nonionic Surfactants. Physical Chemistry*, M. J. Schick, (Ed.), Marcel Dekker, New York, 1987, pp. 195–196

General Comments on Amphiphile Self-Assembly

In water (and other polar solvents), aggregation results from the insolubility of the non-polar parts in water. The packing of the hydrocarbon chains results from the drive to minimize contact with water. Aggregation is opposed by the hydrophilic interaction, giving a repulsion between the polar head group on the micelle surface. The head groups will arrange themselves to minimize the unfavourable repulsions.

The self-assembly of an amphiphile depends on the strength of the opposing force. It must be strong enough to compete with one alternative, which is macroscopic phase separation, but must also be limited in magnitude since otherwise the unimeric state will be the most stable one. Examples of head groups giving amphiphiles that are too weak are hydroxyl, aldehyde, ketone and amine. For a long-chain alcohol, macroscopic phase separation results rather than micelle formation.

For ionic surfactants, the counterion dissociation plays a major role. Because of the counterions, macroscopic phase separation becomes entropically very unfavourable and there is a strong tendency to form small micelles.

Cosolutes may affect amphiphile self-assembly in many different ways. They can, for example, stabilize the micelles by reducing the polar interactions. For ionic surfactants, we can neutralize the charges by adding an oppositely charged surfactant, we can 'screen' the repulsions between head groups, or rather even out the uneven counterion distribution by adding electrolyte, or we can dilute the charges by introducing a non-ionic amphiphile, like a long-chain alcohol. In all cases, we observe a marked reduction of the CMC and, as discussed in Chapter 3, an increase in micelle size.

Self-assembly and micelle formation have, however, a broader significance. Mixed polymer-surfactant solutions have many applications and it has become more and more evident that an important role of the polymer chains in many systems is to promote micelle formation. A macromolecular cosolute will be much more effective in reducing the CMC than a low molecular weight one. This is discussed in some detail in Chapter 13.

Surfactant adsorption at hydrophilic solid surfaces is often pictured as leading to surfactant monolayers and bilayers (Figure 2.17), but it has become

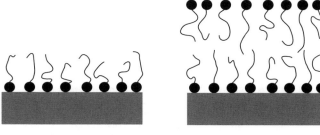

Monolayer adsorption   Bilayer adsorption

Figure 2.17 Conventional pictures of surfactant adsorption at hydrophilic solid surfaces involve monolayer and bilayer structures. Monolayer adsorption is not applicable while bilayers occur as a limiting case. Typically, there are indeed discrete aggregates, as depicted below in Figure 2.18

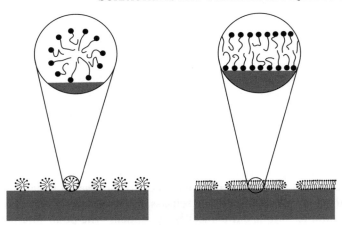

Figure 2.18 In the absence of strong specific interactions between a hydrophilic surface and the surfactant head groups, surfactant molecules will self-assemble into discrete micelles at the surface

increasingly clear that also this process is best regarded as a process of surfactant self-assembly. In such cases, the picture of continuous surfactant layers must be replaced by one with discrete surfactant aggregates, i.e. surface micelles, as illustrated in Figure 2.18.

Bibliography

Evans, D. F. and H. Wennerström, *The Colloidal Domain. Where Physics, Chemistry, Biology and Technology Meet*, Wiley-VCH, New York, 2nd Edn, 1999, Chs 1 and 4.

Friberg, S. E. and B. Lindman (Eds), *Organized Solutions*, Surfactant Science Series, 44, Marcel Dekker, New York, 1992.

Israelachvili, J., *Intermolecular and Surface Forces*, Academic Press, London, 1991.

Lindman, B. and H. Wennerström, *Micelles: Topics in Current Chemistry*, Vol. 87, Springer-Verlag, Berlin, 1980.

Lindman, B., O. Söderman and H. Wennerström, NMR of surfactant systems, in *Surfactant Solutions. New Methods of Investigation*, R. Zana (Ed.), Marcel Dekker, New York, 1987, Ch. 6.

Lindman, B., U. Olsson and O. Söderman, Surfactant solutions: aggregation phenomenon and microheterogeneity, in *Dynamics of Solutions and Fluid Mixtures by NMR*, J-J. Delpuech (Ed.), Wiley, New York, 1995, Ch. 6.

Shah, D. O. (Ed.), Micelles, Microemulsions and Monolayers, Marcel Dekker, New York, New York, 1998.

Shinoda, K., *Principles of Solution and Solubility*, Marcel Dekker, New York, 1978.

Tanford, C., *The Hydrophobic Effect. Formation of Micelles and Biological Membranes*, Wiley, New York, 1980.

3 PHASE BEHAVIOUR OF CONCENTRATED SURFACTANT SYSTEMS

Micelle Type and Size Vary with Concentration

The spherical micelle discussed in Chapter 2 is but one possibility of an amphiphile self-assembly. The spherical micelle does not form at all for many amphiphiles and for others it occurs only in a limited range of concentration and temperature. In general, we can distinguish between three types of behaviour of a surfactant or a polar lipid as the concentration is varied:

(1) The surfactant has a high solubility in water and the physicochemical properties (viscosity, scattering, spectroscopy, etc.) vary in a smooth way from the CMC region up to saturation. This suggests that there are no major changes in micelle structure, but the micelles remain small and do not depart much from a spherical shape.

(2) The surfactant has a high solubility in water but as the concentration is increased there are quite dramatic changes in certain properties. This indicates that there are marked changes in the self-assembly structure.

(3) The surfactant has a low solubility and there is a phase separation at low concentrations.

The three cases are characterized by different ranges of existence of the isotropic solution phase. In either case the new phase formed above saturation may be:

- a liquid crystalline phase

- a solid phase of (hydrated) surfactant, or

- a second, more concentrated, surfactant solution

Different phase structures give very different physicochemical properties and, therefore, in any practical use of surfactants it is mandatory to have control over phase structure. The regions of existence of different phases and the equilibria between different phases are described by phase diagrams. These are significant not only as a basis of applications but also for our general understanding of surfactant self-assembly.

For a relatively short-chain surfactant, like C_8 or C_{10}, one usually observes a slow and regular variation of relevant properties, and no phase separation, up to high concentrations, say 10–40 wt% (Figure 3.1). The viscosity, which is an important property for uses of surfactants, varies smoothly and approximately as predicted for a dispersion of spherical particles up to high concentrations. By scattering experiments and by NMR spectroscopy, direct evidence is obtained for closely spherical aggregates up to the approach of phase separation. For some surfactants it may be that only at micelle volume fractions of the order of 0.3 are appreciable deformations of the micelles seen.

A frequently encountered behaviour for longer-chain surfactants, say C_{14} or above, is that at low or intermediate concentrations the viscosity starts to increase rapidly with concentration. This is exemplified in a plot of the (zero shear) viscosity versus concentration in Figure 3.2. Here the micelles grow with increasing concentration, at first to short prolates or cylinders and then to long cylindrical or thread-like micelles (Figure 3.3).

A third, less common, behaviour is the growth to very long thread-like micelles, already at very low concentrations, sometimes just above the CMC. Growth of micelles is generally a one-dimensional process leading to aggregates with a circular cross-section. The hydrophobic core has a radius which, as for

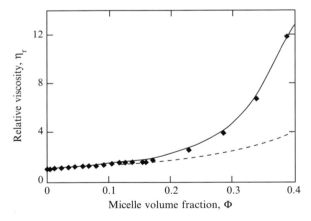

Figure 3.1 Relative viscosity as a function of micelle volume fraction for solutions of spherical micelles. Dashed and solid curves give theoretical predictions for two models of spherical particles, in the latter case taking into account particle–particle interactions. The system exemplified is that of $C_{12}E_5$ micelles with an equal weight of solubilized decane. Reprinted with permission from M. S. Leaver and U. Olsson, *Langmuir*, **10** (1994) 3449. Copyright (1994) American Chemical Society

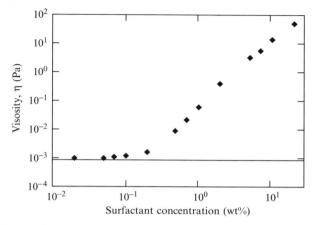

Figure 3.2 For surfactants forming large micelles, the viscosity starts to increase rapidly with surfactant concentration. The increase in zero shear viscosity is plotted as a function of surfactant concentration for $C_{16}E_6$. By courtesy of U. Olsson, M. Malmsten and F. Tiberg

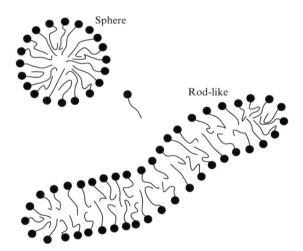

Figure 3.3 At higher concentrations the micelles often grow from spherical aggregates to long rod- or thread-like micelles

the spherical micelles and for the same reasons, equals the length of the extended alkyl chain of the surfactant. The linear length of the rod-like micelles can vary over wide ranges, from well below 10 nm to many hundred nm.

Micellar Growth is Different for Different Systems

Micellar growth is a very common phenomenon and for ionic surfactants the following factors (some of which are exemplified in Figures 3.4 and 3.5) influence the growth:

(1) The tendency to grow increases strongly with the alkyl chain length and there is no growth for shorter chains.

(2) Micellar growth is strongly dependent on temperature and is strongly promoted by a decrease in temperature. For example, for hexadecyltrimethylammonium bromide, there is micellar growth at 30°C but not at 50°C.

(3) While the CMC is only slightly dependent on the counterion within a given class, micellar growth displays a strong variation. The dependence of growth on counterions, however, is very different for different surfactant head groups. For example, for hexadecyltrimethylammonium bromide, there is major micelle growth while there is no growth with chloride as the counterion. For alkali dodecyl sulfates, growth is insignificant with Li^+ as the counterion, moderate with Na^+, but quite dramatic with K^+ or Cs^+. With carboxylate as the head group, the opposite variation along the series

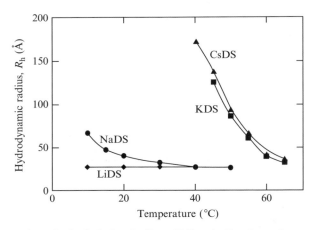

Figure 3.4 The size of alkali dodecylsulfate (DS) micelles, here characterized by the hydrodynamic radius, decreases with increasing temperature and is very sensitive to the choice of counterion. Concentrations are: LiDS, 20 g/l + 1 M LiCl; NaDS, 20 g/l + 0.45 M NaCl; KDS, 5 g/l + 0.45 M KCl; CsDS 5 g/l + 0.45 M CsCl. As seen, KDS and CsDS give much more pronounced growth in spite of lower concentrations of surfactant. Reproduced from K. L. Mittal and E. J. Fendler (Eds), *Solution Behaviour of Surfactants*, Vol. 1, 1982, p. 373, 'Thermodynamics of the sphere-to-rod transition in alkyl sulfate micelles', P. J. Missel, N. A. Mazer, M. C. Carey and G. B. Benedek, Figure 5, with kind permission of Kluwer Academic Publishers

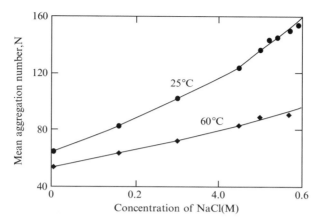

Figure 3.5 The aggregation number of sodium dodecyl sulfate (SDS) micelles increases with decreasing temperature and increasing salt concentration. Reprinted with permission from N. J. Turro and A. Yekta, *J. Am. Chem. Soc.*, **100** (1978) 5951. Copyright (1978) American Chemical Society

of alkali ions is observed. Organic counterions may induce dramatic growth at low concentrations, as exemplified by salicylate in the presence of long-chain cationic surfactants.

(4) In the case of micellar growth, micellar size increases strongly with surfactant concentration.

(5) Large micelles are, contrary to the small spherical micelles, very polydisperse.

(6) Micelle size is very sensitive to cosolutes. Addition of salt promotes micelle growth. Solubilized molecules can have very different effects depending on the surfactant system. However, in general non-polar solubilizates, like alkanes, which are located in the micellar core, tend to inhibit micellar growth, while alcohols or aromatic compounds, which are located in the outer part of the micelles, tend to strongly induce growth. For example, for hexadecyltrimethylammonium bromide there is no growth on adding cyclohexane while hexanol and benzene give dramatic growth.

For other classes of surfactants there are different characteristics of micellar growth. Non-ionic surfactants of the polyoxyethylene type give growth with increasing concentration which is much more marked the shorter the polar group. With four to six oxyethylene units there is dramatic growth, while with eight or more oxyethylenes there is negligible growth under any conditions. These surfactants show a micellar growth that is much more pronounced at a higher temperature, i.e. opposite to other classes of surfactants. This and other unusual features of non-ionics are treated in Chapter 4.

Solutions of large micelles show many parallels with solutions of linear polymers and the micelles have been denoted as 'living polymers'. Because of the polymer-like behaviour, concepts and theories developed for polymer solutions have been successfully applied. Differences from polymers, which complicate the comparison, include the strong dependence of the 'degree of polymerization' on the conditions (surfactant concentration, temperature, etc.). Furthermore, under certain conditions, such as very high concentrations, growth may lead to branched structures.

The large micelles can differ strongly in flexibility and may be referred to as rigid rods, or semi-flexible or highly flexible micelles. As for polymers, they can be characterized by a persistence length. The flexibility of ionic micelles is strongly dependent on electrolyte addition. Thus, salt addition can induce a change from rigid rods to very flexible micelles.

In dilute solutions, where the micelles do not overlap, they behave as independent entities (Figure 3.6). After the overlap volume fraction, $\phi*$, in the so-called semi-dilute concentration regime, the micelles are entangled and there is a transient network characterized by a correlation length; the correlation length is independent of micelle size and polydispersity. The overlap concentration of the system presented in Figure 3.2 is *ca.* 0.1%. The viscosity of solutions of long linear micelles can be analysed in terms of the motion of the micelles, e.g. using the reptation model of polymer systems. In this, the micelles creep like a snake through tubes in a porous structure given by the other micelles. The zero shear

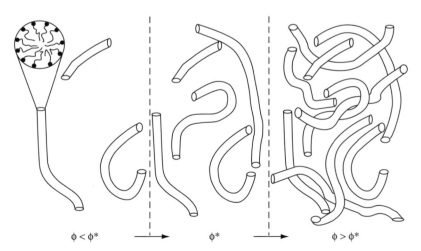

$$\phi < \phi* \quad\longrightarrow\quad \phi* \quad\longrightarrow\quad \phi > \phi*$$

Figure 3.6 There is a close analogy between solutions of long micelles and polymer solutions, including transient networks. The figure illustrates the transition from dilute to semi-dilute solutions and the overlap concentration; $\phi*$ is the overlap volume fraction

viscosity, η, depends on micelle size (aggregation number, N) and volume fraction, ϕ, according to:

$$\eta = \text{constant } N^3 \phi^{3.75}$$

As also observed experimentally, the viscosity increases very strongly with both micelle size and surfactant concentration.

The linear growth of micelles is the strongly dominating type of growth. Disc-like or plate-like structures may also form, but these micelles are quite small and exist only in a narrow range of conditions (concentration, etc.). The linear growth can, as mentioned, lead to branched structures, which at high enough concentrations may lead to the transition into a surfactant micellar structure, which is completely connected so that the concept of distinct micelles loses its meaning (Figure 3.7). In such a case, we use the term bicontinuous structure, since the solutions are continuous not only in the solvent but also in the surfactant.

As we will see below, bicontinuous structures are very significant in many contexts of amphiphile self-assembly. Another type of bicontinuous structure in simple surfactant–water solutions is the 'sponge phase', formed also in quite

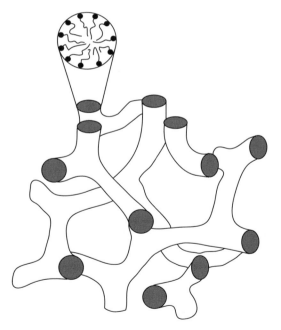

Figure 3.7 Branched micelles

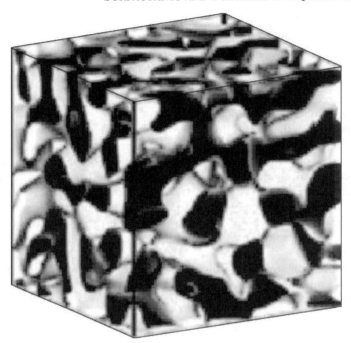

Figure 3.8 The sponge phase. For many surfactants, there is an isotropic solution phase where the surfactant forms a connected three-dimensional network. Since both water and the hydrophobic regions are connected over macroscopic distances, such structures are termed bicontinuous. Reprinted with permission from P. Pieruschka and S. Marcelja, *Langmuir*, **2** (1994) 345. Copyright (1994) American Chemical Society

dilute surfactant solutions (Figure 3.8). This structure forms for all classes of surfactants but in particular for non-ionics; we will consider it in more detail in our discussion of these surfactants (Chapter 4). We will also mention that the structure of the sponge phase is related to that of many microemulsions and will, therefore, be further discussed in this context (Chapter 6).

Surfactant Phases are Built Up by Discrete or Infinite Self-Assemblies

Surfactant micelles and bilayers are the building blocks of most self-assembly structures and it is natural to dwell on the distinction between these and give some further examples. As schematized in Figure 3.9, we can divide phase structures into two groups, i.e. those that are built up of limited or discrete self-assemblies, which may be characterized roughly as spherical, prolate, oblate or cylindrical (more or less flexible), or infinite or unlimited self-assemblies. In the latter case, the surfactant aggregate is connected over macroscopic

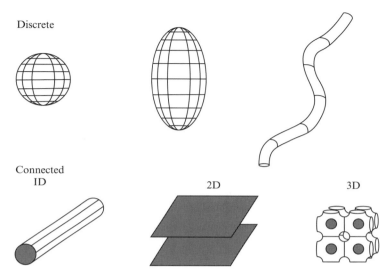

Figure 3.9 Amphiphile self-assembly structures can be divided into discrete micellar-type and connected forms. There may be connectivity in one, two or three dimensions

distances in one, two or three dimensions. The hexagonal phases are examples of one-dimensional continuity, the lamellar phase of two-dimensional continuity, while three-dimensional continuity is found for the bicontinuous cubic phases, for the 'sponge' phase and, as we shall see in Chapter 6, for many microemulsions.

Phases built up of discrete aggregates include the normal and reversed micellar solutions, micellar-type microemulsions and certain (micellar-type) normal and reversed cubic phases. However, discrete self-assemblies are also important in other contexts. Adsorbed surfactant layers at solid or liquid surfaces may involve micellar-type structures and the same applies to mixed polymer-surfactant solutions. We will return to these matters later.

Bilayer structures are common both in nature and in applications and have a broad significance. Cooling a lamellar liquid crystalline phase typically leads to a so-called gel phase (with characteristics different to polymer-based gels) or a crystal with a bilayer structure. Depending on the packing conditions, different arrangements can be found, as illustrated in Figure 3.10. In the so-called gel phases, water between the bilayers is in a liquid-like state. There may also be some mobility of the surfactant molecules.

Other examples of bilayer structures already mentioned are the 'sponge' phase and bicontinuous cubic phase. The sponge phase has been most studied for non-ionic surfactants and is related to common microemulsions. Therefore, we will examine the sponge phase further in Chapters 4 and 6. Bilayers may also easily close in on themselves to form discrete entities, including unilamellar

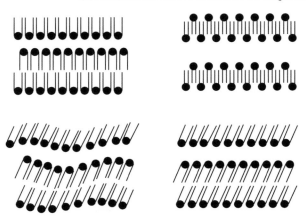

Figure 3.10   Bilayer structures in gel phases. Reproduced by permission of John Wiley & Sons, Inc., from D. F. Evans and H. Wennerström, *The Colloidal Domain. Where Physics, Chemistry, Biology, and Technology Meet*, VCH, New York, 1994, p. 247

vesicles and multilamellar liposomes. Vesicles are of interest because of the division into inner and outer aqueous domains separated by the bilayers. Vesicles and liposomes are normally not thermodynamically stable, although there are exceptions, and tend to phase-separate into the lamellar phase and a dilute aqueous solution. Lipid bilayers are important constituents of living organisms and form membranes that act as barriers between different compartments. Certain surfactants and lipids may form reversed vesicles, i.e. vesicles with inner and outer oleic domains separated by a (reversed) amphiphile bilayer; the bilayer may or may not contain some water.

Micellar Solutions can Reach Saturation

Having considered the different micellar solutions, two questions arise naturally: what limits the range of existence of micelles at higher surfactant concentrations and what happens to the self-assembly structure at saturation? We will now consider the second question and return to the first one later.

   Micellar growth to rods can be considered to arise from two mechanisms. In one, there is an internal driving force to form large aggregates with another geometry and then micelles may grow also at low concentrations. In another, micellar growth is induced by intermicelle repulsions to allow a better packing of the micelles. This will occur at quite high concentrations, when the micelles come in direct close contact. Often we see a combination of the two factors.

   Another way to solve the problem of too strong a crowding of the micelles is a structural transition into an ordered phase, which may be a solid but is more frequently a liquid crystalline phase. We will now discuss:

- the structures formed, and

- phase diagrams of binary surfactant–water mixtures

There is no generally accepted notation for the different phases occurring in surfactant systems. In Table 3.1 we give the most common symbols of liquid crystalline phases.

Structures of Liquid Crystalline Phases

Micellar Cubic Phase

As can be seen in Figure 3.11, this phase is built up of regular packing of small micelles, which have similar properties to small micelles in the solution phase. The micelles are short prolates (axial ratio, 1–2) rather than spheres, since this allows a better packing. This phase is highly viscous.

Hexagonal Phase

This phase is built up of (infinitely) long cylindrical micelles arranged in a hexagonal pattern, with each micelle being surrounded by six others (Figures 3.12 and 3.13). The radius of the circular cross-section (which may be somewhat deformed) is again close to that of the surfactant molecule length.

Lamellar Phase

This phase is built up of bilayers of surfactant molecules alternating with water layers (Figures 3.12 and 3.14). The thickness of the bilayers is somewhat

Table 3.1 Different systems of notation for the most common liquid crystalline and other phases

Phase structure	Our notation	Other notations
Lamellar	lam	L_α, D, G, neat
Hexagonal	hex	H_1, E, M_1, middle
Reversed hexagonal	rev hex	H_2, F, M_2
Cubic (normal micellar)	cub (cub_m)	I_1, S_{1c}
Cubic (reversed micellar)	cub (cub_m)	I_2
Cubic (normal bicontinuous)	cub (cub_b)	I_1, V_1
Cubic (reversed bicontinuous)	cub (cub_b)	I_2, V_2
Gel	gel $\cdot$	L_β
Micellar	mic	L_1, S
Reversed micellar	rev mic	L_2, S
Sponge phase (reversed)	spo	L_3 (normal), L_4
Microemulsion	μem	L, S, μE
Vesicular	ves	—

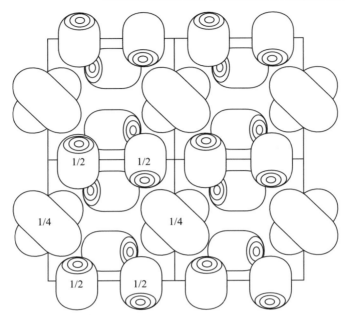

Figure 3.11 A cubic phase built up of discrete micelles. Reproduced from K. Fontell, C. Fox and E. Hansson, *Mol. Cryst. Liquid Cryst.*, **1** (1985) 9, copyright Taylor & Francis, Ltd., http://www.tandf.co.uk/journals

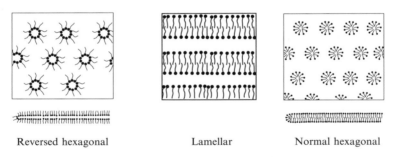

Reversed hexagonal   Lamellar   Normal hexagonal

Figure 3.12 Schematic pictures of anisotropic liquid crystalline phases

less than twice the surfactant molecule length. The thickness of the water layer can vary (depending of surfactant) over wide ranges. The surfactant bilayer can range from being stiff and planar to being very flexible and undulating.

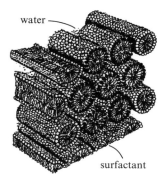

Figure 3.13   The hexagonal phase. Reproduced from K. Fontell, *Mol. Cryst. Liquid Cryst.*, **63** (1981) 59, copyright Taylor & Francis, Ltd., http://www.tandf.co.uk/journals

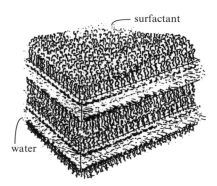

Figure 3.14   The lamellar phase. Reproduced from K. Fontell, *Mol. Cryst. Liquid Cryst.*, **63** (1981) 59, copyright Taylor & Francis, Ltd., http://www.tandf.co.uk/journals

Bicontinuous Cubic Phases

There can be a number of different structures, where the surfactant molecules form aggregates that penetrate space, forming a porous connected structure in three dimensions. We can consider structures as either formed by connecting rod-like micelles, similar to branched micelles, or bilayer structures as visualized in Figure 2.1.

Reversed Structures

Except for the lamellar phase, which is symmetrical around the middle of the bilayer, the different structures have a reversed counterpart in which the polar and non-polar parts have changed roles (Figure 3.15). For example, a reverse

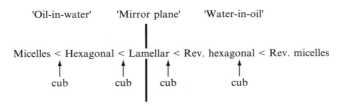

Figure 3.15 The Fontell scheme for the dependence of surfactant liquid crystal structure on composition, illustrating the symmetry of curvature of the surfactant aggregates around the lamellar phase structure and the four principal locations of cubic phases. From K. Fontell, *Colloid Polym. Sci.*, **268** (1990) 264

hexagonal phase (Figure 3.12) is built up of hexagonally packed water cylinders surrounded by the polar heads of surfactant molecules and a continuum of the hydrophobic parts. Reversed (micellar-type) cubic phases and reversed micelles analogously consist of globular water cores surrounded by the surfactant molecules. The radii of these water droplets are typically in the range 20–100 Å.

The phases described are commonly occurring ones, but in addition there are other phases of less importance. Some of these involve discrete aggregates of different shapes and different types of mutual organization and some are similar to the hexagonal phase but with alternative arrangements of the cylinders or non-circular cross-sections of the aggregates.

How to Determine Phase Diagrams

In a phase diagram we can read out how many phases are formed, which are the phases and what are the compositions of the phases. The determination of a complete phase diagram involves considerable work and skill and strongly increases in difficulty as the number of components increases. The distinction between solution and liquid crystalline phases is best made from studies of diffraction properties, either light, neutron or X-ray. Liquid crystalline phases have a repetitive arrangement of aggregates and the observation of a diffraction pattern can firstly give evidence for a long-range order and, secondly, distinguish between alternative structures. However, due to the low intensities often encountered this is not always unambiguous.

The scattering of normal and polarized light is very useful for identification of different structures. Isotropic phases, i.e. solutions and cubic liquid crystals, are clear and transparent while the anisotropic liquid crystalline phases scatter light and appear more or less cloudy. Using polarized light and viewing samples through crossed polarizers give a black picture for isotropic phases, while anisotropic ones give bright images. The patterns in a polarization microscope are distinctly different for different anisotropic phases and can, therefore, be used to identify phases, e.g. to distinguish between hexagonal and lamellar phases. Another very useful technique is NMR spectroscopy, especially observations of quadrupole splittings in deuterium NMR. Different patterns are observed for different phases and allow a direct identification (Figure 3.16).

The viscosity varies strongly between different phases, but since it is also strongly dependent on concentration and surfactant it does not allow an unambiguous determination of the phase present in a sample. Characteristically, however, cubic phases are very viscous and often quite stiff, so the clear gel-like appearance of a phase can normally be used to identify a cubic phase. Hexagonal phases, although less viscous than cubic phases, are also of high viscosity and much more viscous than the lamellar phases.

The distinction between normal and reversed phases can typically be made from sample compositions, since normal phases occur on the 'water-rich' side

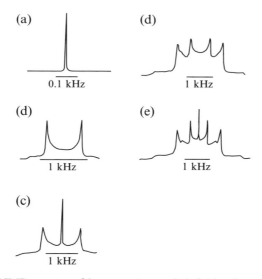

Figure 3.16 ^{2}H NMR spectra of heavy water are helpful in phase diagram determinations. (a) For isotropic phases (micellar solution, sponge phase, cubic phase or microemulsion) a narrow singlet is observed. (b) If there is a single anisotropic liquid crystalline phase, a doublet is obtained. The magnitude of the 'splitting' depends on the degree of anisotropy and is, for example, twice as large for a lamellar phase as for a hexagonal one under comparable conditions. In a two-phase region, with coexistence of two phases, we observe specta of the two phases superimposed; (c) for one isotropic and one anisotropic phase, we observe one singlet and one doublet, and (d) for two anisotropic phases (lamellar and hexagonal) we observe two doublets. (e) In a three-phase region with two anisotropic and one isotropic phase coexisting, we observe two doublets and one singlet. One narrow and one broad singlet may be observed for the case of one anisotropic and one isotropic phase before equilibrium has been reached and the anisotropic phase is dispersed as small microcrystallites. From A. Khan, K. Fontell, G. Lindblom and B. Lindman, *J. Phys. Chem.*, **86** (1982) 4266

of the lamellar phase and the reversed one on the 'water-poor' side. Conductivity is most helpful since the conductivity of a solution of water in closed domains is several orders of magnitude lower than if there are aqueous domains connected over macroscopic distances.

In the determination of phase diagrams, we can distinguish between a first step involving the establishment of which phases are forming and their sequence of appearance as a function of concentration, and a second involving the quantitative determination of the regions of existence of different phases. The first step can, for example, involve macroscopic observations of the effect of the penetration of water into the surfactant crystals. The complete phase diagram determination involves making up a large number of samples with different compositions and firstly establishing whether there is a single homogeneous phase or if there is more than one phase. In the latter case, the next step is to separate the phases, which often is possible by centrifugation; however, the separation can be very difficult or impossible due to high viscosities of phases or a small difference in density. (The NMR method mentioned above circumvents this problem since phase identification can be made without macroscopic separation of phases.) Finally, the different phases are identified and the compositions of phases and of multiphase samples are determined by some form of chemical analysis.

Binary and Ternary Phase Diagrams are Useful Tools: Two Components

In Figures 3.17 and 3.18, two phase diagrams for binary surfactant–water systems are presented, both referring to single C_{12}-chain ionic surfactants. Binary phase diagrams are presented with the temperature along the ordinate and the composition along the abscissa. The composition may be expressed in different ways: mole ratio, mole fraction, weight fraction or weight per cent. Traditionally, compositions on weight rather than mole basis have been used. This has the disadvantage that molar compositions cannot be directly read out, but since stoichiometric complexes are rare this is not a major disadvantage. Weight-based scales give a more suitable division of diagram area into stability regions of different phases and is the only possibility if the surfactant composition is not exactly known.

The phase diagram of dodecyltrimethylammonium chloride (Figure 3.17) represents a case with a relatively low Krafft point and then solid phases have very little role to play. We can see that the isotropic micellar phase exists at room temperature up to high concentrations (*ca.* 40%). The next phase is a cubic phase built up of discrete globular micelles. Between the two phases there is a two-phase region where the two phases coexist. Due to the impossibility of packing globular micelles at high volume fractions, the micelles deform and become elongated to form finally a hexagonal phase. Thereafter, there is a

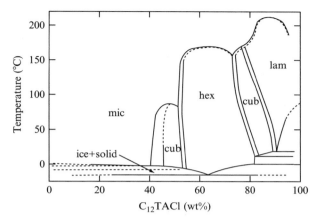

Figure 3.17 Binary phase diagram of dodecyltrimethylammonium chloride–water. Isotropic solutions (mic) exist at lower surfactant concentrations and higher temperatures. In addition, there are liquid crystalline phases (cubic, hexagonal and lamellar) and, at lower temperatures, crystalline phases. From R. R. Balmbra, J. S. Clunie and J. F. Goodman, *Nature*, **222** (1969) 1159

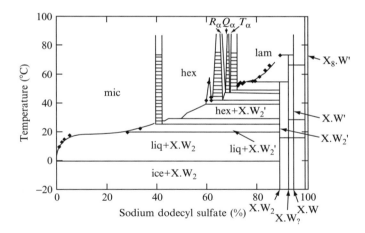

Figure 3.18 Binary phase diagram of sodium dodecyl sulfate (X)–water. The principle liquid crystalline phases are the hexagonal and lamellar ones. The other phase notations refer to less common intermediate liquid crystalline phases and to crystalline hydrates. From R. G. Laughlin, *The Aqueous Phase Behaviour of Surfactants*, Academic Press, London, 1994, p. 111

transformation to a second cubic phase, now of the bicontinuous type. Then we find the lamellar phase and finally solid hydrated surfactant. The sequence of phases found in this case is typical of rather hydrophilic surfactants.

At higher temperatures, the stability relations between the different phases change. The first cubic phase disappears and at very high temperatures the lamellar phase is the only liquid crystalline phase, forming also for the neat surfactant. At the very highest temperatures, there are only isotropic solutions, of course with major changes in aggregate structures as the composition changes.

The phase diagram for SDS (Figure 3.18) shows many similarities. Because of the higher Krafft point, different solid phases play a much more important role at ambient temperature. Except for the hexagonal and lamellar phases we find a number of 'intermediate' liquid crystalline phases of narrow regions of existence.

For double-chain surfactants, the phase diagram looks typically quite different, as exemplified in Figure 3.19 for the case of sodium bis(2-ethylhexyl)-sulfosccinate (Aerosol OT, or AOT) and in Figure 3.20 for dioctadecyldimethy-lammonium chloride (DODMAC). As for single-chain surfactants, the chain melting temperature depends strongly on chain length. In the case of DODMAC, the chains are long, so melting occurs at a high temperature. The most important feature of phase diagrams of this class of surfactant is the low solubility (often extremely low) in water, pointing to an inability to form micelles. Rather, it is typical with a very stable lamellar phase and a broad two-phase region of a dilute solution and a lamellar phase. In addition, for this type of surfactant there is often a formation, at higher surfactant concentrations, of a bicontinuous cubic phase and a reversed hexagonal phase.

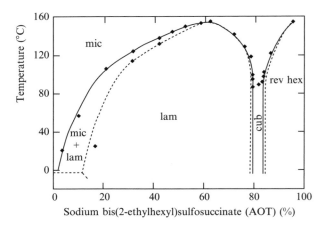

Figure 3.19 The binary phase diagram of sodium bis(2-ethylhexyl)sulfosuccinate (AOT)–water is dominated by a lamellar phase. In addition, there is an isotropic solution phase, a bicontinuous cubic phase and a reversed hexagonal phase. Reproduced by permission of Academic Press from J. Rogers and P. A. Winsor, *J. Colloid Interface Sci.*, **30** (1969) 247

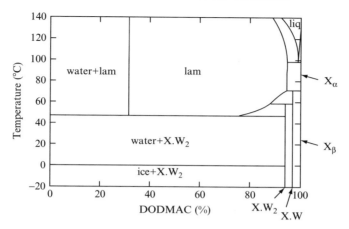

Figure 3.20 The binary phase diagram of dioctadecyldimethylammonium chloride (DODMAC)–water is dominated by a lamellar phase and crystalline hydrates, as well as two-phase regions. From R. G. Laughlin, *The Aqueous Phase Behaviour of Surfactants*, Academic Press, London, 1994, p. 29

In Figure 3.18 are indicated tie-lines, i.e. horizontal (constant temperature) lines connecting two coexisting phases. The end-points of a tie-line give the compositions of the two phases in equilibrium. The relative amounts of the two phases can be obtained by using the 'lever rule'. Transitions between phases in surfactant systems are generally of the first-order type, implying that there is a (smaller or larger) two-phase region between two single-phase regions; sometimes the two-phase region is omitted for simplicity but its existence should not be forgotten. Macroscopic properties are frequently very different in the two-phase region from those in the two single phases.

Binary and Ternary Phase Diagrams are Useful Tools: Three Components

Fundamental to our understanding of the phase diagrams is the phase rule:

$$P + F = C + 2$$

relating, for any point in the phase diagram, the number of phases coexisting (P), the number of components that make up the system (C) and the number of degrees of freedom (F); the latter are temperature, pressure and composition variables. For a two-component system ($C = 2$), $P + F = 4$. Fixing the pressure, we reduce the number of degrees of freedom by one. We then have at most two degrees of freedom ($F = 2$), temperature and surfactant concentration, which occurs with a single phase ($P = 1$). Therefore, the phase diagram can be presented in two dimensions. With two phases present, $F = 1$ and in a two-phase region only one of temperature or concentration can be varied.

With three components ($C = 3$), at constant pressure we have a maximum of three degrees of freedom, i.e. temperature and two concentration variables. In order to make a two-dimensional representation, different choices can be made (depending on the system and the purpose), namely fixing the temperature, fixing one concentration or fixing the ratio of the amounts of two components. It is most common to consider the phase behaviour at constant temperature and to use the Gibbs' triangle (Figure 3.21). Here the apices represent the three pure components, while the three sides of the triangle represent the three two-component systems. Inside the triangle, all three components are present and the amounts can be read as indicated. To illustrate the temperature dependence, a stack of isothermal phase diagrams can be given in a triangular prism (Figure 3.22).

Two three-component phase diagrams are illustrated in Figures 3.23 and 3.24. In the first there is an aqueous mixture of a single-chain ionic surfactant and a weakly polar amphiphile. The phase diagram is built up of six different phases: two isotropic solution phases, three liquid crystalline phases and solid surfactant. We note that in binary mixtures with water, the surfactant forms only micelles and a hexagonal phase. Addition of octanol induces a transition to the lamellar phase and, at higher alcohol concentrations, to reversed structures,

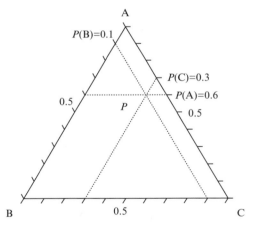

Figure 3.21 The phase diagram of three components A, B and C at a fixed temperature is presented in a triangle. The amounts of the different components at a point P are given by the distance from the opposite base. A sample represented by the point P is composed of 60% of A, 10% of B and 30% of C. Reproduced by permission of John Wiley & Sons, Inc., from D. F. Evans and H. Wennerström, *The Colloidal Domain. Where Physics, Chemistry, Biology, and Technology Meet*, VCH, New York, 1994, p. 421

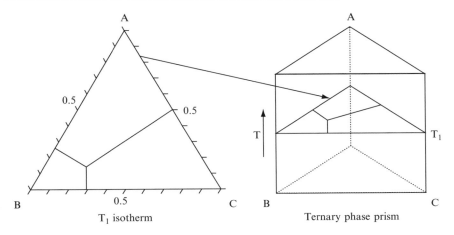

Figure 3.22 A three-dimensional representation of the phase diagram of a ternary system including temperature as the variable. From R. G. Laughlin, *The Aqueous Phase Behaviour of Surfactants*, Academic Press, London, 1994, p. 369

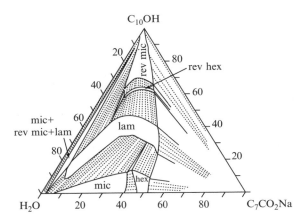

Figure 3.23 Ternary phase diagram of the sodium octanoate–decanol–water system at 25°C. There are two isotropic solution phases, micellar and reversed micellar, and three liquid crystalline phases, hexagonal, lamellar and reversed hexagonal. From R. G. Laughlin, *The Aqueous Phase Behaviour of Surfactants*, Academic Press, London, 1994, p. 397

i.e. reversed hexagonal and reversed micellar. In addition to the single-phase regions, there are a large number of two-phase and three-phase regions describing the coexistence of the different phases.

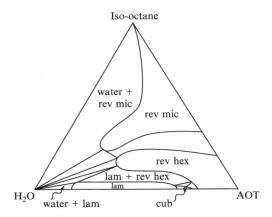

Figure 3.24 Ternary phase diagram of a double-chain anionic surfactant, Aerosol OT, mixed with isooctane and water at 25°C. There are two isotropic solution phases, water (with a low surfactant concentration) and reversed micellar, and three liquid crystalline phases, lamellar, bicontinuous cubic and reversed hexagonal (K. Fontell, unpublished). Reproduced by permission of John Wiley & Sons, Inc., from D. F. Evans and H. Wennerström, *The Colloidal Domain. Where Physics, Chemistry, Biology, and Technology Meet*, VCH, New York, 1994, p. 471

The ternary system in Figure 3.24 looks very different. Here we consider a double-chain ionic surfactant, which in mixtures with water gives an extensive lamellar phase as well as a bicontinuous cubic phase and a reversed hexagonal phase; the solubility in water is low. On addition of the oil, there is a transition from the lamellar to reversed hexagonal phases and then to the reversed micellar phase.

The sequence in which phases occur as a function of surfactant concentration or addition of a less polar component generally follows a simple sequence summarized in the Fontell scheme (Figure 3.15). Depending on the system and on temperature, etc., a smaller or larger number of the different phases actually occurs. For example, for single-chain surfactants we generally encounter structures from the mirror plane and to the left, while double-chain surfactants give the phases from the mirror plane and to the right.

The phase diagram in Figure 3.25 for a mixture of a cationic and an anionic surfactant shows an additional feature, namely of regions of stable vesicle solutions. Such systems show for stoichiometric mixtures the precipitation of a solid crystal. In addition, there are lamellar and vesicle regions forming with some excess of one surfactant.

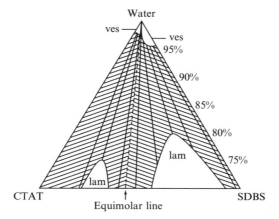

Figure 3.25 The dilute part of the ternary phase diagram for an aqueous mixture of a cationic (hexadecyltrimetylammonium tosylate, CTAT) and an anionic surfactant (sodium dodecylbenzene sulfonate, SDBS). Reprinted with permission from E. W. Kaler, A. Kamalakara, B. E. Rodriguez and J. A. N. Zasadzinski, *Science*, **245** (1989) 1371. Copyright (1989) American Association for the Advancement of Science

Surfactant Geometry and Packing Determine Aggregate Structure: Packing Parameter and Spontaneous Curvature of the Surfactant Film are Useful Concepts

We have noted the, quite general, rule that single-chain surfactants tend to form micelles and other 'normal' structures, while double-chain surfactants prefer to form lamellar phases and reversed structures. We can easily understand this if we try to pack space-filling models of surfactants into different aggregate shapes. We will, for example, find that due to the bulkiness of the hydrophobic part, double-chain molecules cannot be packed into spherical micelles. We also noted in our simple geometrical characterization of chain packing in spherical micelles that the quantity $v/(l_{max}a)$ can be at most 1/3 for a spherical micelle (p. 61). If we double v by adding a second alkyl chain, keeping the other factors constant, it can easily be understood that structures of other geometries may become favoured. We can alternatively see the critical packing parameter as the ratio between the cross-sectional area of the hydrocarbon part and that of the head group.

While relating chemical structure to aggregate structure is much more complex than such a simple geometrical analysis, this is an illustrative and useful starting point, in particular in analysing trends in phase behaviour. Basically, the structure formed is a result of the balance between the polar and the non-polar parts of a surfactant molecule; this explains the interest in (mainly empirical) scales of hydrophilic–lipophilic balances (HLBs), which are discussed in Chapter 21. These are useful classifications of surfactants, helpful *inter alia* in selecting a surfactant for a certain application, but they do not allow a deeper analysis or understanding. More recent approaches to the problem are based on the indicated concepts of surfactant packing and the spontaneous curvature of the surfactant film.

The definition of the critical packing parameter or the surfactant number, introduced in Chapter 2, is illustrated in Figure 3.26. Analyses of different geometrical shapes of aggregates, like the one made above for spheres, lead to the simple rules illustrated in Figure 3.27.

There are two factors that are not taken into account in the simple geometrical model but have a great influence on the aggregate structure. The first of these is the interaction between the head groups in the aggregates. Clearly a strongly repulsive interaction between head groups will drive aggregates to the left in the Fontell scheme while the opposite applies for attractive interactions. This problem can be circumvented by estimating an 'effective' head group area. For example, for ionic surfactants, the head group interactions will be strongly affected by the electrolyte concentration so that a decreases on addition of electrolytes. Electrostatic calculations of distances between head groups can be made to estimate a CPP value or surfactant number in this case. For non-ionics, temperature rather than electrolyte concentration is very important for interactions between head groups and is decisive for aggregate structure. We will return to this question in Chapter 4.

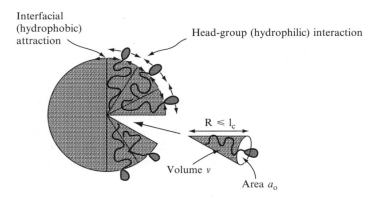

Figure 3.26 The critical packing parameter (CPP) or surfactant number relates the head group area, the extended length and the volume of the hydrophobic part of a surfactant molecule into a dimensionless number CPP $= v/(l_{max}a)$. From J. N. Israelachvili, *Intermolecular and Surface Forces, with Applications to Colloidal and Biological Systems*, Academic Press, London, 1985, p. 247

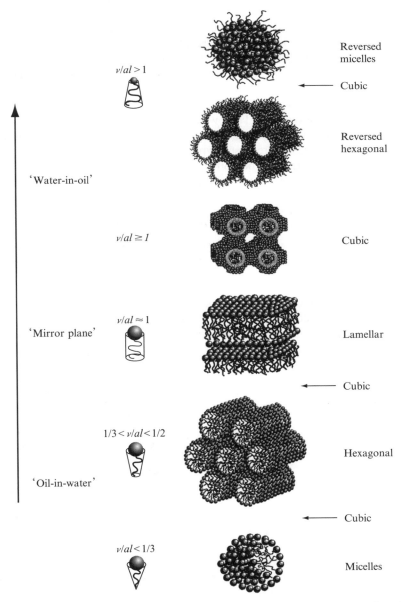

Figure 3.27 Critical packing parameters (CPPs) of surfactant molecules and preferred aggregate structures for geometrical packing reasons

Considering what surfactants fall into the different categories of Figure 3.27, we have already noted that CPP < 1/3 and thus spherical micelles are found for single-chain surfactants with a strongly polar head group, like an ionic head in the absence of an electrolyte. Here we also find non-ionics with large head groups. The range of CPP values of 1/3–1/2 and rod-like structures are characteristic of single-chain ionics with an added electrolyte, single-chain ionics with a strongly bound counterion or non-ionics with an intermediate head group size. Higher CPP values are characteristic of double-chain amphiphiles or non-ionics with short head groups. Here we find many membrane lipids. For ionics, electrolyte addition again increases the CPP value and may drive aggregate structures from bilayers to reversed type structures.

Surfactant aggregates can be considered to be built up of surfactant films and, depending on the curvature of the films, different structures result. (The curvature is the inverse of the radius of curvature.) We may define the spontaneous curvature to be positive if the film is curved around the hydrophobic part and negative if it is curved towards the polar part (Figure 3.28). A normal micelle has thus a positive film curvature, while a reversed one has a negative one. Planar films, like in a lamellar phase, have zero curvature. Many bicontinuous structures are characterized by a more complex 'saddle-shaped' geometry with two principal radii of curvature with opposite signs. An important case is that of minimal surfaces, where the mean curvature is zero.

The spontaneous curvature can vary essentially from the inverse surfactant molecule length to a similar negative value. The spontaneous curvature decreases

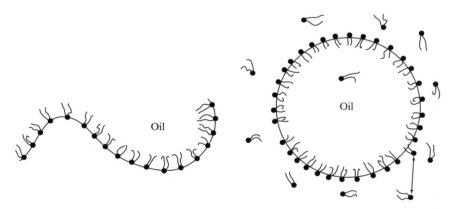

Figure 3.28 A micelle with solubilized oil is characterized by a positive spontaneous curvature. In general, a surfactant film between oil and water is flexible with both regions of positive (towards oil) and negative curvature (towards water)) curvatures. Reproduced by permission of John Wiley & Sons, Inc., from D. F. Evans and H. Wennerström, *The Colloidal Domain. Where Physics, Chemistry, Biology, and Technology Meet*, VCH, New York, 1994, pp. 456–457

on adding a second chain to a surfactant and on decreasing head group repulsions, e.g. by adding an electrolyte to an ionic surfactant.

Polar Lipids Show the same Phase Behaviour as other Amphiphiles

Polar lipids obey the same rules as surfactants and show the same type of phase behaviour. In Figure 1.3 are given the chemical structures of a number of different lipids and we can easily realize the very different tendencies of associating in water of, for example, a very non-polar triglyceride and a highly polar bile salt. In fact, triglycerides and diglycerides are generally not amphiphilic enough to be soluble in water and show self-assembly. The other extreme, the bile salt, tends to form only small aggregates, with a large positive spontaneous curvature. Other lipids fall in between and show different types of self-assembly structures, as exemplified in a number of phase diagrams. (In the literature, these phase diagrams are generally presented with amphiphile and water reversed, compared to surfactant phase diagrams, i.e. with the water content increasing from left to right.)

Phospholipids, like lecithin, behave similarly to other double-chain surfactants in showing a strong preference to form a lamellar phase (Figure 3.29). The lamellar phase can take up water and swell to *ca.* 45% water. At higher water contents, there is a coexistence of the lamellar phase and a very dilute aqueous solution. Monoolein (Figure 3.30) also gives a lamellar phase, but only in a limited range of concentration and temperature. This system is rather unusual in giving a bicontinuous cubic phase which firstly is stable over wide ranges of

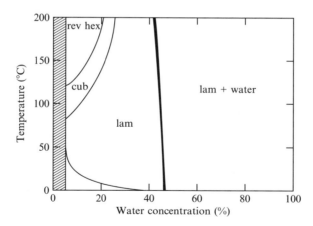

Figure 3.29 The phase diagram of lecithin (from egg yolk) and water is dominated by a lamellar phase which at higher water contents is in equilibrium with a very dilute aqueous solution. From K. Larsson, *Lipids—Molecular Organization, Physical Functions and Technical Applications*, The Oily Press Ltd, Dundee, Scotland, UK, 1994, p. 64

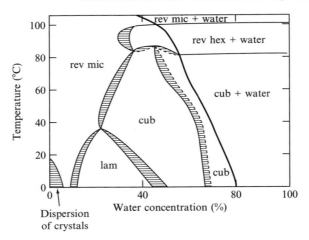

Figure 3.30 The phase diagram of monoolein and water is characterized by a large region of bicontinuous cubic phases. The cubic phase can be in equilibrium with excess water. In addition, there is a lamellar phase and a reversed micellar phase. From K. Larsson, *Lipids—Molecular Organization, Physical Functions and Technical Applications*, The Oily Press Ltd, Dundee, Scotland, UK, 1994, p. 61

concentration and temperature and secondly coexists with excess water. In the two-phase region there is a formation of Cubosomes®, dispersed cubic phase.

Bile salts are characterized by very low surfactant numbers or large positive curvatures of the surfactant films. They give, therefore, only micellar solutions in mixtures with water and generally no liquid crystalline phases. If we add a bile salt to a less polar lipid, we get the expected changes in aggregate structure. For lecithin, the lamellar phase incorporates some bile salt but is then transformed to first a hexagonal phase and then to a micellar phase. For monoolein, the cubic phase is transformed to the lamellar phase, which at a higher content is transformed to a micellar phase.

Liquid Crystalline Phases may Form in Solvents other than Water

We have noted that surfactants have a much stronger tendency to self-assembly to micelles in water than in other polar solvents and this applies in general to all types of self-assembly aggregates. This is exemplified in Figure 3.31 for hexadecyltrimethylammonium bromide in different polar solvents. As can be seen, the isotropic solution phase becomes much more significant with other solvents than with water, whereas the liquid crystalline phases have lower stability ranges.

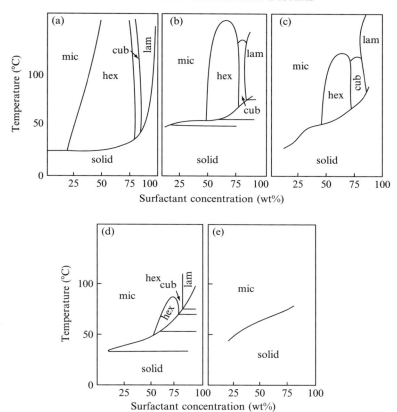

Figure 3.31 Binary phase diagrams of hexadecyltrimethylammonium bromide with different polar solvents: (a) water; (b) glycerol; (c) formamide; (d) ethylene glycol; (e) *N*-methyl formamide. With increasing concentration of surfactant, we find the isotropic (micellar) solution phase and hexagonal, cubic and lamellar liquid crystalline phases. From M. Sjöberg, *Surfactant Aggregation in Nonaqueous Polar Solvents*, Ph. D. Thesis, Department of Physical Chemistry, The Royal Institute of Technology, Stockholm, Sweden, 1992

Bibliography

Evans, D. F. and H. Wennerström, *The Colloidal Domain. Where Physics, Chemistry, Biology and Technology Meet*, Wiley-VCH, New York, 2nd Edn, Chs 6 and 10.

Fontell, K., Some aspects on the cubic phases in surfactant and surfactant-like lipid systems, *Adv. Colloid Interface Sci.*, **41** (1992) 127.

Larsson, K., *Lipids—Molecular Organization, Physical Functions and Technical Applications*, The Oily Press Ltd, Dundee, Scotland, UK, 1994.

Laughlin, R. G., *The Aqueous Phase Behaviour of Surfactants*, Academic Press, London, 1994.

Robb, I. D. (Ed.), *Specialist Surfactants*, Blackie Academic and Professional, London, 1996.

Tiddy, G. J. T., Surfactant–water liquid crystal phases, *Phys. Rep.*, **57** (1980) 1.

4 PHYSICOCHEMICAL PROPERTIES OF SURFACTANTS AND POLYMERS CONTAINING OXYETHYLENE GROUPS

Polyoxyethylene Chains make up the Hydrophilic Part of many Surfactants and Polymers

For ionic surfactants, the electrostatic interactions are, as seen above, decisive for the properties of simple and complex systems. Non-ionic surfactants are controlled by very different hydrophilic interactions. The most important type of non-ionic surfactant (see Figure 1.7) is that with an oligo(oxyethylene) group as the polar head. Denoting an oxyethylene group by E, simple non-ionics can be abbreviated as C_mE_n if we have an alkyl chain as the lipophilic part. We can also have a more complex hydrophobic part with a branched and/or unsaturated group; many non-ionics contain aromatic groups.

For typical ionic surfactants, as well as many other surfactants, the volume of the polar group is much smaller than that of the non-polar part. For the polyoxyethylene surfactant, the situation is different in that the volumes of the two parts are of similar size; typically the polar part is larger than the non-polar one. It is, therefore, appropriate and fruitful to consider a non-ionic surfactant as a short AB block copolymer.

One conspicuous feature of a non-ionic surfactant is the temperature dependence of physicochemical properties. This may be problematic in applications but can also be turned into an advantage, since temperature triggered systems can be designed. To master and understand the special temperature-dependent (effective) interactions between the polar solvent and the polyoxyethylene chains proves to be most essential for non-ionics. These interactions are obviously not unique for surfactant systems, but have a general bearing also for polymers containing oxyethylene groups, some of which are pictured in Figure 4.1.

It is thus logical to treat oxyethylene-based surfactants and polymers, homo, graft and block copolymers, in a single context. In this chapter we will first

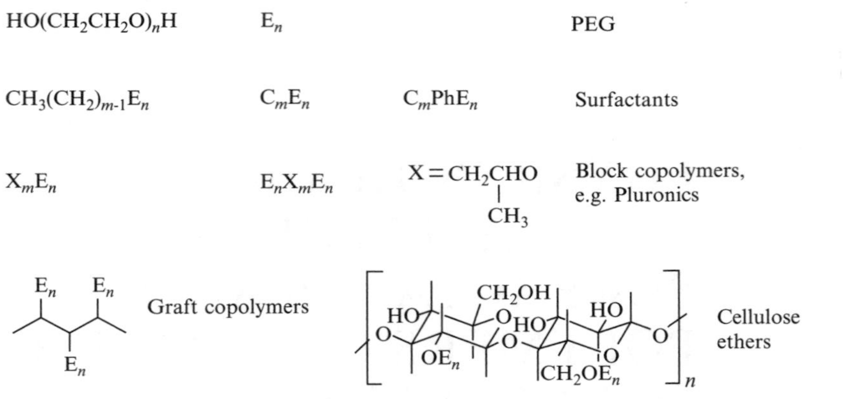
HO(CH2CH2O)nH
En
PEG
CH3(CH2)m-1En
CmEn
CmPhEn
Surfactants
XmEn
EnXmEn
X = CH2CHO
CH3
Block copolymers, e.g. Pluronics
En
En
En
Graft copolymers
CH2OH
HO
O
HO
HO
O
O
O
OEn
O
CH2OEn
n
Cellulose ethers

Figure 4.1 Chemical structure of some oxyethylene (EO)-based polymers and surfactants

review the general properties of non-ionic surfactants and, in comparison with other types of surfactants, identify their particular behaviour. We will then consider the unusual temperature effects in a broader context and review our present understanding of mechanisms in terms of intermolecular interactions. For the polymer solutions, we will here focus on the special temperature-dependent properties and treat the more general polymer aspects in Chapters 9 and 12.

Surfactants in applications are rarely chemically homogenous. For all types of surfactants there may be a considerable distribution in the non-polar part, different alkyl chain length, partial unsaturation, etc., resulting from the fatty raw material used. This has normally only minor consequences for our analysis of physicochemical observations and our modelling. For non-ionics, we have typically, in addition, a marked heterogeneity in the polar head groups (Figure 4.2), as discussed in Chapter 1. It is, therefore, necessary to treat the non-ionic surfactant systems used in commercial applications as mixed surfactant systems. However, at a quite high price, well-defined homogeneous non-ionics are available. The significance of those for scientific work and for our improved understanding cannot be overestimated. Below we will mainly refer to results obtained in work on well-defined surfactants.

CMC and Micellar Size of Polyoxyethylene-Based Surfactants are Strongly Temperature Dependent

The CMC of non-ionic polyoxyethylene surfactants, C_mE_n, varies strongly with m, while the variation with n is quite weak, with a slight increase as the EO chain becomes longer (Table 2.3). A plot of the logarithm of the CMC as a

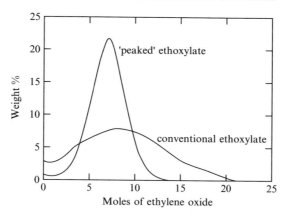

Figure 4.2 Conventional polyoxyethylene surfactants have a much broader distribution of head group sizes than recently developed ('peaked') ones

function of the number of carbons in the alkyl chain (Figure 2.5) gives, as noted above, a straight line with a slope that is considerably larger than for ionics. Lengthening of the alkyl chain by one methylene unit gives a lowering of the CMC by a factor of three rather than two. As noted above, non-ionics show a behaviour paralleling the solubility of hydrocarbons in water, while the variation is weaker for ionics because of an opposing electrostatic effect, which becomes more important the lower the CMC.

The temperature dependence of the CMC (Figures 4.3 and 2.6) differs from that of ionics in two respects: it is markedly stronger and there is typically a monotonic decrease with increasing temperature rather than an increase at higher temperatures. A non-ionic micelle (Figure 4.4) has a thick interfacial layer of polar head groups rather than the quite sharp transition from the hydrophobic micellar interior to the aqueous bulk of ionics. We will see that changes in the intermolecular interactions in the polar layer account for the special temperature-dependent behaviour of non-ionics.

The spherical micelle pictured in Figure 4.4 is typical for surfactants with long polyoxyethylene chains, in particular at low temperatures and concentrations. As for ionics, micellar growth may occur but the conditions are different for non-ionics. In particular, the temperature dependence of micelle size is opposite to that of ionics.

The most significant features are illustrated in Figure 4.5, where the hydrodynamic radii of the micelles are given as a function of temperature for three relatively dilute surfactants. For a surfactant like $C_{12}E_8$ there is insignificant or moderate growth up to high temperatures; a longer polyoxyethylene chain would give even less growth. For a short oxyethylene chain, illustrated here by $C_{12}E_5$, there is a dramatic growth with increasing temperature.

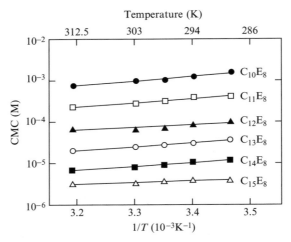

Figure 4.3 The logarithm of the CMC plotted against the inverse absolute temperature for non-ionics with eight oxyethylenes in the head group. From top to bottom, the numbers of carbon atoms in the alkyl chain are 10, 11, 12, 13, 14 and 15. From K. Meguro, M. Ueno and K. Esumi, in *Nonionic Surfactants: Physical Chemistry*, M. J. Schick (Ed.), Marcel Dekker, New York, 1987, p. 136

In general, we can summarize micellar size effects for non-ionics as follows:

1. The polyoxyethylene chain length is the prime factor in determining growth. The shorter the chain, then the larger the tendency for growth, both with temperature and with concentration.

2. As for all types of surfactants, micellar growth is promoted by an increased alkyl chain length. For example, C_{16} surfactants give much stronger growth than C_{12} ones.

3. Cosolutes influence growth differently than for ionics. Salting-out (see below) electrolytes tend to promote growth and salting-in ones to inhibit growth. Ionic surfactants have a strong tendency to reduce growth and even quite small amounts may inhibit growth.

Temperature Dependence can be Studied using Phase Diagrams

Phase diagrams of non-ionic surfactant–water systems show many resemblances to those of ionics, with those of longer polyoxyethylene chains being similar to single-chain ionics and those of short oxyethylene chains being more similar to double-chain ionics. However, there are also principal differences. We exemplify this by using the phase diagram of $C_{12}E_5$ (Figure 4.6). At low temperatures, we have the previously encountered sequence of micellar solution

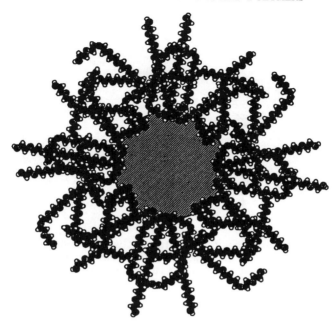

Figure 4.4 A schematic picture of a non-ionic micelle. Reprinted with permission from M. Jonströmer, B. Jönsson and B. Lindman, *J. Phys. Chem.*, **95** (1991) 3293. Copyright (1991) American Chemical Society

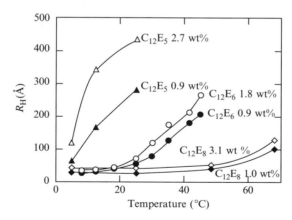

Figure 4.5 Non-ionic micelles grow with increasing temperature for surfactants with short polar heads, while growth is weak or insignificant for larger head groups. The size is characterized by the hydrodynamic radius, R_H. From B. Lindman and M. Jonströmer, in *Springer Proceedings in Physics: Physics of Amphiphilic Layers*, J. Meunier, D. Langevin and N. Boccara (Eds), Springer-Verlag, Berlin, 1987, p. 235

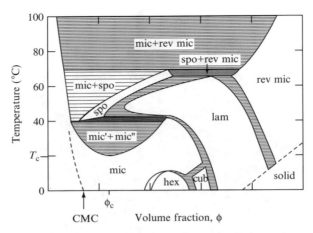

Figure 4.6 Phase diagram of a non-ionic surfactant–water system given as a function of the volume fraction of surfactant on a logarithmic scale; mic, rev mic and spo denote isotropic solution phases and hex, lam and cub denote hexagonal, lamellar and cubic liquid crystalline phases, respectively. On heating a micellar solution, there is a phase separation into two isotropic solutions. The boundary of the two-phase region is a so-called lower consolute curve. The minimum of the lower consolute curve is a critical point, defined by a critical temperature (T_c) and composition (ϕ_c or c_c). From R. Strey, R. Schomäcker, D. Roux, F. Nallet and U. Olsson, *J. Chem. Soc., Faraday Trans.*, **86** (1990) 2253. Reproduced by permission of The Royal Society of Chemistry

phase, normal hexagonal, bicontinuous cubic, lamellar and solid surfactant, with an increasing percentage of surfactant. In the micellar region, there is a very strong micellar growth, both with increasing temperature and with increasing concentration. A major difference when compared to typical ionics is the splitting of the water-rich solution phase into two liquid phases at higher temperature and also the appearance of the sponge phase, a new solution phase, at still higher temperatures. We also note that the lamellar phase and the reversed micellar phase, the liquid surfactant with solubilized water, gain in importance at higher temperatures at the expense of the micellar, hexagonal and cubic phases.

Comparing the phase behaviour of a sequence of surfactants with the same alkyl chain but different polyoxyethylene chain lengths (Figure 4.7), we note a dramatic variation. With a long polyoxyethylene chain, the micellar phase (with a strong preference for small, roughly spherical, micelles), a cubic phase of discrete micelles, and the normal hexagonal phase dominate. Shortening the oxyethylene chain is accompanied by a larger tendency to micellar growth and by the progressive disappearance of the phases mentioned. Instead, in particular the lamellar phase and the surfactant-rich solution phase, and at higher temperatures also the sponge phase, increase their stability. For a surfactant like

$C_{12}E_3$, no micelles form on addition of surfactant to water, but there is directly a two-phase region of the lamellar phase and a very dilute solution, a situation reminding us of that for lecithin. Besides the lamellar phase, the dominating phase is the reversed micellar, which has a narrow channel extending into the water-rich part of the diagram.

The L_3 or 'Sponge' Phase

This phase was originally discovered in phase diagram studies of the system $C_{10}E_6$, and was denoted 'anomalous' since it was unexpected to encounter a second micellar solution region at higher temperatures. Now we know that it is of the bicontinuous type, with surfactant bilayer films connected over macroscopic distances. The surfactant bilayers form zero or low mean curvature surfaces separating different water channels. We encountered this structure previously in Chapters 2 and 3.

The L_3 or 'sponge' phase is common for non-ionic surfactants and has been mainly studied for these. In recent years, we have also learnt that it occurs for all classes of surfactants, provided there is a suitable balance between the hydrophilic and lipophilic parts. For ionics, the formation of the sponge phase can be induced by the addition of electrolyte.

Sequence of Self-Assembly Structures as a Function of Temperature

If we summarize a large number of observations of phase diagrams, we may obtain a quite general sequence of phase structures as a function of temperature, which is spherical micelles, discrete cubic phase, elongated micelles, hexagonal phase, lamellar phase, bicontinuous cubic phase, sponge phase and reversed micellar solution. Selected structures are illustrated in Figure 4.8. We should note that temperature and polyoxyethylene chain length have opposing effects and that for a given surfactant only a few of the structures are formed within the accessible range of temperatures.

The Critical Packing Parameter and the Spontaneous Curvature Concepts are Useful Tools

These results can now be examined in the light of the principles of surfactant self-assembly described in Chapter 3. We recall that the dimensionless critical packing parameter or surfactant number is given by the volume of surfactant hydrophobe divided by the product of the area of the polar head group and the extended length of the surfactant molecule and that the spontaneous curvature of a surfactant film is taken to be positive if it is curved towards the hydrophobic part.

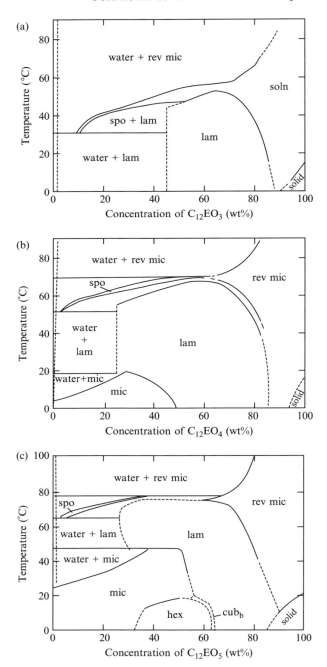

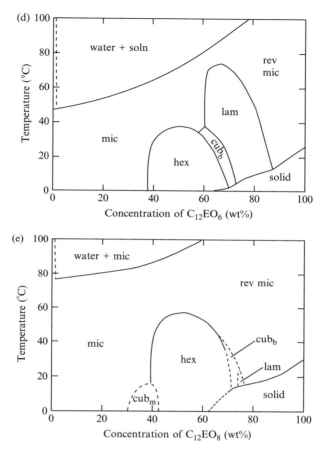

Figure 4.7 Phase diagrams (somewhat simplified) of non-ionic surfactant–water systems for C_{12} surfactants with 3, 4, 5, 6 or 8 oxyethylenes in the polar head; water, mic, rev mic and spo denote isotropic solution phases and hex, lam, cub_m and cub_b denote hexagonal, lamellar and discrete (micellar) and bicontinuous cubic liquid crystal-line phases, respectively. From D. J. Mitchell, G. J. T. Tiddy, L. Waring, T. Bostock and M. P. MacDonald, *J. Chem. Soc., Faraday Trans. 1*, **79** (1983) 975. Reproduced by permission of the Royal Society of Chemistry

The sequence of self-assembly structures that we have described indicates that the surfactant number of non-ionics increases progressively with increasing temperature. The only way in which we can understand this is in terms of a decreasing area per polar head group at higher temperatures. Therefore, it appears that there is a closer packing of polar head groups at the aggregate surface at higher temperatures. This is consistent with observations of a

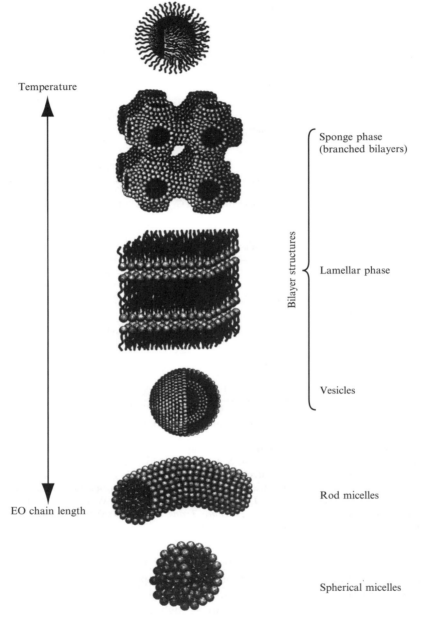

Figure 4.8 The surfactant self-assembly structures of non-ionic surfactants are mainly governed by temperature and the number of oxyethylene groups. Only selected structures are represented

decreased hydration number—or a decreased concentration of water in the head group area—at higher temperatures, as observed by spectroscopy or water self-diffusion. The spontaneous curvature is similarly inferred to change to lower values at higher temperatures (Figure 4.9).

Since unusual temperature dependences are ubiquitous for oxyethylene-containing systems we are looking for some general underlying mechanism. As indicated, there is a weakened interaction between the oxyethylene groups and the water solvent with increasing temperature. There is a smoothly decaying hydration with increasing temperature which is strikingly similar for quite dissimilar systems. In fact, the apparent number of hydration per oxyethylene group is sensitive to temperature but closely the same for different systems.

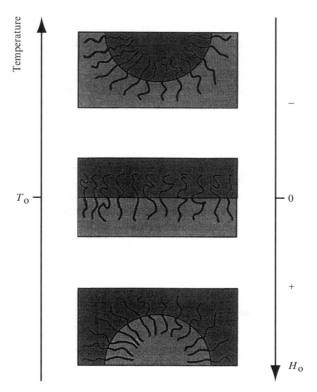

Figure 4.9 For non-ionic surfactants, best considered as short AB block copolymers, the spontaneous curvature, H, changes from positive at low temperatures to negative at high temperatures. At some intermediate temperature, T_0, the spontaneous curvature is zero and the surfactant is termed 'balanced'. The inferred change in spontaneous curvature with temperature suggests that while water is a good solvent for the head groups at low temperature, the oxyethylene–water interactions become less favourable at higher temperatures

The lower consolute curve is typical of phase diagrams of oxyethylene-based surfactants and polymers. This implies that the effective solute–solute interactions have a significant temperature dependence and changes from repulsive to attractive with increasing temperature. In turn, this can reflect changes in either solute–solute, solute–solvent or solvent–solvent interactions, or a combination of those. All alternatives have been suggested and there is as yet no consensus about the dominating effect. However, a water–water interaction mechanism, relating to a temperature-dependent structuring of water around oxyethylene groups, is less likely in view of analogous observations in solvents other than water. Hydrogen bonding between water molecules and ether oxygen is another model that has been analysed. Here we describe a model based on temperature-dependent solute conformational effects which has a strong predictive power.

A polyoxyethylene chain may exist in a large number of conformations, which have different energies. The conformation of an oxyethylene group which is gauche around the C—C bond and anti around the C—O bond (Figure 4.10, top) has the lowest energy of all conformers. This low-energy conformation, which will dominate at low temperatures, has a large dipole moment. On the other hand, it has a low statistical weight. With increasing temperature, other conformations of higher statistical weight will become increasingly more important. These have smaller or no dipole moments, like the anti–anti–anti conformation (Figure 4.10, bottom).

The conformational changes will consequently make the polyoxyethylene chains progressively less polar as the temperature is increased. Becoming less polar they will interact less favourably with water, leading to reduced hydration, and more favourably among themselves, leading to a closer packing of head groups in surfactant self-assemblies, as well as to an increased tendency to separate into a more concentrated phase. The succession of self-assembly structures with increasing temperature also follows logically from the

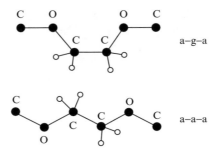

Figure 4.10 Different conformations of an oxyethylene group have different stabilities and polarities. The upper anti–gauche–anti conformation has a lower energy and is more polar than the lower anti–anti–anti conformation

decreased polarity. The same applies to many other observations, to be given below, such as the increased adsorption of homopolymers, copolymers and surfactants at higher temperatures, resulting from the worsened solvency conditions. Irrespective of the model adopted, the temperature effects of non-ionic polymers and surfactants are best analysed in terms of water being a good solvent for the oxyethylene groups at low temperatures, while at high temperatures it is a bad solvent. We can effectively tune the solute–solvent interaction with temperature.

Clouding is a Characteristic Feature of Polyoxyethylene-Based Surfactants and Polymers

A common observation for a solution of a non-ionic surfactant is that on heating the solution may start to scatter light strongly in a well-defined temperature range. It becomes 'cloudy'. This is a consequence of one feature of the phase diagram. The isotropic solution region is bordered towards higher temperatures by a lower consolute curve, above which there is a phase separation into one surfactant-rich and one surfactant-poor solution. The onset of phase separation is manifested by a cloudiness of the solutions. The minimum in the lower consolute curve is a critical point. The approach of this is accompanied by strong scattering of light due to critical fluctuations.

The clouding temperature, or the cloud point, depends strongly on the polyoxyethylene chain length but is less influenced by the hydrophobe size. Normally, the cloud point is recorded for a certain fixed solute concentration (say 1% by weight). We infer from the phase diagrams given above that the cloud point of $C_{12}E_8$ is around 80°C, while it is $ca.$ 50 and 10°C for $C_{12}E_6$ and $C_{12}E_4$, respectively. For still shorter polyoxyethylene chains, the surfactant is insoluble, even at the freezing temperature of water, so the cloud point is below 0°C. In Figure 4.11, the cloud point at a fixed surfactant weight concentration is plotted as a function of the number of oxyethylene units for C_{12} surfactants.

The clouding phenomenon occurs for many systems and is a common feature for a large class of solutions with solutes containing oxyethylene groups. For poly(ethylene glycol) or poly(ethylene oxide) (PEO), there is a simple closed-loop two-phase region. As can be seen in Figure 4.12, the two-phase region grows strongly in all directions as the molecular weight of the polymer is increased. The basic features of the closed-loop appearance of the phase diagram are the same for the surfactant systems, but these are more complex due to self-assembly, leading to additional phases. Furthermore, the cloud point is strongly dependent on micelle size, which is very different for different surfactants. A high micellar aggregation number will have an effect analogous to a high degree of polymerization of a polymer and thus give a low cloud point and a critical point at a low concentration.

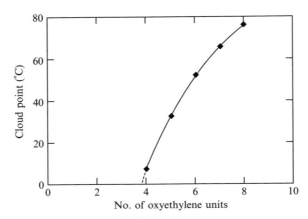

Figure 4.11 The cloud point as a function of the oxyethylene chain length for C_{12} surfactants

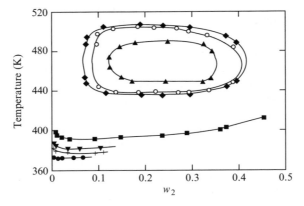

Figure 4.12 Phase diagrams for poly(ethylene glycol)–water for different polymer molecular weights. From the bottom, the mean molecular weights are 10^6, 2×10^4, 1.4×10^4, 8×10^3, 2270, 2250 and 2160. Reprinted from *Polymer*, **17**, S. Saeki, N. Kuwahara, M. Nakata and M. Kaneko, 'Upper and lower critical solution temperature in poly(ethylene glycol) solutions', 685–689, Copyright (1976), with permission from Elsevier Science

Clouding is strongly dependent on cosolutes. Electrolytes may either increase or decrease the cloud point (as illustrated in Figure 4.13) and may be termed salting-in or salting-out, respectively. This may be understood from the interaction between the polymer or surfactant and the cosolute. The effect is dominated by the anions. Some anions, like SCN^-, show a preference for the polymer or surfactant relative to the bulk solvent and are enriched in the vicinity of the oxyethylene groups. Others, like Cl^-, do not show such

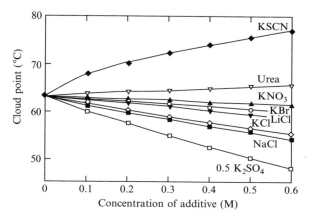

Figure 4.13   Effect of additives on the cloud point of 5 mM $C_{12}E_7$. From K. Meguro, M. Ueno and K. Esumi, in *Nonionic Surfactants: Physical Chemistry*, M. J. Schick (Ed.), Marcel Dekker, New York, 1987, p. 151

preference and are depleted in the vicinity of the oxyethylene groups. In the former case, increased solubility and an increased cloud point results, while the opposite is observed in the latter case.

Very low concentrations of an ionic surfactant strongly increase solubility and thus also the cloud point. This is due to the formation of mixed micelles in the case of a clouding surfactant and to an association of the surfactant to the polymer (see Chapter 13) in the case of a clouding polymer. This results in charged aggregates which are much more difficult to concentrate in one of the phases due to the unfavourable electrostatic interactions arising from the entropy of the counterion distribution.

Physicochemical Properties of Block Copolymers Containing Polyoxyethylene Segments Resemble those of Polyoxyethylene-Based Surfactants

A related behaviour to that of non-ionic surfactant and poly(ethylene glycol) is shown by random and block copolymers containing oxyethylene groups. (A list of examples is given in Figure 4.14.) Random copolymers, containing, in addition to oxyethylene, less polar groups, such as oxypropylene (PO), mainly show a similar behaviour to poly(ethylene glycol) but with lower solubility and lower cloud points.

Many polyoxyethylene block copolymers are known and have found important applications. They range from poly(ethylene glycol) 'end-capped' by alkyl chains to complex structures of a 'star-type' geometry. Triblock systems, with poly(propylene glycol) (PPO) or poly(butylene glycol) (PBO), have attracted

$(EO)_n (PO)_m (EO)_n$ 'normal Pluronics'

$(PO)_n (EO)_m (PO)_n$ 'inverse Pluronics'

$RO(EO)_n (PO)_m$ R is C_{12} to C_{18}

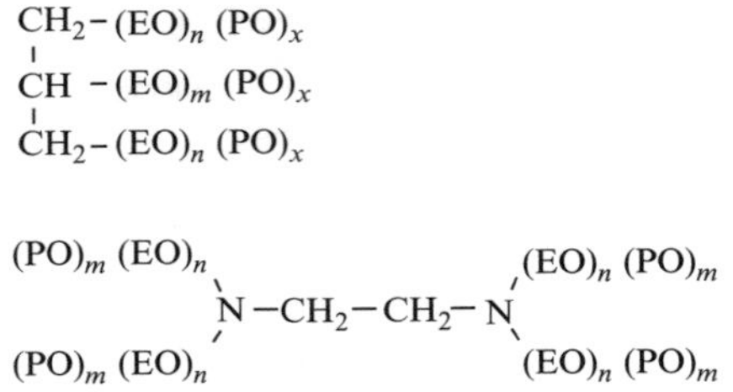

$CH_2-(EO)_n (PO)_x$
|
$CH\ -(EO)_m (PO)_x$
|
$CH_2-(EO)_n (PO)_x$

$(PO)_m (EO)_n$ \qquad\qquad\qquad\qquad $(EO)_n (PO)_m$
$\searrow$ \qquad\qquad\qquad\qquad $\nearrow$
N $-CH_2-CH_2-$ N
$(PO)_m (EO)_n$ $\nearrow$ \qquad\qquad\qquad\quad $\searrow$ $(EO)_n (PO)_m$

Figure 4.14   Structures of some oxyethylene–oxypropylene copolymers

special interest. They show many analogies with surfactant systems since they are also characterized by a considerable self-assembly. However, because of the higher molecular weight, temperature-dependent effects become much more pronounced. One well-known example is the transition of a low-viscous solution of a PEO–PPO–PEO polymer into a high viscous clear 'gel' at a moderate temperature increase (Figure 4.15). This gelation is in fact the result of formation of a cubic liquid crystalline phase.

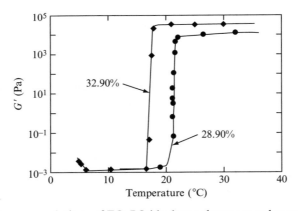

Figure 4.15   Aqueous solutions of EO–PO block copolymers may become very viscous over a narrow temperature range. Here the storage modulus, G', which characterizes the solid-like or elastic properties of the system, is shown as a function of temperature for two weight percentages of a Pluronic copolymer. Reprinted with permission from P. Bahadur and K. Pandya, *Langmuir*, **8** (1992) 2666. Copyright (1992) American Chemical Society

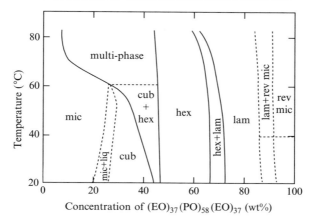

Figure 4.16 Phase diagram of a triblock EO–PO–EO copolymer-water system; mic and rev mic are isotropic solution phases and cub, hex and lam denote cubic, hexagonal and lamellar liquid crystalline phases, respectively. Reprinted with permission from P. Alexandridis, D. Zhou and A. Khan, *Langmuir*, **12** (1996) 2690. Copyright (1996) American Chemical Society

A phase diagram of a typical two-component system of such a triblock copolymer is represented in Figure 4.16. Here we find the same type of phases and phase sequences as for surfactant systems. However, there are often a larger number of liquid crystalline phases for the block copolymer systems. The phases appearing can be controlled by varying the ratio between the more polar oxyethylene groups and the less polar oxypropylene groups, so that with a high proportion of oxyethylene we find mainly the phases to the left in the Fontell scheme (Chapter 3) and conversely with a high proportion of oxypropylene the phases to the right dominate. In contrast to typical surfactants, these block copolymers can form structures with a wide range of spontaneous curvatures and surfactant numbers. This is even more striking for ternary systems, including an oil component as illustrated later in Figure 12.11.

A CMC, defining the onset of self-assembly, can be defined equally well for the block copolymers as for simple surfactants. For the block copolymers, there is a dramatic decrease in CMC with increasing temperature (Figure 4.17). Clouding is also observed for the block copolymers and, as expected, the cloud point increases with the proportion of oxyethylene groups and decreases with increasing molecular weight.

Temperature Anomalies of Oxyethylene-Based Surfactants and Polymers are Ubiquitous

We have already encountered a number of novel temperature effects for the different types of oxyethylene-containing solutes, i.e. surfactants, homopolymers

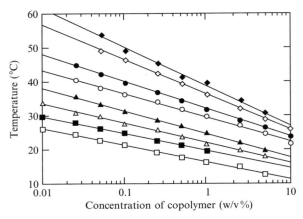

Figure 4.17 The transition from free polymer to micelles in EO–PO block copolymer solutions takes place at much lower concentrations as the temperature is increased. This feature is illustrated for EO–PO block copolymers of different polarities and molecular weights from the Pluronics family. Reprinted from *Colloid Surf.* **96**, P. Alexandridis and T. A. Hatton, 'Poly(ethylene oxide) – poly(propylene oxide) – poly(ethylene oxide) block copolymer surfactants in aqueous solution and at interfaces: thermodynamics, structure, dynamics and modeling', 1–46, Copyright (1995), with permission from Elsevier Science

and copolymers. In fact, these and other temperature 'anomalies' have a much broader significance and we will find them for a broad range of systems.

Clouding and phase behaviour are examples of such unusual temperature effects. We expect, in general, two substances to become more miscible at higher temperatures due to the more significant contribution from entropy to the free energy of mixing (see Chapter 10). For a few systems, there is an inverse effect of temperature on solubility, as we have noted for the oxyethylene-containing solutes. Then, we have a lower consolute curve and a lower critical point. We expect then that if temperature can be raised to sufficiently high values we will retrieve complete miscibility so that we have a closed-loop type of phase behaviour.

Clouding is one temperature anomaly displayed by these systems, but there are numerous others. A closely related observation is that if we have a two-phase system of oil and water and add surfactant, the surfactant will be distributed into the lower aqueous phase at low temperatures but into the upper oil phase at higher temperatures. The surfactant not only decreases its solubility in water with increasing temperature but also increases its solubility in oil. This behaviour is opposite to that of other surfactants, such as ionics. A related observation is the strong decrease of the CMC of a non-ionic surfactant with an increase in temperature, again different from what is observed with other surfactants. The CMC in oil shows an opposite behaviour to that in water.

Micelle size and shape are strongly dependent on temperature, with a growth from spheres to rods and to bilayer structures. Again, this is the opposite to what is seen for other surfactant classes.

As described in Chapter 6, the microstructure of non-ionic microemulsions shows a dramatic variation with temperature. At low temperatures, a structure of oil droplets is formed, while at higher temperatures water droplets form. Again this is opposite to what we are used to for other surfactants.

Emulsion stability due to non-ionic surfactants is strongly temperature dependent as well (Figure 4.18). At low temperatures, oil-in-water emulsions are stable but at high temperatures water-in-oil emulsions are stable (Figure 4.19). In an intermediate temperature range it is not possible to form stable emulsions. The temperature of transition between the two types of emulsions is quite well defined and is denoted as the phase inversion temperature (PIT). The latter depends on the hydrocarbon used but there is a good correlation between the PIT and the cloud point of the surfactant, as is further discussed in Chapters 21 and 22.

In detergency, we observe unusual maxima as a function of temperature, with the maximum occurring at a higher temperature the longer the EO chain of the surfactant (see further in Chapter 22).

The interactions between aggregates or between surfaces change from being repulsive at lower temperatures to becoming attractive at higher temperatures.

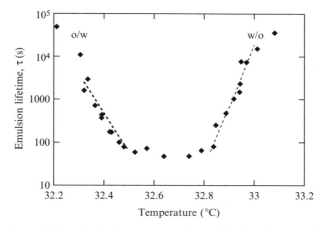

Figure 4.18 On emulsification in the presence of a non-ionic surfactant, the droplets are small in the vicinity of the phase inversion temperature (PIT) due to a low interfacial tension. Emulsion stability is high well away from the PIT but very low close to the PIT. The logarithm of the emulsion lifetime is plotted as a function of temperature for the system $C_{12}E_5$−n-octane–water. We will return to the mechanisms in Chapter 21. Reprinted with permission from A. Kabalnov and J. Weers, *Langmuir*, **12** (1996) 1931. Copyright (1996) American Chemical Society

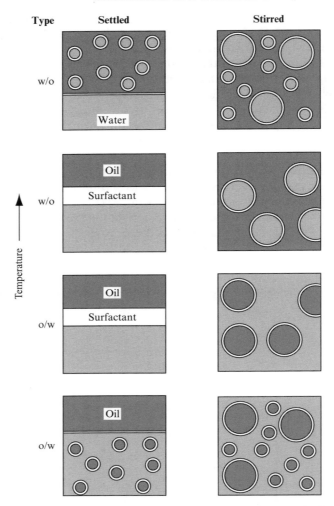

Figure 4.19 Schematic illustration of the state of dispersion in non-ionic surfactant–oil–water systems as a function of temperature. Small circles indicate swollen micelles (microemulsion droplets) and large circles emulsion drops

This observation applies to self-assembly structures like micelles, to particles stabilized by surfactants or polymers, or macroscopic surfaces covered by a surfactant or a polymer. This is *inter alia* seen in the temperature dependence of the stability of dispersions. Polymer and surfactant adsorption increase in magnitude with increasing temperature (see Chapter 17). At higher temperatures the adsorbed layers become more compact.

In mixed systems (polymer-polymer or polymer-surfactant), a number of unusual temperature dependencies are noted, for example, with regard to incompatibility and association.

Temperature Anomalies are Present in Solvents other than Water

It appears that these effects are not confined to water as a solvent. They are also found for some other polar solvents although the effects are weaker. As an example, we may consider non-ionic surfactants in formamide. Here there is also the formation of micelles and different liquid crystalline phases. However, the CMCs are much higher and the ranges of liquid crystal stability lower. The solubility is much higher than in water, which corresponds to a higher cloud point. As illustrated in Figure 4.20, the cloud point of $C_{12}E_3$ in formamide is a

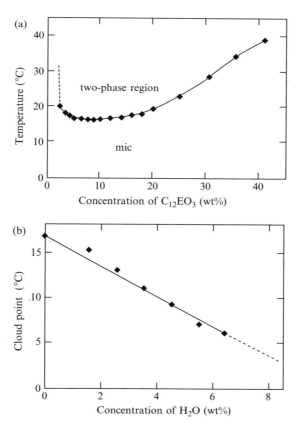

Figure 4.20 Non-ionic surfactants in formamide behave qualitatively as in water. They give a cloud point (as here for $C_{12}E_3$), which decreases as water is added. From T. Wärnheim, J. Bokström and Y. Williams, *Colloid Polym. Sci.*, **266** (1988) 562

little higher than that of $C_{12}E_4$ in water and the cloud point decreases progressively as water is added.

Bibliography

Alexandridis, P. and B. Lindman (Eds), *Amphiphilic Block Copolymers*, Elsevier, Amsterdam, 2000.

Bailey Jr, F. E. and J. V. Koleske, *Alkylene Oxides and Their Polymers*, Marcel Dekker, New York, 1991.

Laughlin, R. G., *The Aqueous Phase Behaviour of Surfactants*, Academic Press, London, 1994.

Nace, U. M. (Ed.), *Polyoxyalkylene Block Copolymers*, Marcel Dekker, New York, 1996.

Olsson, U. and H. Wennerström, Globular and bicontinuous phases of nonionic surfactant films, *Adv. Colloid Interface Sci.*, **49** (1994) 113.

Piirma, I., *Polymeric Surfactants*, Marcel Dekker, New York, 1992.

Schick, M. J. (Ed.), *Nonionic Surfactants: Physical Chemistry*, Marcel Dekker, New York, 1987.

5 MIXED MICELLES

Technical uses of surfactants mostly involve more than one surfactant species. Surfactant mixtures form micelles that include all of the surfactant species present. This chapter deals with mixed micelles and how to understand and predict the critical micelle concentration (CMC) of surfactant mixtures as well as the molecular composition of the mixed micelles. Mixed micelles have received increasing attention since the early 1980s. The derivations of the equations presented in the following are given in the Appendix at the end of the chapter.

Systems of Surfactants with Similar Head Groups Require no Net Interaction

We will start the treatment of mixed micelles with the simplest case, i.e. when there is no net interaction between the surfactant species that are mixed. Such is the case when mixing two surfactants with the same head group but with different chain lengths, for example. Note that, of course, there is an interaction between the head groups of the surfactants in the mixed micelle, but since the head groups are of the same kind this interaction is not different for the different surfactant species and hence the net interaction is zero.

Comparing with the hydrophilic–lipophilic balance (HLB) of a mixture, which is calculated as the weight average of the HLB of the single surfactants, it is reasonable to assume that the CMC of a surfactant mixture is an average of the CMCs of the single surfactants, i.e.

$$CMC = x_1 CMC_1 + x_2 CMC_2 \tag{5.1}$$

where CMC is the critical micelle concentration of the surfactant mixture and the CMC_i are the critical micelle concentrations of the single surfactant species, and x_1 and x_2 are the mole fractions of the respective surfactants in the system. Equation (5.1) leads to erroneous results, however, if x_1 is meant to represent the fraction of surfactant 1 in the whole system, i.e. where:

$$x_1 = \frac{C_1}{C_1 + C_2} \tag{5.2}$$

where C_1 and C_2 are the molar surfactant concentrations of the respective species. However, looking inside the micelles, equation (5.1) makes sense, viz. when x_1 represents the mole fraction of surfactant 1 in the micelle itself and not the whole system, i.e. x_1^m. Thus, the CMC of a surfactant mixture is written as follows:

$$CMC = x_1^m CMC_1 + x_2^m CMC_2 \qquad (5.3)$$

where x_1^m is the mole fraction of surfactant 1 in the micelles. Equation (5.3) is of no use for the prediction of the CMC of a mixture, however, since the surfactant composition in the micelles, x_1^m, is not known *a priori*. It can be shown (see the Appendix) that if x_1 is the solution composition, the expression for the CMC of the surfactant mixture can be written as follows:

$$\frac{1}{CMC} = \frac{x_1}{CMC_1} + \frac{x_2}{CMC_2} \qquad (5.4)$$

This equation is easily extended to a mixture of three, or more, surfactants:

$$\frac{1}{CMC} = \frac{x_1}{CMC_1} + \frac{x_2}{CMC_2} + \frac{x_3}{CMC_3}, \text{ etc.} \qquad (5.5)$$

The molar composition of the mixed micelle is given by the following:

$$x_1^m = \frac{x_1 CMC_2}{x_1 CMC_2 + x_2 CMC_1} \qquad (5.6)$$

Figures 5.1(a) and 5.1(b) show respectively the calculated CMC and the micelle composition as a function of the solution composition using equations (5.4) and (5.6) for three cases where $CMC_2/CMC_1 = 1$, 0.1 or 0.01. As can be seen in these figures, the CMC and also the micellar composition change dramatically with solution composition when the CMCs of the surfactants differ considerably, i.e. when the ratio of the CMCs is far from 1. Thus, a small amount of a very hydrophobic surfactant will change the solution properties of a system dramatically. This fact is used when preparing microemulsions, for example, where the addition of long-chain alcohols will change the properties considerably.

If we for the moment assume that component 2 represents a very surface active species (i.e. CMC_2/CMC_1 is a small number, e.g. of the order of 0.01) and that it is present in low concentrations (x_2 is of the order of 0.01) we find from equation (5.6) that $x_1^m \approx x_2^m \approx 0.5$, i.e. at the CMC of the system the micelles are up to 50% composed of component 2. It is therefore not surprising

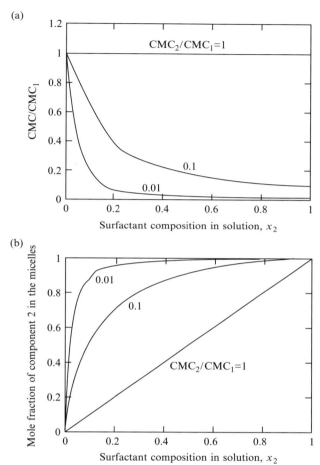

Figure 5.1 (a) The calculated CMC and (b) the micellar composition as a function of the solution composition for three systems where $CMC_2/CMC_1 = 1$, 0.1 and 0.01, respectively

that surface active contaminants sometimes play an important role in surface chemistry. An example of such a contaminant is dodecyl alcohol in sodium dodecyl sulfate (SDS), which is formed from hydrolysis of the surfactant.

The next example is shown in Figure 5.2, where the CMC as a function of the molar composition for a mixture of SDS and NP-E$_{10}$ (ethoxylated nonylphenol with 10 oxyethylene units) is drawn. Here, the CMC as functions of both the molar composition of the solution and the molar composition in the micelles are plotted. This figure shows that if the molar composition of the micelles is

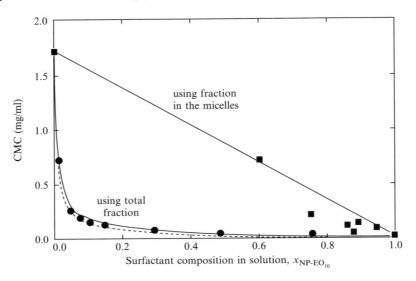

Figure 5.2 The critical micelle concentration as a function of surfactant composition, x_1, or the micellar surfactant composition, x_1^m, for the system SDS + NP-E$_{10}$, showing the validity of equation (5.3)

used as the x-axis the CMC is more or less the arithmetic mean of the CMCs of the two surfactants, as predicted by equation (5.3). If, on the other hand, the molar composition in the solution is used as the x-axis (which at the CMC is equal to the total molar composition), then the CMC of the mixture shows a dramatic decrease at low fractions of NP-E$_{10}$. This decrease is due to preferential absorption of NP-E$_{10}$ in the micelle. This is illustrated further in Figure 5.3, which shows the mole fraction of non-ionic surfactant in the micelles as a function of the mole fraction of non-ionic surfactant in the solution. This figure clearly illustrates that the non-ionic surfactant is preferentially absorbed in the micelle. This is due to the higher hydrophobicity of this non-ionic surfactant, as revealed by its lower CMC when compared to the anionic SDS. (The present sample of SDS has a CMC of *ca.* 1.75 mg/ml while the CMC of the NP-E$_{10}$ is *ca.* 0.05 mg/ml). The dotted line in the figure is calculated from equation (5.6). We note that this prediction does not perfectly follow the experimental results, with the reason being that in this system there is a small net interaction between the surfactants. This will be discussed further below.

Figures 5.4(a) and 5.4(b) show examples where mixtures of potassium soaps with 8, 10 and 14 carbons (KC$_8$, KC$_{10}$ and KC$_{14}$) have been investigated. The resemblance with Figure 5.1 is obvious, thus confirming the dominant role of the more hydrophobic surfactant.

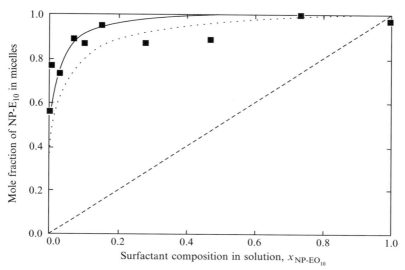

Figure 5.3 The surfactant composition in the micelles, $x_{NP\text{-}E_{10}}^{m}$, as a function of the surfactant composition in the bulk solution, $x_{NP\text{-}E_{10}}$ for the system SDS + NP-E$_{10}$. The dashed line shows the same composition in the micelles and bulk, while the dotted line is the calculated composition assuming no interactions (equation (5.6)). The full drawn line is calculated from equation (5.9). Reprinted with permission from B. Kronberg, M. Lindström and P. Stenius, *Phenomena in Mixed Surfactant Systems*, J. F. Scamehorn (Ed.), ACS Symposium Series, No. 311, American Chemical Society, Washington, DC, 1986, p. 225. Copyright (1986) American Chemical Society

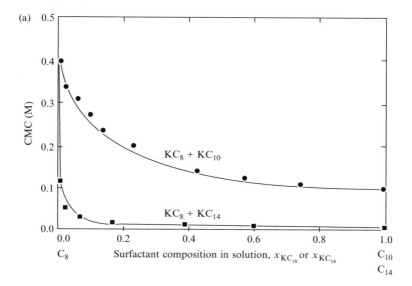

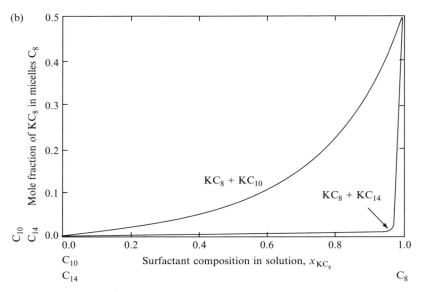

Figure 5.4 (a) The CMC and (b) surfactant composition of $KC_8 + KC_{10}$ and $KC_8 + KC_{14}$ mixtures. The continuous lines are calculated from equations (5.3) and (5.6), respectively. Reprinted with permission from R. Nagarajan, *Mixed Surfactant Systems*, P. M. Holland and D. N Rubingh (Eds.), ACS Symposium Series, No. 501, American Chemical Society, Washington, DC, 1992, p. 54. Copyright (1992) American Chemical Society

General Treatment of Surfactant Mixtures Requires a Net Interaction

In many systems, surfactants of different kinds are mixed, such as anionics and non-ionics. Here the non-ionic surfactants shield the repulsion between the head groups of the anionic surfactants in the micelle and hence there is a net interaction between the two types of surfactant. Another example is, of course, mixtures of anionic and cationic surfactants, where indeed a very strong interaction between the surfactants is found. In these cases, a more elaborate analysis has to be performed in order to gain any molecular insight. Equation (5.3) is then rewritten to give:

$$CMC = x_1^m f_1^m CMC_1 + x_2^m f_2^m CMC_2 \qquad (5.7)$$

where f_1^m and f_2^m are the activity coefficients of the surfactants in the micelle. An expression for the activity coefficient can be found by using the regular solution theory, i.e.

$$\ln f_1^m = (x_2^m)^2 \beta \qquad (5.8a)$$

and:

$$\ln f_2{}^m = (x_1{}^m)^2 \beta \qquad (5.8b)$$

where β is an interaction parameter, quantifying the net interaction between the surfactant species in the micelle. In Chapter 10, we give a thorough description of the regular solution theory, where the χ parameter is identical to the β parameter used here. We use the notation β in this present chapter since this is the common notation in the literature treating mixed micelles. Positive β values imply that there is a net repulsion between the two surfactant components, while negative β values imply a net attraction. If β is zero, the activity coefficients will be unity and equation (5.7) reverts to equation (5.3). Negative β values are most commonly found, significant for a net attraction between the surfactant species. Positive β values can, however, exist in, for example, mixtures of normal hydrocarbon-based surfactants with fluorinated ones.

In terms of the solution composition we have the following:

$$\frac{1}{CMC} = \frac{x_1}{f_1{}^m CMC_1} + \frac{x_2}{f_2{}^m CMC_2} \qquad (5.9)$$

and:

$$x_1{}^m = \frac{x_1 f_2{}^m CMC_2}{x_1 f_2{}^m CMC_2 + x_2 f_1{}^m CMC_1} \qquad (5.10)$$

Figures 5.5(a) and 5.5(b) show the effects of increasingly negative β parameters on the CMC and micellar composition, respectively. Here, a ratio of the two CMCs of 0.1 has been used. Figures 5.5(a) shows that when the attraction between the surfactants increases, i.e. when the β parameter becomes increasingly more negative, the CMC of the mixture decreases. Here, β values of around -2 are typical for mixtures of anionic and non-ionic surfactants. Values in the -10 to -20 range are typical for mixtures of anionic and cationic surfactants. (Note that in the $\beta = -20$ case the CMC curve almost follows the X-axis on this scale in the figure). Figure 5.5(b) shows that with an increasingly negative value of the β parameter the mixed micelles tend towards a mixing ratio of 50:50, which is a reflection of the mutual electrostatic attraction between the surfactant components becoming increasingly dominating.

Both the predicted mixed CMC and the micellar composition have a variable sensitivity to the value assigned to the β parameter. This sensitivity depends on the ratio between the CMCs of the single surfactants. When the CMCs of the single surfactants are similar, the predicted value of the mixed CMC, or the micellar composition, is very sensitive to small variations in the β parameter.

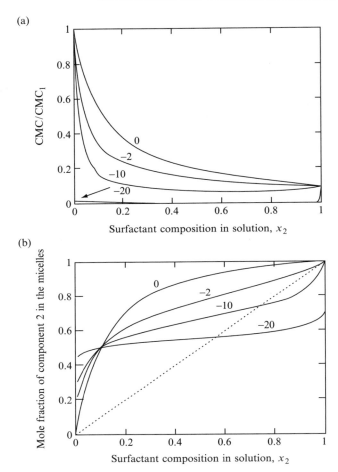

Figure 5.5 (a) Calculated CMC and (b) micellar composition for the indicated values of the β parameter for systems where $CMC_2/CMC_1 = 0.1$

On the other hand, when the ratio between the CMCs of the single surfactants is large, the predicted value of the mixed CMC, or the micellar composition, is insensitive to variations in the β parameter. This is realized upon a comparison of the results shown in Figures 5.2 and 5.6. The dashed line given in Figure 5.2 is calculated with β = 0, while the continuous line is calculated with β = −2. We note that there is not a tremendous difference between the two curves and the dashed line (β = 0) could well serve as a good approximation for the critical micelle concentration of the mixtures. Figure 5.6, on the other hand, shows the CMC of a mixture of SDS and C_8E_4, where the surfactants have similar CMCs.

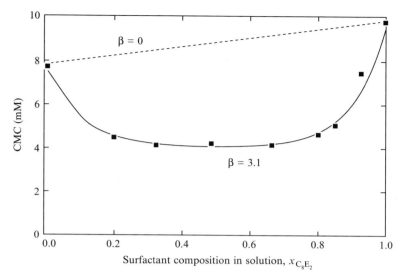

Figure 5.6 Variation in CMC for SDS and C_8E_4 mixtures . The dashed line is calculated from equation (5.3), i.e. the ideal case when $\beta = 0$, and the full drawn line from equation (5.9) with $\beta = -3.1$, illustrating that when the CMCs of the two single surfactants are close to each other the predicted CMC is very sensitive to the value of the β parameter. Reprinted with permission from P. Holland, *Mixed Surfactant Systems*, P. M. Holland and D. N Rubingh, (Eds), ACS Symposium Series, No. 501, American Chemical Society, Washington, DC, 1992, p. 31. Copyright (1992) American Chemical Society

Here the dashed line, corresponding to $\beta = 0$, differs considerably from the continuous curve, calculated with $\beta = -3.1$.

A system with a large net attraction between the surfactants is shown in Figures 5.7(a) and 5.7(b) for a mixture of sodium decyl sulfate, $(C_{10}SO_4Na)$, and decyltrimethylammonium bromide, $(C_{10}TBr)$. The dashed line in Figure 5.7(a) is calculated with $\beta = 0$, while the continuous curve corresponds to $\beta = -13.2$. We note that a very large synergetic effect is obtained when there is a large net attraction between the surfactants. This figure is also an illustration of the fact that it is indeed possible to lower the CMC of an anionic surfactant by adding a cationic surfactant without causing any precipitation. A requirement is that the hydrocarbon chains of the surfactants are short; in this example, each surfactant contains only 10 carbon atoms. Figure 5.7(b) shows that the micellar composition is constant at almost all surfactant compositions in this system, illustrating that the electrostatic attraction between the two surfactant species is the dominating driving force.

In Table 5.1, some β values for typical surfactant mixtures are given. We can note some trends, e.g. for mixtures of non-ionic and ionic surfactants the β

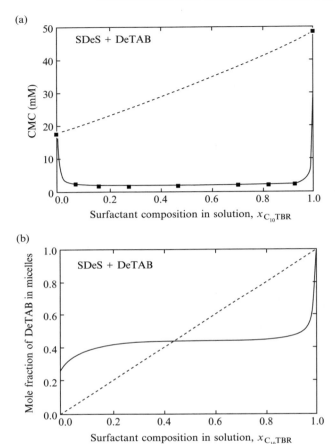

Figure 5.7 (a) CMC and (b) the micellar composition for mixtures of sodium decyl sulfate, (SDeS) and decyltrimethylammonium bromide, (DeTAB), illustrating that an anionic and a cationic surfactant can be used in combination in order to obtain a very high surface activity. Reprinted with permission from R. Nagarajan, *Mixed Surfactant Systems*, P. M. Holland and D. N Rubingh, (Eds), ACS Symposium Series, No. 501, American Chemical Society, Washington, DC, 1992, p. 54. Copyright (1992) American Chemical Society

value decreases in magnitude with increasing salt concentration. This is due to shielding of the electrostatic repulsion of the salt. We also note that for mixtures of $C_{12}\phi SO_3$ and NP-E$_{10}$ the β parameter decreases in magnitude, i.e. the net attraction decreases, with increasing temperature.

It is possible to calculate the optimum composition, i.e. the composition at which the CMC has a minimum value, in a surfactant mixture. There are two conditions for a minimum in CMC to appear, namely (a) β is negative, and (b)

Table 5.1 Typical β-values obtained for various surfactant mixtures. Reprinted with permission from P. Holland, *Mixed Surfactant Systems*, P. M. Holland and D. N. Rubingh (Eds), ACS Symposium Series, No. 501, American Chemical Society, Washington, DC, 1992, p. 31. Copyright (1992) American Chemical Society

β	Type[a]	Binary mixture	Medium	T (°C)
−3.9	AN	$C_{12}SO_4Na/C_{12}E_8$	Water	25
−2.6	AN	$C_{12}SO_4Na/C_{12}E_8$	0.5 M NaCl	25
−3.6	AN	$C_{12}SO_4Na/C_{10}E_8$	0.5 mM Na_2CO_3	23
−4.1	AN	$C_{12}SO_4Na/C_8E_{12}$	Water	25
−3.4	AN	$C_{12}SO_4Na/C_8E_8$	Water	25
−3.1	AN	$C_{12}SO_4Na/C_8E_4$	Water	25
−1.6	AN	$C_{12}E_2SO_4Na/C_8E_4$	Water	25
−4.3	AN	$C_{15}SO_4Na/C_{10}E_6$	Water	25
−3.4	AN	$C_{12}SO_3Na/C_{12}E_8$	Water	25
−4.4	AN	$C_{12}SO_4Na/C_{12}AO$	0.5 mM Na_2CO_3	23
−3.7	AN	$C_{12}SO_4Na/C_{10}PO$	1.0 mM Na_2CO_3	24
−2.4	AN	$C_{12}SO_4Na/C_{10}MSO$	1.0 mM Na_2CO_3	24
−1.6	AN	$C_{12}SO_4Na/NPE_{10}$	0.4 M NaCl	30
−3.2	AN	$C_{12}\phi SO_3Na/NPE_{10}$	0.15 M NaCl	30
−2.8	AN	$C_{12}\phi SO_3Na/NPE_{10}$	0.15 M NaCl	38
−2.1	AN	$C_{12}\phi SO_3Na/NPE_{10}$	0.15 M NaCl	46
−1.8	AN	$C_{12}\phi SO_3Na/NPE_{10}$	0.15 M NaCl	54
−0.0	AN	$C_{10}PO/C_{10}MSO$	1.0 mM Na_2CO_3	24
−0.1	NN	$C_{10}E_3/C_{10}MSO$	Water	25
−0.4	NN	$C_{12}E_3/C_{12}E_8$	Water	25
−0.8	NN	$C_{12}AO/C_{10}E_4$	0.5 mM Na_2CO_3	23
−16.5	AC	$C_{12}SO_4K/C_{12}AO$	Water	25
−10.5	AC	C_8SO_4Na/C_8TABr	Water	25
−18.5	AC	$C_{10}SO_4Na/C_{10}TABr$	Water	25
−13.2	AC	$C_{10}SO_4Na/C_{10}TABr$	0.05 M NaBr	23
−25.5	AC	$C_{12}SO_4Na/C_{12}TABr$	Water	25
−1.8	CN	$C_{10}TABr/C_8E_4$	0.05 M NaBr	23
−1.5	CN	$C_{14}TABr/C_{10}E_5$	Water	23
−2.4	CN	$C_{16}TACl/C_{12}E_5$	Water	23
−3.1	CN	$C_{16}TACl/C_{12}E_8$	0.1 M NaCl	25
−2.6	CN	$C_{18}TACl/C_{12}E_5$	2.4 M NaCl	25
−4.6	CN	$C_{20}TACl/C_{12}E_8$	Water	25
−2.7	CN	$C_{12}PyrCl/C_{12}E_8$	Water	25
−1.4	CN	$C_{12}PyrCl/C_{12}E_8$	0.1 M NaCl	25
−1.0	CN	$C_{12}PyrCl/C_{12}E_8$	0.5 M NaCl	25
−1.3	CN	$C_{12}PyrCl/NPE_{10}$	0.03 M NaCl	30
−1.7	CN	$C_{16}TACl/NPE_{10}$	0.03 M NaCl	30
−0.2	CC	$C_{16}TACl/C_{12}PyrCl$	0.15 M NaCl	30
−10.6	ZA	$C_{12}Np/C_{10}SO_4Na$	Water	30
−14.1	ZA	$C_{12}Np/C_{12}SO_4Na$	Water	30
−15.5	ZA	$C_{12}Np/C_{14}SO_4Na$	Water	30
−5.0	ZA	$C_{12}BMG/C_{12}SO_4Na$	Water	25
−1.2	ZC	$C_{12}BMG/C_{12}TABr$	Water	25
−0.9	ZN	$C_{12}BMG/C_{12}E_8$	Water	25

[a]A, anionic; C, cationi; Z, zwitterionic; N, nonionic.

$|\ln(CMC_2/CMC_1)| < |\beta|$. The optimum surfactant composition, $x_2(\text{min})$, is obtained from the following:

$$x_2(\text{min}) = \frac{\ln(CMC_2/CMC_1) + \beta}{2\beta} \qquad (5.11)$$

Figure 5.8 shows the optimum composition as a function of the β parameter for three different values of the CMC_2/CMC_1 ratio. The optimum composition is at 50% when the CMCs of the two surfactants are the same. For other cases, the optimum composition is shifted towards the surfactant that is more hydrophobic.

The Concept of Mixed Micelles can also be Applied to Amphiphiles not Forming Micelles

There are amphiphilic molecules that are unable to form micelles and instead phase-separate at higher concentrations. A well-known example is the class of medium-chain alcohols. The concentration at which phase separation occurs is analogous to the critical micelle concentration of micelle-forming surfactants. Hence, the solubility limit of medium-chain alcohols can be used as a 'CMC' of the alcohol in water in order to predict the CMC of a mixed system where the medium-chain alcohol is mixed with an ordinary surfactant. Figure 5.9 shows such a set of results. Here, the CMC of an anionic surfactant is plotted versus the alcohol concentration. The solid lines are calculated from equation (5.4), where the solubility limit of the alcohol has been used as the CMC.

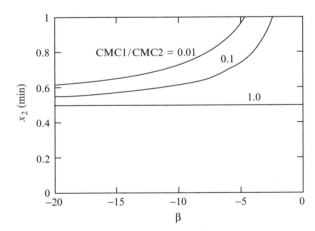

Figure 5.8 The optimum composition, giving a minimum in the CMC, for surfactant systems with $CMC_2/CMC_1 = 1$, 0.1 and 0.01, is very dependent on the β parameter

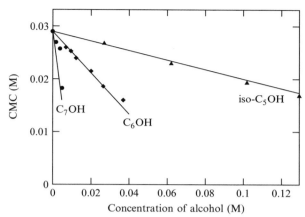

Figure 5.9 The mixed micelle concept can be applied to mixtures of surfactants with medium-chain alcohols, not forming micelles by themselves. The continuous lines are calculated from equation (5.4), where the solubility limit of the alcohol has been used as the CMC. The surfactant used is potassium dodecanoate (cf. Figure 5.8)

Mixed Surfactant Systems at Higher Concentrations Show Interesting Features

The phase behaviour of mixed surfactant systems may vary dramatically with the mixing ratio at higher surfactant concentrations. In applications of mixed surfactant systems, either synergistic or antagonistic effects may be obtained depending on the situation. These may be understood from observing mixed micellar behaviour, but perhaps even more clearly by using phase diagrams of mixed surfactant systems.

Mixtures of two similarly charged surfactants—anionic, cationic or non-ionic—show homogeneous phases, which depend on the ratio of surfactants and on the phase behaviour of the individual surfactants in a straightforward way. Figure 5.10 shows the phase diagram of an aqueous mixture of two cationic surfactants, i.e. one single-chain micelle-forming and one double-chain bilayer-forming. In such a case we can understand the phase behaviour in terms of an average surfactant critical packing parameter (CPP value) or spontaneous curvature (see Chapter 3). If, for example, the total surfactant concentration in the figure is kept constant at 50% and movement is from the double-chain surfactant (high CPP) towards the single-chain surfactant (low CPP), a hex-agonal liquid crystalline phase will form at the expense of the lamellar liquid crystalline phase.

Mixtures of two oppositely charged surfactants show a much richer phase behaviour. Precipitation of the crystalline salt of the two amphiphilic ions is a common phenomenon. If the stability of the crystalline state is lowered, e.g. by

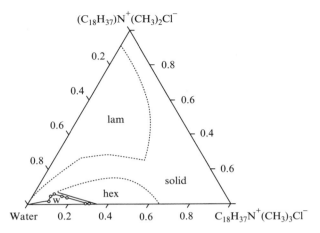

Figure 5.10 Phase diagram for a mixture of two similarly charged surfactants. Reprinted with permission from H. Kunieda and K. Shinoda, *J. Phys Chem.*, **82** (1978) 1710. Copyright (1978) American Chemical Society

using shorter alkyl chains, precipitation becomes less predominant or is absent. Then, a rich phase behaviour results with a large number of liquid crystalline phases. One conspicuous feature of these systems is the regions of thermodynamically stable vesicle solutions (see Figure 5.11).

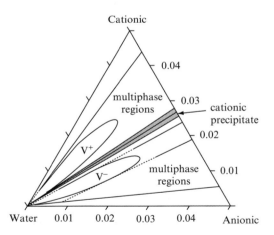

Figure 5.11 In a mixture of a cationic and an anionic surfactant there are two regions of thermodynamically stable vesicles, V^+ and V^-, respectively. Reprinted from A. Kahn, *Curr. Opinion Colloid Interface Sci.*, **1** (1996), 'Phase science of surfactants', 614–623, Copyright (1996), with permission from Elsevier Science

The normal self-assembly of a single ionic surfactant is counteracted by the lowering of the entropy arising from the counterion condensation (see Chapter 8). A large entropy increase results, however, when mixed surfactant aggregates, from cationic and anionic surfactants, are formed. All counterions, from both surfactants, are released from the aggregate surfaces. Thus, the entropy is not lowered as observed for the association of a single surfactant. This is, in fact, the driving force for the association of a cationic and an anionic surfactant mixture. For these mixed systems, some average surfactant number or CPP value is completely misleading. Rather, there is a non-monotonic change in the CPP, with a pronounced maximum as the mixing ratio is varied.

On removing the counterions from a stoichiometric mixture of oppositely charged surfactants, we obtain a so-called 'cat-an-ionic' surfactant, composed of two oppositely charged amphiphilic ions. A catanionic amphiphile behaves very much like a double-chain non-ionic one. In particular, we may compare it with a double-chain zwitterionic amphiphile. The electrostatic interactions residing in the counterion distribution are responsible for the major differences between charged and uncharged surfactants. One example is the very large swelling of a lamellar phase of an ionic surfactant and the small one of a non-ionic surfactant. However, even small additions of an ionic surfactants to a non-ionic, zwitterionic or catanionic surfactant can introduce very extensive swelling. This is illustrated for the lamellar phase of lecithin in Figure 5.12. The stability and extensive swelling of the lamellar phase in a mixture of oppositely charged surfactants on a slight departure from charge stoichiometry are closely related phenomena.

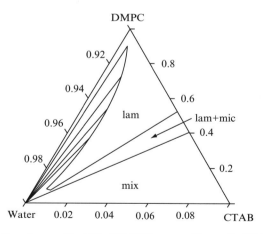

Figure 5.12 Lamellar phases of lecithin (DMPC) are readily swelled if the bilayers attain a charge from the addition of a surfactants. (CTAB) Reproduced by permission of Academic Press from L. Rydhag, P. Stenius and L. Ödberg, *J. Colloid Interface Sci.*, **86** (1982) 274

Mixed Surfactant Systems are used Technically

Figure 5.13 shows the soil removal from wool by a mixture of an alkyl ether sulfate and a linear alkyl sulfate. This figure shows an enhanced soil removal at compositions of *ca.* 10–20% of the alkyl ether sulfate. This is most likely due to a higher solubilizing power at this composition, which in turn is due to the lower CMC. The results presented in Figure 5.13 were obtained at a constant total surfactant concentration and thus the maximum number of micelles exists at concentrations close to the minimum in the CMC of the mixture.

Figures 5.14(a–c) show the foam height, dispersion ability of MnO_2 and zeolite, and wetting time of cotton, respectively, of a surfactant system consisting of SDS and $C_{16}E_3$. All of the three figures show an optimum at low fractions of the non-ionic surfactant. Hence, the surface activity has been enhanced by the addition of the more surface active component, $C_{16}E_3$. This figure is an illustration of the power of mixing surfactants in order to enhance the surface activity, and hence the technical performance.

Yet another use of mixing surfactants is to lower the Krafft point, or the Krafft temperature. The Krafft temperature is the melting point of the hydrated surfactant and at temperatures below the Krafft point the surfactant is not usable due to the low solution concentration (see Chapter 2). Certain systems show an eutecticum with respect to the Krafft temperature, as is illustrated in Figure 5.15 for a mixture of two alkylbenzene sulfonates, thus enabling the use of these surfactants at lower temperatures than the Krafft points of the single species.

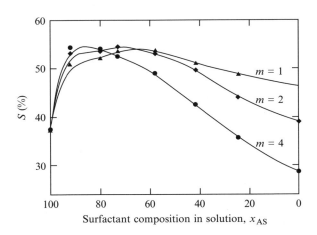

Figure 5.13   Soil removal from wool by mixtures of an alkyl ether sulfate and an alkyl sulfate (AS)

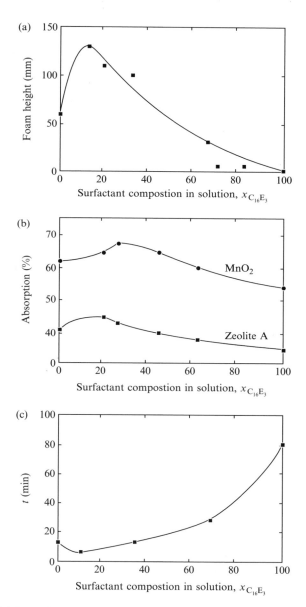

Figure 5.14 (a) Foam height, (b) dispersion ability of MnO_2 and zeolite, and (c) wetting time of cotton of a surfactant system with SDS and $C_{16}E_3$. From F. Jost, H. Leiter and M. Schwuger, *Colloid Polym. Sci.*, **266** (1988) 554

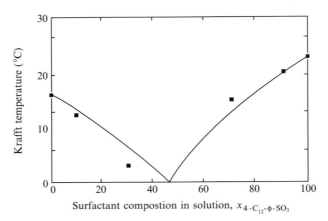

Figure 5.15 The Krafft temperature for a mixture of 4-dodecylbenzene sulfonate, $4\text{-}C_{12}\text{-}\phi\text{-}SO_3^-$ and 3-decylbenzene sulfonate, $3\text{-}C_{10}\text{-}\phi\text{-}SO_3^-$. Reprinted with permission from J. Scamehorn, *Mixed Surfactant Systems*, P. M. Holland and D. N Rubingh (Eds) ACS Symposium Series, No. 501, American Chemical Society, Washington, DC, 1992, p. 392. Copyright (1992) American Chemical Society

Finally, the cleaning efficiency of a surfactant pair, where none of the single surfactants are able to remove lipid from a surface while the mixture is able to remove up to ca. 55% of the lipid, is shown later in Figure 22.19. This optimum in cleaning efficiency is a reflection of a combination of (a) an optimum in critical packing parameter of the surfactant mixture and (b) an optimum in the number of micelles that are able to solubilize the lipid.

Appendix

Derivation of Equations (5.3) and (5.5)

The chemical potential of a surfactant species i in solution can be written as

$$\mu_i = \mu_i^0 + RT \ln c_i \qquad (5.12)$$

where c_i is the surfactant concentration, assuming that the solution is ideal such that the activity coefficient is unity. Here, μ_i^0 is the chemical potential in a reference state. At the critical micelle concentration, CMC, we therefore have the following:

$$\mu_i^{CMC} = \mu_i^0 + RT\ln CMC_i \qquad (5.13)$$

In a solution of a surfactant mixture at the CMC, the solution concentration of species i, i.e. the concentration of i in the aqueous phase is $c_i = x_i$ CMC, where CMC is the critical micelle concentration of the surfactant mixture. Hence:

$$\mu_i = \mu_i^0 + RT\ln (x_i CMC) \qquad (5.14)$$

The chemical potential of component i in the mixed micelle is given by the following:

$$\mu_i^{\text{mixed micelle}} = \mu_i^{\bullet} + RT\ln (f_i^m x_i^m) \qquad (5.15)$$

where x_i^m and f_i^m are, respectively, the mole fraction and the activity coefficient of the ith component in the micelle and $\mu_i^{\bullet}$ is the chemical potential of component i in a reference state, which is the neat micelle of component i. When $x_i^m = 1$, i.e. for the single surfactant species in solution, we have the following:

$$\mu_i = \mu_i^{\bullet} = \mu_i^{CMC} = \mu_i^0 + RT\ln CMC_i \qquad (5.16)$$

since $f_i^m = 1$. At equilibrium, the chemical potential of component i in the solution equals that in the micelle and so we have the following:

$$\mu_i^{\text{mixed micelle}} = \mu_i^0 + RT\ln CMC_i + RT\ln (f_1^m x_i^m)$$
$$= \mu_i^0 + RT\ln (x_i CMC) \qquad (5.17)$$

Hence:

$$CMC x_i = CMC_i x_i^m f_i^m \qquad (5.18)$$

Since the sum of the mole fractions of the different components is equal to unity we find that, assuming ideal mixing of the surfactants in the micelles, i.e. $f_i^m = 1$, the following:

$$CMC = \Sigma CMC_i x_i^m \qquad (5.19)$$

which is identical to equation (5.3) for a two-component system. Rewriting equation (5.18) as follows:

$$x_i^m = x_i \frac{CMC}{f_i^m CMC_i} \qquad (5.20)$$

and again using the fact that $\Sigma x_i^m = 1$, we find that:

$$\frac{1}{CMC} = \Sigma \frac{x_i}{f_i^m CMC_i} \qquad (5.21)$$

which is identical to equation (5.5) when $f_i^m = 1$.

Calculating the β Parameter

In calculating this parameter from experimental data for the CMC of a surfactant mixture, we use expressions (5.8) and (5.18). First eliminating the β parameter, we find the following:

$$F = (x_1^m)^2 \ln (x_1 CMC) - (x_1^m)^2 \ln (x_1^m CMC_1) - (x_2^m)^2 \ln (x_2 CMC)$$

$$+(x_2^m)^2 \ln (x_2^m CMC_2) = 0 \tag{5.22}$$

Here, the only unknown is x_1^m. We now use the Newton–Raphson method to iterate the equation in order to get x_1^m. Hence, the first derivative of F with respect to x_1^m is as follows:

$$F' = \frac{dF}{dx_1^m} = 1 + 2x_2^m[\ln (x_2 CMC) - \ln (x_2^m CMC_2)]$$

$$+2x_1[\ln (x_1\ CMC) - \ln (x_1^m CMC_1)] \tag{5.23}$$

Assigning a first value to x_1^m (here we give this value the symbol $x_1^m(1)$), the next value of $x_1^m(= x_1^m(2))$ is given by the following:

$$x_1^m(2) = x_1^m(1) - \frac{F(x_1^m(1))}{F'(x_1^m(1))} \tag{5.24}$$

This is repeated until two consecutive values are identical within the desired accuracy. Then, the β parameter is calculated via equations (5.8) and (5.18), as follows:

$$\beta = \frac{1}{(x_1^m)^2} \ln \left(\frac{x_2 CMC}{x_2^m CMC_2} \right) = \frac{1}{(x_2^m)^2} \ln \left(\frac{x_1 CMC}{x_1^m CMC_1} \right) \tag{5.25}$$

Bibliography

Holland, P. M. and D. N. Rubingh (Eds), *Mixed Surfactant Systems*, ACS Symposium Series, No. 501, American Chemical Society, Washington, DC, 1992.

Hua, X. Y. and M. Rosen, *J. Colloid Interface Sci.*, **90**, 212 (1982), and subsequent papers in the series of Rosen.

Ogiono, K. and M. Abe, *Mixed Surfactant Systems*, Surfactant Science Series 46, Marcel Dekker, New York, 1993.

6 MICROEMULSIONS

The Term Microemulsion is Misleading

Microemulsions are macroscopically homogeneous mixtures of oil, water and surfactant, which on the microscopic level consist of individual domains of oil and water separated by a monolayer of amphiphile. Microemulsions should not be regarded as emulsions with very small droplet size; micro- and macroemulsions are fundamentally different (Figure 6.1). Whereas macroemulsions are inherently unstable systems in which the droplets eventually will undergo coalescence, microemulsions are thermodynamically stable with a very high degree of dynamics with regard to the internal structure. As a thermodynamically stable phase based on surfactant self-assemblies, it has much in common with other surfactant phases, micellar solutions and liquid crystalline phases.

Emulsion	Microemulsion
Unstable, will eventually separate	Thermodynamically stable
Relatively large droplets (1-10 μm)	Small aggregates (~10 nm)
Relatively static system	Highly dynamic system
Moderately large internal surface, moderate amount of surfactant needed	High internal surface, high amount of surfactant needed
Small oil/water curvature	The oil/water interfacial film can be highly curved

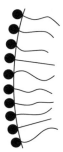 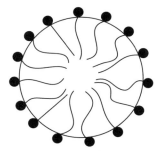

Figure 6.1 Characteristic differences between emulsions and microemulsions

Microemulsions can form in mixtures of surfactant, oil and water alone, but in many cases, a second surfactant, or a so-called cosurfactant, such as a medium-chain alcohol, is required. In the case of other surfactants, the addition of salt or a cosolvent can lead to the formation of microemulsions.

Phase Behaviour of Oil–Water–Surfactant Systems can be Illustrated by Phase Diagrams

The application potential of microemulsions was recognized at an early stage and has triggered a buildup of knowledge about the phase behaviour of oil-water-surfactant systems. The phase behaviour of a three-component system can, at fixed temperature and pressure, best be represented by a phase diagram, as shown in Figure 6.2. At low surfactant concentrations, there is a sequence of equilibria between phases, commonly referred to as Winsor phases. A microemulsion phase may be in equilibrium with excess oil (Winsor I, or lower phase microemulsion), with excess water (Winsor II, or upper phase microemulsion) or with both excess phases (Winsor III, or middle phase microemulsion). For non-ionic surfactants, the I → III → II transition (from the upper left to the lower right in Figure 6.2) may occur by raising the temperature while for ionic surfactant systems containing an electrolyte, i.e. a quaternary system, the transition may be induced by increasing salinity. System (a), which represents a composition based on non-ionic surfactant at a low temperature, is indicative of the phase behaviour of a hydrophilic surfactant.

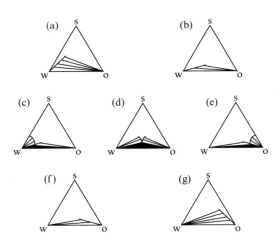

Figure 6.2 A series of phase diagrams of a ternary system transformed from Winsor I (a, b) via Winsor III (c to e) to Winsor II (f, g). The dark triangles are three-phase regions. Tie-lines indicate the compositions of the equilibrium phases of the two-phase regions: S, W and O stand for surfactant, water and oil, respectively

Only small quantities of oil can be solubilized into the o/w microemulsion which is in equilibrium with almost pure oil, as indicated by the tie-lines. On raising the temperature the surfactant becomes less hydrophilic and more oil can be solubilized into the microemulsion, but the system is still of the Winsor I type (b). Systems (c), (d) and (e) illustrate Winsor III systems with a three-phase triangle surrounded by two-phase regions. On increasing the temperature, the microemulsion apex moves from left to right. When it is in a central position (d), i.e. at the point where the microemulsion contains equal amounts of oil and water, the system is referred to as balanced. The height of the microemulsion triangle at the point where the system is balanced can be seen as a measure of the surfactant efficiency. With a very efficient amphiphile the microemulsion apex may appear at only a few per cent surfactant, and the importance of such systems is discussed in Chapter 22 in relation to enhanced oil recovery. Systems (f) and (g) may be seen as mirror images of (b) and (a). The phase behaviour depicted in the diagrams of Figure 6.2 can be visualized by test tube experiments, as shown in Figure 6.3.

In the triangular diagrams of Figures 6.2 and 6.3, temperature is not a variable. In order to illustrate the effect of temperature on the phase behaviour of oil–water–surfactant systems, a phase prism, as shown in Figure 6.4, can be used. Characterization of an entire prism is tedious, however, and in order to simplify the work, the number of degrees of freedom is often reduced by one, either by keeping the oil-to-water ratio constant, usually at 1:1 (Figure 6.4(a)), or by using a constant surfactant concentration (Figure 6.4(b)).

Figure 6.5 illustrates a section through the phase prism for an oil–water–non-ionic surfactant system at a 1:1 oil-to-water ratio, equivalent to the plane cut out of Figure 6.4(a). The three-phase region, consisting of a microemulsion in equilibrium with excess oil and water, exists between temperatures T_1 and T_2 and the temperature range is very dependent on surfactant concentration. At a surfactant concentration C^* and at the balanced temperature T^*, the three-phase region meets the one-phase microemulsion (the microemulsion apex of the three-phase region). At higher surfactant concentrations, the microemulsion is in equilibrium with the lamellar phase.

The section through the phase prism representing constant surfactant concentration, equivalent to the plane cut out of Figure 6.4(b), is also a useful tool for studying the phase behaviour of non-ionic systems. A typical example is shown in Figure 6.6. The diagram illustrates the relationship between temperature and relative amounts of oil and water. (Note that by convention the weight fraction of oil rather than the oil-to-water ratio is given on the x-axis.)

The phase diagram shows an isotropic solution phase forming a narrow channel which connects the surfactant in water at lower temperatures with the surfactant in oil at higher temperatures. This diagram illustrates the limited region of existence of the microemulsion phase which is typical of systems based on non-ionic surfactants. The system is balanced at around 28°C. At

142

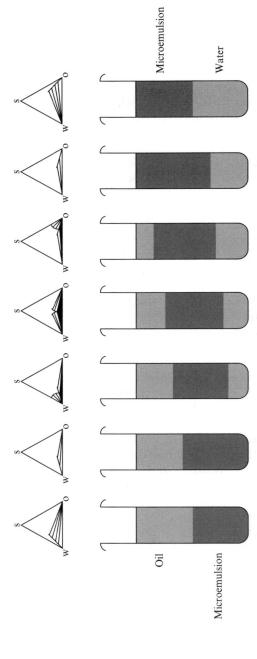

Figure 6.3 Phase changes of a system containing equal amounts of oil and water and a given (low) amount of surfactant. For a non-ionic surfactant system the left-to-right transition may be induced by increasing the temperature

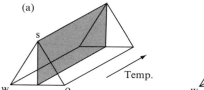 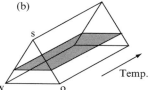

Figure 6.4 Phase prisms illustrating cuts at (a) a constant oil-to-water ratio, and (b) a constant surfactant concentration. From U. Olsson and H. Wennerström, *Adv. Colloid Interface Sci.*, **49** (1994) 113

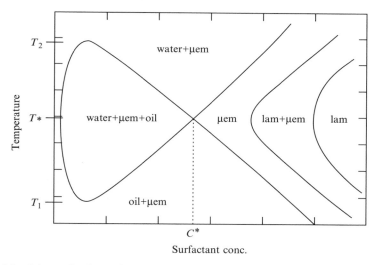

Figure 6.5 Schematic phase diagram of a ternary system based on non-ionic surfactant; 'lam' denotes a lamellar liquid crystalline phase. The oil-to-water ratio is kept constant and the surfactant concentration is varied. From U. Olsson and H. Wennerström, *Adv. Colloid Interface Sci.*, **49** (1994) 113

higher temperatures, the surfactant is too oil soluble and an aqueous phase separates out. At lower temperatures, the surfactant is too hydrophilic and oil separates out. The diagram also shows that a lamellar liquid crystalline phase forms at intermediate temperatures, both at high and low weight fractions of oil.

The Choice of Surfactant is Decisive

The curvature of the oil-water interface of a microemulsion may vary from highly curved towards oil, to zero mean curvature, and to highly curved towards water. Contrary to emulsions, the curvature of microemulsions can

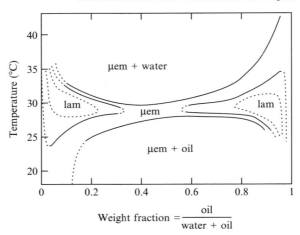

Figure 6.6 Phase diagram of the system $C_{12}E_5/H_2O$/cyclohexane-hexadecane at a constant surfactant concentration of 7.0 wt%. Reprinted with permission from U. Olsson, K. Nagai and H. Wennerström, *J. Phys. Chem.*, **92** (1988) 6675. Copyright (1988) American Chemical Society

be considerable at the scale of the surfactant. This implies that not only the hydrophile-lipophile balance, but also the molecular geometry of the surfactant, is an important factor in finding the optimum microemulsion surfactant.

A popular way of dealing with surfactant geometry is to use the packing parameter concept, introduced in Chapter 2. The geometric or packing properties of surfactants depend on their optimal head group area, a, as well as on the hydrocarbon volume, v, and the extended length of the surfactant hydrophobe chain, l_{max}. The value of a is governed by repulsive forces acting between the head groups and attractive hydrophobic forces between the hydrocarbon chains. Steric chain-chain and oil penetration interactions determine v and l_{max}. As described earlier, the value of the dimensionless critical packing parameter (CPP), $v/(l_{max}a)$, can be used to determine what type of aggregate will spontaneously form in solution.

From geometrical considerations it may be stated that surfactants with moderately long, straight-chain aliphatic hydrocarbon tails are best suited to prepare o/w microemulsions, surfactants with rather bulky hydrophobes are good for bicontinuous microemulsions (see below) and surfactants with highly branched hydrophobic tails should be used for w/o microemulsions. Indeed, this has been found to be the case. Often a combination of surfactants is used in the formulation. The surfactant geometry will then be the mean geometry of the species involved. Consequently, a combination of a surfactant with a single straight-chain tail and one with two branched tails, such as the top and bottom compounds of Figure 6.7, may constitute an ideal mixture to formulate a bicontinuous microemulsion.

For some microemulsion applications, in particular in enhanced oil recovery (Chapter 22), there is a need for surfactants which by themselves, i.e. without cosurfactant or cosolvent, can solubilize large amounts of oil and water. By proper optimization of the surfactant geometry, including the surfactant molecular weight, molecules with extreme solubilizing capacities have been obtained. An example is shown in Figure 6.8. The relatively simple twin-tailed sulfate surfactant used in 1.54 wt% can solubilize 49.2% aqueous NaCl solution and 49.2% hexane, representing 32 times as much water and hexane as surfactant. Even higher solubilizations have been obtained with somewhat larger molecules with tailor-made branching of the hydrophobic chain.

The surfactant of Figure 6.8 has a very low solubility in both oil and water. This is an important characteristic of a surfactant to be used with high efficiency in Winsor III systems. Very low saturation concentrations in both

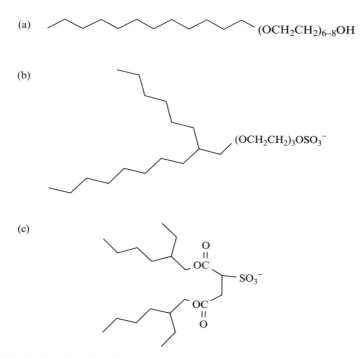

Figure 6.7 Examples of surfactants for different types of microemulsions: (a) an alcohol ethoxylate (used at temperatures well below the cloud point) for o/w microemulsions: (b) a branched-tail ether sulfate for bicontinuous microemulsions; (c) a double-tail sulfonate for w/o microemulsions

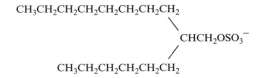

Figure 6.8 Structure of a surfactant, sodium 2-hexyldecylsulfate, with a very high solubilizing capacity for both oil and water

excess phases are needed in order to have the surfactant confined to the location where it exerts its action, i.e. at the oil–water interface.

Ternary Phase Diagrams can be Complex

For an amphiphile of less pronounced amphiphilic character, like a short-chain surfactant, the immiscibility of oil and water is overcome only at rather high concentrations of amphiphile. For a stronger amphiphile, such as a long-chain surfactant, lower amounts of surfactant are needed to form a microemulsion and the phase diagram may be more complex, with a three-phase triangle in addition to two-phase regions. Some different appearances of the phase diagram of ternary surfactant–oil–water systems were illustrated in Figure 6.2. (Liquid crystalline phases were omitted there.) The number of phases may vary, as exemplified in Figure 6.3. For a strong or efficient surfactant, i.e. one with a very pronounced amphiphilic character, which produces microemulsions at low surfactant contents and a lamellar phase at higher concentrations, the phase diagram is more complex, as exemplified in Figure 6.9. The example is taken for an exactly balanced surfactant (such as $C_{12}E_5$ with decane at $38.2°C$), i.e. the microemulsion that is formed at the lowest surfactant concentration has equal volume fractions of oil and water. The phase diagram is then symmetric around this solvent ratio.

We will now be concerned with the microstructure of a microemulsion system as a function of a relevant system parameter which may be temperature, electrolyte concentration or surfactant composition of a mixture. It will then be illustrative to present a phase diagram at a fixed surfactant concentration and varying compositions of the solvent mixture. For non-ionics, we consider a Shinoda cut, as illustrated in Figure 6.6. Here, the microemulsion channel goes from the lower left to the upper right; it is possible to form microemulsions at all mixing ratios between oil and water by adjusting the temperature to a proper value.

How to Approach Microstructure?

An efficient surfactant, i.e. one forming a microemulsion at low concentration, is characterized by a low concentration of surfactant unimers in both solvents.

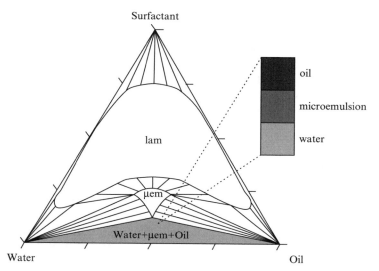

Figure 6.9 Phase behaviour under balanced conditions of a surfactant–water–oil system; μem denotes the microemulsion region and lam the lamellar phase region. From A. Kabalnov, B. Lindman, U. Olsson, L. Piculell, K. Thuresson and H. Wennerström, *Colloid Polym. Sci.*, **274** (1996) 297

There is then a very strong segregation into three types of domains: oil and water domains and surfactant monolayer films.

This segregation into domains, or pseudo-phases, is the first question relating to microemulsion structures. The second question to address concerns, in analogy with our general discussion of surfactant self-assembly structures (Chapter 3), whether or not the surfactant films give discrete or connected structures. In the early literature, discrete, i.e. droplet-type, structures were stressed but it was realized during the later 1970s that this picture is inconsistent with several observations, such as the stability over the entire range of solvent mixing ratios.

Having decided whether we have in a particular situation a discrete or bicontinuous type structure, we address the subsequent question which relates to the shapes of droplets and type of bicontinuous structure.

Molecular Self-Diffusion can be Measured

A general and reliable approach for probing microstructure in terms of connectivity is by monitoring the molecular self-diffusion coefficients. This approach is not limited to microemulsions, but is more generally applicable and has been instrumental in increasing our understanding of cubic liquid crystals and other isotropic solution phases as well.

In self-diffusion experiments, we investigate the molecular displacements over large distances, from several microns and upwards. This means that molecular motions within discrete aggregates, like micelles and microemulsion droplets, will not contribute to the experimental observations.

Self-diffusion of an entity (molecule, aggregate, particle) depends on a number of factors: size and shape of the diffusing entity, friction and obstruction. At low concentrations in a homogeneous medium, the diffusion coefficient of a spherical entity (radius R) is given by the Stokes–Einstein relationship, as follows:

$$D = kT/(6\pi\eta R) \qquad (6.1)$$

With a typical viscosity, η, of the order of 1 mPas (1 cP in older units), D will be *ca.* $2 \times 10^{-9}/R$, if R is expressed in Å and D in m^2/s. This means that a small molecule (R of the order of 1 Å) will have a diffusion coefficient of the order of 10^{-9} m^2/s. This will apply to the local (short-range) motion. For the case where there is no confinement into closed domains and the molecular species can translate freely, this value will apply also to the long-range motion, which is what is probed in the experiments. For large particles, the diffusion coefficient will be much lower and, as we will see, for the case of microemulsions with discrete droplets, the droplet size is typically of the order of 100 Å; the self-diffusion coefficient will then be of the order of 10^{-11} m^2/s.

There are several different ways in which molecular self-diffusion coefficients may be obtained. The most popular ones have been techniques based on radioactive labelling and NMR spectroscopy. In the former, diffusion is monitored over distances from a few to several mm. In the latter, which is the most versatile and convenient, the Fourier-transform pulsed-gradient spin-echo nuclear magnetic resonance (FT PGSE NMR) technique is used to monitor molecular displacements. Here, the motion of the nuclear spins in molecules is monitored in a spatially varying magnetic field. It is then possible to monitor diffusion over distances from a few μm to several μm. The obtained NMR signal (the 'spin-echo') contains contributions from the different molecules in the sample containing the spin probe (normally ^{1}H) but by Fourier transformation these different contributions can be resolved. It is generally possible to obtain the self-diffusion coefficients for all components in a complex liquid mixture with good precision within a few minutes.

Confinement, Obstruction and Solvation Determine Solvent Self-Diffusion in Microemulsions

Figure 6.10 illustrates how the self-diffusion of oil and water can be used to distinguish between structures with discrete droplets or micelles and structures in which both solvents form domains that are connected over macroscopic

Figure 6.10 Solvent molecules in domains, which are connected over macroscopic distances, can translate rapidly over long distances while solvent molecules, which are trapped in discrete or closed domains, diffuse very slowly. (a) An oil-swollen micelle or an oil-in-water microemulsion and (b) a bicontinuous structure. (We note that bicontinuity can only occur in three dimensions so this is not easily illustrated in two dimensions.) The points represent labelled molecules at the initial time (left figures) and after a certain time (right figures)

distances. A solvent confined to discrete particles will have a D-value orders of magnitude lower than that of a solvent forming connected domains.

The predicted self-diffusion behaviour of the principle structures can be summarized as follows:

1. A water-in-oil droplet structure, i.e. with discrete water domains in an oil continuum, will have a water diffusion that is much slower than that of oil; oil will have a D-value of the same order of magnitude as that of neat oil. Both water and surfactant diffusion will correspond to the diffusion of the droplets, with D-values of the order of $10^{-11}\,\mathrm{m}^2/\mathrm{s}$ or below.

2. An oil-in-water situation will have the opposite relation between the diffusion rates of the two solvents but will otherwise have analogous self-diffusion characteristics.

3. A bicontinuous microemulsion, where both solvents form domains extending over macroscopic distances, will be characterized by high D-values of both solvents. Surfactant diffusion is expected to be of the order

of $10^{-10}\,\mathrm{m^2/s}$ (as in a lamellar phase), i.e. much more rapid than for a droplet-type microemulsion, since surfactant diffusion is uninhibited, but significantly slower than in a simple molecule-dispersed solution.

4. For the molecule-dispersed or structureless situation we expect all components to have rapid diffusion (around $10^{-9}\,\mathrm{m^2/s}$ or slightly below).

These simple arguments apply perfectly to a case of complete segregation into oil and water domains and surfactant films and are appropriate for an efficient surfactant, i.e. a surfactant with low solubility in both solvents and which can mix large amounts of oil and water in homogeneous solution. If confinement is less pronounced, the difference between the D-values will be reduced. In the extreme case of a molecule-dispersed solution, with no aggregation, all components will be characterized by rapid diffusion.

Both single molecule translation and droplet translation will be affected by the obstruction due to domains that are inaccessible for that particular species. Droplet diffusion is retarded in proportion to the volume fraction, ϕ, of the droplets. For the case of spheres, the relative reduction is approximately given by the following:

$$D/D_0 = 1 - 2\phi \tag{6.2}$$

Solvent molecules in a continuous medium containing spherical droplets are retarded only moderately and the diffusion coefficient relative to the droplet-free case is given by:

$$D/D_0 = 1/(1 + \phi/2) \tag{6.3}$$

Also for elongated droplets, such as prolate or cylindrical shapes, the obstruction effect is moderate. This is illustrated in Figure 6.11, where we also see that oblate or disc-shaped droplets retard solvent diffusion very strongly, even at low volume fractions. Indeed, the presence of larger planar obstructing surfaces will reduce the D-value to two-thirds at low volume fractions. The same obstruction effect will apply for the zero mean curvature structures, already referred to in conjunction with sponge and cubic liquid crystalline phases (Chapter 3). The noted differences in obstruction effects between different obstructing geometries will be useful in assigning the microstructures of microemulsions.

The surfactant molecules in the surfactant films will be solvated, which will retard the solvent diffusion further. This retardation will be proportional to the surfactant concentration and can be accounted for.

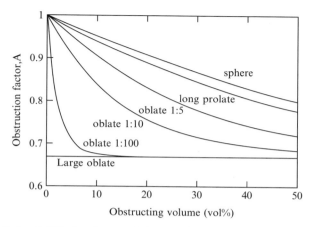

Figure 6.11 Solvent diffusion is retarded strongly by the presence of large oblates or discs, but only moderately by spheres or prolates. The obstruction factor, i.e. the ratio between the self-diffusion coefficient in the presence of obstructing particles and that of the neat solvent, is given as a function of the particle volume fraction for spheres, prolates and oblates, with different axial ratios. From B. Jönsson, H. Wennerström, P.-G. Nilsson and P. Linse, *Colloid Polym. Sci.*, **264** (1986) 77

Self-Diffusion Gives Evidence for a Bicontinuous Structure at Balanced Conditions

Essentially all types of surfactants can form microemulsions provided the conditions are appropriate. Furthermore, the solvent composition of the microemulsion can be varied over wide ranges by varying a suitable parameter, which may be temperature, salinity, cosurfactant or cosolvent concentration or the mixing ratio between two different surfactants. Stressing that the microemulsion behaviour is general, we choose a non-ionic surfactant system as an example. Non-ionic surfactant microemulsions offer the easiest illustration since they may be prepared from only three components and since the spontaneous curvature of non-ionic surfactant films can be varied by temperature rather than by varying the composition.

The self-diffusion behaviour at different temperatures in the micoemulsion channel of non-ionics is illustrated in Figure 6.12. The relationship between the diffusion coefficients of the two solvents is very different at different temperatures. At low temperatures, water diffusion is close to that of neat water while oil diffusion is strongly reduced and gives evidence for oil confinement into closed domains. At high temperatures, the converse situation applies with uninhibited oil diffusion and confinement of water into droplets.

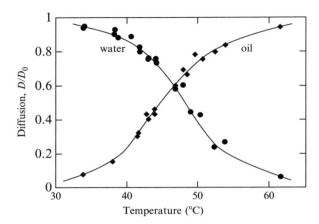

Figure 6.12 As temperature is increased through the microemulsion channel of a non-ionic system, water diffusion is reduced from high to very low values, while the opposite applies to oil. The ordinate gives the relative diffusion coefficient, i.e. the measured diffusion coefficient, D, over that of the neat solvent at the temperature in question. Reprinted with permission from U. Olsson, K. Shinoda and B. Lindman, *J. Phys. Chem.*, **90** (1986) 4083. Copyright (1986) American Chemical Society

At intermediate temperatures, the relative diffusion coefficients of both solvents are high, which gives evidence for a bicontinuous microstructure, i.e. with both oil and water domains being connected over macroscopic distances. It is observed for all types of surfactants that at the crossing point of the water and oil curves we have D/D_0-values which are *ca.* 0.6. This is close to the theoretical maximum of two-thirds for D/D_0 at the crossing point. Such a situation applies when the obstructing surface is effectively planar, allowing translation only in two directions out of three, or has a zero mean curvature.

Our structural description of bicontinuous microemulsions derives from the minimal surface-type structures, often found for cubic liquid crystalline phases (Chapter 3), but since the microemulsion is a solution without long-range order, a perturbed version of the regular minimal surface structure applies. In fact, the microstructure is close to that of the sponge phase illustrated in Figure 3.8. The main differences are that the sponge phase has bilayer surfactant films and the bicontinuous microemulsion monolayer surfactant films, and that in the sponge phase all channels are water filled while in the bicontinuous microemulsion every second one is filled with oil.

The Microstructure is Governed by Surfactant Properties

As discussed above, a value of the critical packing parameter below unity will give an o/w structure and a value above unity a w/o structure. For a value around one, we can form either a lamellar liquid crystalline phase or a bicontinuous microemulsion. The competition between these two alternatives is general and is decided by the flexibility of the surfactant film, so that a more flexible film favours the microemulsion.

We can alternatively discuss microstructure in terms of the spontaneous curvature of the surfactant films (Chapter 3). When H_0 is positive, an o/w structure forms while a w/o structure is preferred when H_0 is negative. When the spontaneous curvature is small, we may again have either a lamellar structure or a bicontinuous microemulsion.

Figure 6.13 summarizes the microstructure for a non-ionic system (cf. Figure 6.6 and the accompanying discussion); the picture would be analogous for

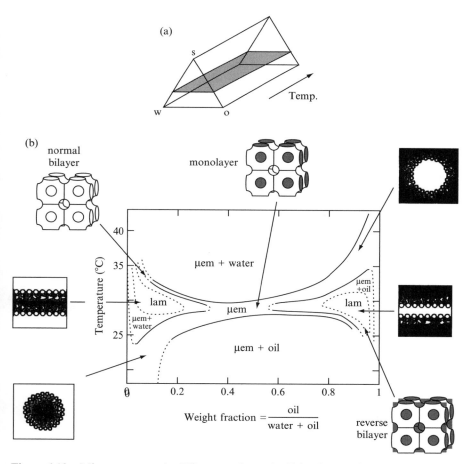

Figure 6.13 Microstructures in different regions of a Shinoda cut of the phase diagram of a non-ionic system. From A. Kabalnov, B. Lindman, U. Olsson, L. Piculell, K. Thuresson and H. Wennerström, *Colloid Polym. Sci.*, **274** (1996) 297

other surfactants, as when temperature is replaced by salinity for an ionic surfactant. Along the microemulsion channel from the lower left to the upper right there is a progressive change in microstructure involving droplet growth, elongation, connection, fully bicontinuous structure, structural inversion, disconnection and disintegration into smaller droplets. In the perpendicular, more narrow, channel there is a bicontinuous structure throughout. We note the proximity of the various bicontinuous structures to lamellar phase. The main factor, here controlled by temperature, in determining microstructure is the spontaneous curvature or the critical packing parameter, while the proportion of the two solvents plays a smaller role.

Figure 6.14 illustrates the microstructure in the alternative cut of the phase diagram (Figure 6.5), i.e. at a fixed volume ratio, in this case 1:1, between the solvents. For a balanced microemulsion system, i.e. one where the microemulsion forms at the lowest surfactant concentration for equal volumes of oil and water, the microstructure is bicontinuous. For such a surfactant system, the three-phase triangle in the phase diagram is symmetrical with respect to the water–oil base. It is generally true for three-phase, so called Winsor III, systems that the middle phase has a bicontinuous structure.

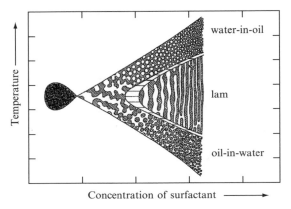

Figure 6.14 An illustration of microstructure in the cut of Figure 6.5. From R. Strey, *Colloid Polym. Sci.*, **272** (1994) 1005

Bibliography

Bourrel, M. and R. S. Schecter, *Microemulsions and Related Systems: Formulation, Solvency, and Physical Properties*, Surfactant Science Series 30, Marcel Dekker, New York, 1988.

Delpuech, J.-J. (Ed.), *Dynamics of Solutions and Fluid Mixtures by NMR*, Wiley, Chichester, UK 1995, Ch. 6.

Friberg, S. E. and P. Bothorel (Eds), *Microemulsions: Structure and Dynamics*, CRC Press, Boca Raton, Florida, CA, 1987.

Friberg, S. E. and B. Lindman (Eds), *Organized Solutions*, Surfactant Science Series 44, Marcel Dekker, New York, 1992.

Kahlweit, M. and R. Lipowsky (Eds), *Microemulsions, Ber. Bunsenges. Phys. Chem.*, **100**(3) (1996).

Kumar, P. and K. L. Mittal (Eds), *Handbook of Microemulsion Science and Technology*, Marcel Dekker, New York, 1999.

Olsson, U. and H. Wennerström, Globular and bicontinuous phases of nonionic surfactant films, *Adv. Colloid Interface Sci.*, **49** (1994) 113.

Shah, D. O. (Ed.), *Micelles, Microemulsions and Monolayers: Science and Technology*, Marcel Dekker, New York, 1999.

Shinoda, K., *Principles of Solution and Solubility*, Marcel Dekker, New York, 1978.

Stilbs, P., Fourier transform pulsed-gradient spin-echo studies of molecular diffusion, *Prog. NMR Spectroscopy*, **19** (1987), 1.

Wennerström, H., O. Söderman, U. Olsson and B. Lindman, in *Encyclopedic Handbook of Emulsion Technology*, J. Sjöblom (Ed.), Marcel Dekker, New York, 2000, Ch. 5.

7 INTERMOLECULAR INTERACTIONS

Pair Potentials Act between Two Molecules in a Vacuum

Intermolecular interactions are of fundamental importance in understanding how atoms and molecules organize in liquids and solids. As an example, consider the formation of a micelle from charged surfactant molecules. Why do micelles form and what are the forces acting between the surfactant molecules? Obviously, it is not sufficient to only consider the interaction between the surfactants, because of the trivial fact that micelles do not form in the gas phase. The solvent plays a crucial role for the micellar aggregation and water is in this respect almost unique. Figure 7.1 demonstrates that micelles cannot form in a solvent with a low dielectric permittivity, resulting in only a weak screening of the electrostatic repulsion between the head groups. Can one understand the formation of a micelle based on the knowledge of how surfactants, counterions and water interact? Yes—it can be understood on a qualitative level, but it is not possible to quantitatively predict it from first principles. The following sections describe in more detail which type of intermolecular interactions operate between these molecules.

In the study of atomic and molecular forces, one can discard forces whose effects do not coincide with molecular dimensions; i.e. gravitational forces are negligible. Only electrostatic forces arising from the interaction between electrons and protons on different molecules are of interest for the present applications. At this point it can be appropriate to recapitulate what are the dimensions of small molecules and how strong (in kJ/mol) is a typical hydrogen bond in, for example, aqueous solution. A rough estimate of the size of a water molecule can be obtained from its density, $1 \, g/cm^3$, at room temperature and normal pressure. This leads to an approximate diameter of 3 Å. The dominant interaction between two water molecules is the dipole–dipole term. With the known dipole moment of water, 1.85 D (debye), and assuming the average separation between two water molecules to be 3 Å, one finds a typical interaction energy to be around 10–20 kJ/mol. Such a calculation is, of course, only a rough estimate, but it will be shown later that it is of the correct order of magnitude. This example is taken to encourage the use of approximate

(a)

(b)

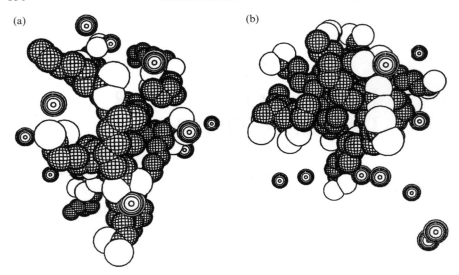

Figure 7.1 Micellar structures as seen in molecular dynamics simulation of sodium octanoate in two different solvents with different dielectric permittivities, ε_r: (a) $\varepsilon_r = 20$ and (b) $\varepsilon_r = 80$. Micelles cannot form in the low-permittivity solvent due to the strong electrostatic head group–head group repulsion. Reproduced by permission of the American Institute of Physics from B. Jönsson *et al., J. Chem. Phys.*, **85** (1986) 2259

estimates when considering intermolecular interactions. The following account is a little bit more stringent and investigates how sophisticated quantum chemical calculations can contribute to the understanding of intermolecular interactions.

Assume that we have two atoms an infinite distance apart; then, the total energy consists of the individual contributions, i.e. the energy of the isolated atoms 1 and 2. When bringing them together they will interact and the total energy is given by the following:

$$E_{tot}(r) = E_1 + E_2 + U(r) \tag{7.1a}$$

where $U(r)$ is the intermolecular potential and is defined as the work required to bring two atoms together from an infinite separation to a distance r:

$$U(r) = -\int_{\infty}^{r} F(s)ds, \text{ i.e. } F(r) = -\frac{dU}{dr} \tag{7.1b}$$

in which $F(r)$ is the force acting between two atoms/molecules. Since the force is the negative derivative of the potential energy, one concludes that a repulsive force is not necessarily associated with a repulsive interaction energy.

Consider now an atom consisting of a heavy, positively charged nucleus surrounded by fast electrons able to instantaneously respond to changes in the nuclear positions. The Born–Oppenheimer approximation then states that the potential energy only depends on the relative positions of the nuclei. With this assumption it is possible to numerically solve the Schrödinger equation. In practice, this approach is limited to molecules including less than approximately 1000 electrons. In order to investigate the interactions of larger molecules, one can try to divide up the total energy into more manageable contributions and treat them separately. Hopefully, the loss in theoretical rigour will be compensated by a deeper insight into the physical nature of the molecular forces.

Note that the discussion initially is limited to the interaction between a pair of atoms/molecules in a vacuum. This is a true pair interaction, which is independent of solvent and temperature. Later in this chapter, the investigation will include the interaction between two molecules in a medium, e.g. two ions in water. Their interaction will depend on the dielectric permittivity of the solvent, which makes the interaction temperature dependent. This type of interaction will be referred to as an effective pair potential. Effective potentials are frequently encountered in chemistry and the hydrophobic interaction and the screened Coulomb interaction are typical examples.

The Intermolecular Interaction can be Partitioned

The short-range interaction, i.e. the repulsive force experienced by molecules as their electron clouds start to overlap, is called the exchange repulsion, U_{exc}, and

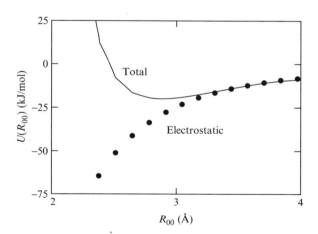

Figure 7.2   The interaction of two water molecules as calculated in an *ab initio* quantum chemical calculation. The continuous line represents the total energy, U, while the symbol (●) shows the direct electrostatic component, U_{ele}. By courtesy of G. Karlström

can be derived quantum mechanically from the Pauli principle. The remaining forces can be interpreted in terms of classical electrostatics. The total energy is given by the following:

$$U = U_{exc} + U_{ele} + U_{ind} + U_{dis} \qquad (7.2)$$

where U_{ele} is the direct electrostatic interaction (see equation (7.3a)) and U_{ind} and U_{dis} are the induction and dispersion energies, respectively.

This energy partitioning is not unique and it also assumes that the interacting species (atoms and/or molecules) retain their identities and individual properties. This is not always the case and a chemical reaction where new covalent bonds are formed is a typical example where this approach fails. Consider, for example, the formation of H_2O from H_2 and O_2; neither of the reactants have any dipole moment, while the product has a very large dipole moment. The partitioning and classification of interaction terms are also to some extent a personal choice. The 'hydrogen bond' is not seen as a separate type of interaction and 'charge transfer complexes' are not mentioned. The hydrogen bond is described by the electrostatic term, U_{ele}, and charge transfer is mainly an induced interaction contained in U_{ind}. There will, of course, be coupling terms between these four types of interaction, but they will for most systems be small and hence can be neglected.

The partitioning can be extended to deal also with intramolecular interactions in large molecules such as polymers or proteins. In this case, one can treat atoms or groups of atoms, which are separated by more than three or four bonds, as belonging to different molecules. It is common practice to separate out the bond, bond angle and dihedral parts of the intramolecular contributions due to the implicit approximation above. Certain intramolecular motion is almost classical, e.g. the rotation around a dihedral angle, whereas the fast vibrations of a bond are strictly quantum mechanical. It is not possible to accurately treat these forces, and in view of the present objectives one can be satisfied with rather crude classical approximations.

Electrostatics

For two arbitrary charge distributions $\rho_a(\mathbf{r}_a)$ and $\rho_b(\mathbf{r}_b)$, the Coulomb energy is given by the following:

$$U_{ele} = 1/(4\pi\varepsilon_0) \int \rho_a(\mathbf{r}_a) \, 1/r_{ab} \, \rho_b(\mathbf{r}_b) \, d\mathbf{r}_a \, d\mathbf{r}_b \qquad (7.3a)$$

The expression for two point charges reduces to:

$$U_{ele} = 1/(4\pi\varepsilon_0) \, q_a q_b / r_{ab} \qquad (7.3b)$$

In general, it is convenient to split the charge distribution into a multipole expansion and talk about ion–ion, ion–dipole, dipole–dipole, etc., interactions. Molecules that deviate from spherical symmetry possess permanent electrostatic moments. The zeroth-order moment is the total charge Q and the first-order moment is the dipole moment, which is a vector:

$$Q = \Sigma q_i, \quad \mu = \Sigma q_i \mathbf{r}_i \tag{7.4}$$

Higher moments, such as quadrupoles, octupoles, etc., are tensor quantities. The electrostatic moments of different molecules interact with each other without distoring their charge distribution.

The dipole moments vary greatly between polar molecules such as H_2O, where $\mu = 1.85 D(6.1 \times 10^{-30}$ Cm) and NH_3, where $\mu = 1.47D$ (4.9×10^{-30} Cm), compared to alkanes, which only possess weak dipole moments and then only for certain conformations. Table 7.1 gives a compilation of dipole moments for some small molecules.

Consider again the interaction between two water molecules; the diameter of a water molecule is roughly 3 Å and assuming that only electrostatic interactions in the form of a dipole–dipole potential are important,

$$U_{dip-dip} = \frac{1}{4\pi\varepsilon_0} \left[\frac{\mu_1 \mu_2}{r^3} - 3\frac{(\mu_1 \mathbf{r})(\mu_2 \mathbf{r})}{r^5} \right] \tag{7.5a}$$

which for a colinear parallel arrangement of the dipoles becomes:

$$U_{dip-dip} = -\frac{2\mu^2}{4\pi\varepsilon_0 r^3} \approx -15 \, \text{kJ/mol} \tag{7.5b}$$

with $\mu = 1.85$ D and $r = 3$ Å. As a comparison, the quantum mechanically calculated interaction energy for the water dimer is around $-25 \, \text{kJ/mol}$.

Table 7.1 Permanent dipole moments of small molecules (in debye units; $1D = 3.336 \times 10^{-30}$ Cm)

H_2O	1.85	$HCONH_2$ (formamide)	3.7
NH_3	1.47	CH_3COCH_3 (acetone)	2.9
CO	0.11	C_2H_4O (ethylene oxide)	1.9
CO_2	0		
CH_3OH (methanol)	1.7	$C_6H_5NO_2$	4.2
CH_3CH_2OH (ethanol)	1.7		
Hexanol	1.7		
C_6H_5OH (phenol)	1.5		

The attractive interaction between two water molecules is dominated by a dipole–dipole interaction, whereas in the carbon dioxide or benzene dimer the electrostatics is characterized by a quadrupole–quadrupole term. For neutral molecules, it is usually the dipole term that gives the most significant contribution to the energy, cf. water. It is important to note that the different moment interactions have different distance dependence; thus the ion–ion interaction goes as r^{-1}, while the dipole–dipole interaction behaves as r^{-3}, and in addition have a strong orientational dependence. A quadrupole–quadrupole interaction decays as r^{-5}, which explains why it only makes a minor contribution in the presence of a non-zero dipole moment. We will later see that the range of interaction has a very profound effect on macroscopically observable quantities.

Induction

Apart from the electrostatic forces it is also possible to treat the induction phenomenon classically. In the case of a point dipole and a neutral molecule without any permanent electrical moments, the electrical field, $\mathbf{E}_{\text{ext}}$, from the dipole induces a small charge redistribution in the form of an oppositely directed dipole moment:

$$\mu_1 = 4\pi\varepsilon_0\alpha\mathbf{E}_{\text{ext}} \tag{7.6}$$

where α is the polarizability. The energy associated with the induced dipole is given by the interaction of the induced dipole moment with the applied field minus the energy it takes to create the dipole in the field. The induction force is a many-body effect in that the induced dipole moment depends on all the permanent as well as all induced dipole moments.

Consider two molecules with permanent dipole moments and polarizabilities. The electric field exerted from molecule 1 at molecule 2 is $\approx \mu_1/r^3$, and the corresponding induced dipole at molecule 2 thus becomes $\mu_{\text{ind}} \approx \alpha\mathbf{E} \approx \alpha\mu_1/r^3$. The permanent dipole at molecule 1 interacts with the induced dipole at molecule 2 according to (leaving out trivial factors) the following:

$$\mu_1\mu_{\text{ind}}/r^3 \approx \mu_1\alpha\mu_1/r^6 \tag{7.7}$$

Comparing with the permanent dipole–dipole interaction one finds that:

$$U_{\text{ind}}/U_{\text{dip-dip}} \approx \alpha/r^3 \tag{7.8}$$

Since α is of the dimension of molecular size we conclude that the induced interactions can be of the same order of magnitude as $U_{\text{dip-dip}}$ at distances

Table 7.2 Polarizability for some small molecules (in units of $4\pi\varepsilon_0 \text{Å}^3$)

H_2O	1.5	He	0.20
CO_2	2.9	Ar	1.63
CO	2.0	Xe	4.0
Glycine	5.9	H_2	0.81
C_6H_6	10.6	Cl_2	4.6
C_2H_4	4.3		

comparable to the molecular size, but the permanent moments dominate at larger separations.

Typical values for the polarizability are given in Table 7.2. Benzene, which has a delocalized π-bonding system, displays a very high polarizability. It is obvious that larger molecules can acquire large induced dipole moments by small polar molecules like water. The dipole moment of a molecule in solution is not strictly a well-defined quantity, but we can make a rough estimate for a water molecule in aqueous solution and find that the dipole moment has increased from 1.85 D in the gas phase to approximately 2.3 D in solution. Such an increase in the 'effective' or total dipole moment has a profound effect on the properties of liquid water. A rough estimate shows that the induced interactions in liquid water increase the boiling point by 100°!

Dispersion

Dispersion forces are of quantum mechanical origin but can be interpreted via more immediate concepts. The electrons in a molecule move fast when compared to the nuclei and the electrical field changes due to the surrounding molecules. However, there are fluctuations in the electron density of a molecule regardless of the surroundings, i.e. fluctuations also occur in vacuum. These give rise to momentary dipoles which in turn generate an electrical field. For two argon atoms, these two fields can, in analogy with the induction term, interact and give rise to an attraction. It is the correlated movement of electrons that is the source of the dispersion energy. It is always present and is responsible for the attraction between noble gas molecules. The magnitude of the dispersion interaction between two molecules is approximately proportional to the product of their polarizabilities.

As in the case with induction, one can introduce higher-order terms and the commonly occurring $1/r^6$ expressions for dispersion energies is only the leading term. The $1/r^6$ dependence is valid at intermediate distances; at larger distances ($> 100\,\text{nm}$), account must be taken of the fact that electrons cannot respond instantaneously to field changes, thus causing a retardation effect, which changes the distance dependence of the first term to $1/r^7$.

The long-range forces, U_{ele}, U_{ind} and U_{dis}, are of varying importance depending on the system of interest. Dispersion energies dominate the interaction between noble gas atoms. This is also true for the CO dimer, where the monomer dipole moment happens to be close to zero. For polar molecules, the electrostatic interaction is stronger, while the induction energy is rather small, as can be seen for water and ammonia in Table 7.3. It is still, however, the dispersion energy that furnishes the main attractive energy in the ammonia dimer.

The dispersion interaction always gives an attractive contribution between two apolar hydrocarbon molecules, as well as between a water molecule and a hydrocarbon. The interfacial tension between water and air is approximately 70 mN/m, but is 50% lower at a water–hydrocarbon interface, mainly due to the dispersion interactions. The difference between cationic and anionic surfactants is also quite often attributed to induction and dispersion forces. Thus, a cationic surfactant with a highly polarizable counterion like Br^- could be expected to show a different behaviour when compared to one with OH^- or Cl^- as the counterion. For example, the CMC for dodecyltrimethylammonium with chloride as the counterion is 20.3 mM, while it decreases to 15.6 mM with a bromide counterion (see Table 2.2). With anionic surfactants and small cations like Li^+, Na^+ and K^+, there is, however, no significant difference in the CMC value (see Table 7.5).

Exchange Repulsion

When two electron clouds approach each other and start to overlap, consideration must be taken of the Pauli principle, which prohibits occupation of an orbital with more than two electrons. Higher-energy orbitals must then be used to accommodate the extra electron density, thus causing an effective repulsion. The repulsive forces can therefore be used to give a 'size' to the molecules. The excluded volume parameter, b, in the van der Waals equation, $(p + n^2a/V)(V - nb) = nRT$, is a measure of this size, e.g. the molecular radii

Table 7.3 Attractive contributions to the dimer energy, at a configuration where the attractive and repulsive energies just balance each other (kJ/mol)

System	U_{ele}	U_{ind}	U_{dis}	U_{bond}
Ar	0	0	−1.2	—
Xe	0	0	−1.9	—
CO	≈0	≈0	−1.4	343
HCl	−0.2	−0.07	−1.8	431
NH_3	−6.3	−0.9	−13.0	389
H_2O	−16.0	−0.9	−5.3	464

of CCl_4 is 2.4 Å, benzene 2.3 Å and ethanol 1.9 Å. When one refers to 'molecular size' or draws a space-filling picture of a molecule, it is effectively the exchange repulsion that is described. The choice of a separation of 3 Å between two water dipoles in a previous example, when trying to estimate the energy of the water dimer, was also guided by the exchange repulsion.

A summary is given in Table 7.4 of the intermolecular force contributions.

The Hydrogen Bond

The hydrogen bond is no unique interaction type but can be understood in terms of the forces presented above. It is special in that the electrostatic contributions dominate the attractive forces. A hydrogen atom bound to a strongly electronegative atom, e.g. oxygen, will cause the electrons associated with the hydrogen to be displaced towards the electronegative atom. As hydrogen lacks lower-lying electron shells its repulsive radii are small and a strong electrostatic attraction can be formed between it and other electronegative atoms/molecules. As seen for the water dimer, the electrostatic attraction is significantly larger than dispersion or induction contributions. At the minimum of the potential energy of the dimer (with the oxygen–oxygen separation $r_{oo} \approx 2.9$ Å):

$$U_{ele} = -31 \, kJ/mol$$

$$U_{ind} = -4 \, kJ/mol$$

$$U_{dis} = -6 \, kJ/mol$$

$$U_{exc} = +20 \, kJ/mol$$

In liquid water at room temperature, the average internal energy per water molecule can also be subdivided into the same components:

$$<U_{ele}> = -53 \, kJ/mol$$

$$<U_{ind}> = -18 \, kJ/mol$$

Table 7.4 Summary of intermolecular force contributions

Type	Range	Sign
Electrostatics	Long	Attractive/repulsive
Induction	Medium	Attractive
Dispersion	Medium	Attractive
Exchange repulsion	Short	Repulsive

$$<U_{dis}> = -17\,kJ/mol$$

$$<U_{exc}> = +47\,kJ/mol$$

where the brackets denote that these are average internal energies.

Theoretical Models and Parameters

In order to treat specific problems, one needs to determine forces either through direct measurement or by using quantum chemical methods. Fitting experimental data can also yield force parameters. The development of explicit analytical potential functions is not trivial. For larger molecules it is virtually impossible to either measure or calculate forces. Instead, one has to use parameters from smaller molecules and transfer them into larger segments, e.g. assuming that the charge distribution in an amide bond in a polypeptide is the same irrespective of the nature of the side chain.

It is only for small noble gas dimers that potential surfaces exist that can accurately reproduce all known experimental observations. For example, the argon–argon potential surface is completely described by using only exchange repulsion and dispersion forces. The accurate functional form of this pair potential is awkward but for a more qualitative description a simple Lennard–Jones potential suffices:

$$u_{LJ}(r) = 4\varepsilon\,[(\sigma/r)^{12} - (\sigma/r)^{6}] \tag{7.9}$$

The physical interpretations of ε and σ are depicted in Figure 7.3. For larger systems, as, for example, amino acids, the commonly applied functional

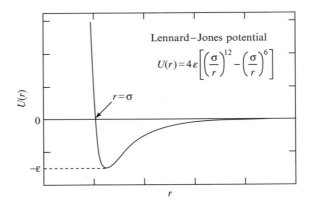

Figure 7.3 Schematic picture of a Lennard–Jones potential, where ε is the depth of the attractive well and σ is the extension of the repulsive part (see equation (7.9))

form contains atom centres interacting via Lennard–Jones and Coulomb interactions:

$$u(r_{ij}) = q_i q_j / (4\pi\varepsilon_0 r_{ij}) + 4\varepsilon_{ij}[(\sigma_{ij}/r_{ij})^{12} - (\sigma_{ij}/r_{ij})^6] \qquad (7.10)$$

where i and j are centres on different molecules and the total interaction is a summation over all centres i and j.

Effective Pair Potentials Act between Two Molecules in a Medium

Water as a Dielectricum

From what has been said above, it is apparent that the electrostatic interactions should play an important role in the formation of a micelle from ionic surfactants. They are both large in magnitude and of long range. Although the direct interaction of two ions in water is very strong it will be attenuated by the water molecules, so that the effective interaction is much weaker—this phenomenon is usually referred to as dielectric screening. Water has a very high dielectric constant and the effective interaction of two ions in water is given by:

$$u^{\mathrm{eff}}(r) = q_i q_j / (4\pi\varepsilon_0 \varepsilon_r r) \qquad (7.11)$$

Thus, to a good approximation, one can treat water as a dielectric continuum and focus on ionic interactions. This approach is usually called the primitive model and has been a cornerstone in electrolyte theory as well as in electric double-layer theory. It will appear in the context of colloid stability and polyelectrolyte solutions in Chapters 8 and 9, respectively. One manifestation of the long-range properties of ionic interactions is the relative insensitivity of CMC of ionic amphiphiles to the type of counterion. For example, changing the counterion from a sodium to a tetraethylammonium ion has only a marginal effect on the CMC (see Table 7.5).

Table 7.5 The critical micelle concentration (CMC) for dodecyl sulfate chains with different counterions. The small variation in CMC is a consequence of the long-range character of the Coulomb interaction

Surfactant	CMC (mM)
$C_{12}SO_4^- Na^+$	8
$C_{12}SO_4^- Li^+$	9
$C_{12}SO_4^- K^+$	8
$C_{12}SO_4^- N(CH_3)_4^+$	6
$C_{12}SO_4^- N(Et)_4^+$	5

Up to this point, focus has been solely on pair potentials, i.e. the interaction between two molecules or atoms independent of all the others. This is the strict mechanical view, but it is quite often possible and profitable to consider effective potentials. This means that one measures the interaction between a pair of molecules in a solution and averages over all possible conformations of the solvent. The resulting force between the two molecules is not a mechanical force but a thermodynamic force, i.e. it will depend on temperature and density. Common examples are the dielectric potential $u^{\text{eff}}(r)$ above, where the effect of the solvent is incorporated into ε_r. The hydrophobic interaction and the screened Coulomb interaction, which will be encountered in the next sections, are other examples of effective potentials.

From the structure of the micelle it is obvious that the electrostatic forces will counteract its formation (see also the simulated structures in Figure 7.1). There is a strong repulsion between the negatively charged carboxylate head groups on the micellar surface. The question then arises: which forces promote the formation of a micellar aggregate? It could, of course, be due to dispersion or induction forces, but these are generally too weak and roughly equal in magnitude between water–water, water–surfactant and surfactant–surfactant. Hence, they could hardly be the dominant contribution to aggregate formation.

Hydrophobic Interaction

The force responsible for the aggregate formation has to be found somewhere else. We have previously seen that the water–water interaction is comparably strong due to the hydrogen bonds formed between water molecules. Hence, introducing a non-polar molecule into water strongly disturbs the hydrogen bond network, with a loss of interaction energy. This loss in energy can be minimized if the water molecules around the solute adjust themselves, but the price has to be paid in lowered entropy. As a consequence, one usually finds that the free energy of transfer of a non-polar molecule into water at room temperature contains a large entropy contribution (see Figure 7.4(a)). When dissolving non-polar molecules (e.g. the hydrocarbon tails of surfactant molecules), they will try to minimize the damage to the water hydrogen bond network by aggregating. Figure 7.4(b) demonstrates the attractive interaction

---

Figure 7.4 Monte Carlo simulations of the hydrophobic effect: (a) the free energy of hydration for a Lennard–Jones particle in water. The ε-parameter is kept constant and equal to 0.62 kJ/mol, while the σ-parameter is varied. The energy and entropy contributions are shown separately ($A_{\text{ex}} = U_{\text{ex}} - TS_{\text{ex}}$). (b) The free energy of interaction between two neon atoms in liquid water. The continuous represents the interaction of two neon atoms in the gas phase. (c) The free energy of interaction of a neon atom with a hydrophobic wall in liquid water (symbols). The continuous line shows the interaction of two neon atoms in liquid water. Reproduced by permission of the American Institute of Physics from J. Forsman and B. Jönsson, *J. Chem. Phys.*, **101** (1994) 5116

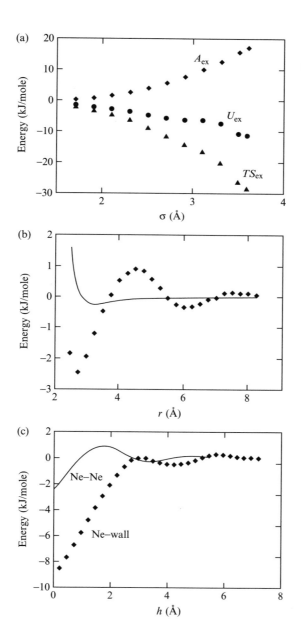

between two neon atoms in water. It is clearly demonstrated that this attraction is a solvation effect, i.e. the neon atoms are pushed together by the solvent (water). The corresponding interaction between two neon atoms in the gas phase has a much weaker attractive minimum. Figure 7.4(c) shows that replacing one of the neon atoms with an infinite hydrophobic wall, not unexpectedly, increases the attraction further (in Chapter 8, we will consider the interaction of two hydrophobic walls immersed in water). Thus, the hydrophobic interaction is the mechanism that promotes the formation of micelles. Table 7.6 shows that the longer the hydrocarbon tail a surfactant molecule carries, then the more easily it will aggregate, as indicated by the lowered CMC. The hydrophobic interaction can be of considerable strength and is a delicate balance of energetic (enthalpic) and entropic terms. It is now also generally accepted to be the main driving force for protein folding. Different arguments can be pursued in favour of one or another explanation of this interaction. One way to understand it is by realizing the strong cohesive energy in the water–water interaction. The interaction is further emphasized by the orientational dependence, and any attempt to break the structure will cost free energy; whether it appears in the form of an entropy or energy term is secondary. For apolar molecules, such as hydrocarbons, this free energy loss cannot be regained through interaction with the solute.

One simple estimate of the free energy cost of transferring a hydrocarbon molecule from its non-polar environment to water is given by the following:

$$G^{\text{transfer}} = \gamma 4\pi R^2 \qquad (7.12)$$

where γ is the interfacial tension and R the solute radius ($\gamma = 0.05\,\text{J/m}^2$). The same type of expression can also be used to estimate the hydrophobic interaction between two non-polar solutes at contact:

$$G \text{ (at contact)} \approx -\gamma 4\pi R r \qquad (7.13)$$

Table 7.6 The critical micelle concentration (CMC) for alkyl sulfate chains with different chain lengths. The decrease in CMC is a consequence of the hydrophobic interaction between the alkyl chains

Surfactant	CMC (mM)
$C_8SO_4^-$	160
$C_{10}SO_4^-$	40
$C_{12}SO_4^-$	10
$C_{14}SO_4^-$	2.5

where r is the radius of a water molecule.

A semi-quantitative measure of the solvent ability to dissolve non-polar molecules is given by the so-called Gordon parameter, i.e. $\gamma V^{-1/3}(\mathrm{J/m^3})$, where γ is the surface tension of the solvent and V its molar volume. Water has a very high Gordon parameter, ≈ 2.7, while hexane has a very low value of ≈ 0.3 (see Figure 7.5). Thus, we see that the hydrophobic interaction is largely due to the high cohesive energy density in water. One can also note that water is actually a 'better solvent' than its high Gordon parameter implies, which means that water to some extent is capable of compensating for the loss in cohesive energy when dissolving a non-polar solute. We will later find that the hydrophobic interaction (or hydrophobic effect) can be qualitatively understood in terms of regular solution theory (the Bragg–Williams model).

Thus, we have a clear qualitative understanding of the hydrophobic interaction in surfactant assemblies but lack a quantitative description of it. This is in contrast to the effective potential between two ions in solution, where equation (7.11) provides an accurate representation of the interaction.

Debye–Hückel Theory

Solutions containing charged species, which could be small atomic ions or highly charged aggregates, have special physicochemical properties. This is

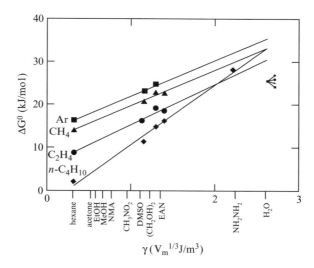

Figure 7.5 The free energy of transfer of a non-polar gas into different solvents as a function of its Gordon parameter. The energy increases approximately linearly with the Gordon parameter for most liquids. Water on the other hand, shows a significant deviation. Reproduced by permission of John Wiley & Sons, Inc., from D. F. Evans and H. Wennerström, *The Colloidal Domain*, VCH, New York, 1994

due to the long-range nature of the Coulomb potential decaying as r^{-1}. Already small concentrations of an electrolyte (<1 mM) will lead to markedly non-ideal behaviour. Addition of an inert salt causes large changes in colloidal stabilities, reaction rates and titration curves. Electrolytes of various kinds have a profound influence on surfactants and polymers in solution. For example, in a solution containing ionic surfactants, the addition of inert electrolyte will significantly lower the CMC, as is demonstrated in Figure 7.6. Thus, electrostatic effects in solution are important, but fortunately they can be qualitatively, and sometimes quantitatively, understood from simple Debye–Hückel (DH) theory.

The electrostatic potential ϕ is related to the charge distribution ρ via Poisson's equation:

$$\varepsilon_0 \varepsilon_r \Delta\phi(x, y, z) = -\rho(x, y, z) \tag{7.14}$$

where Δ is the Laplace operator:

$$\left(\frac{\partial^2}{\partial x^2} + \frac{\partial^2}{\partial y^2} + \frac{\partial^2}{\partial z^2}\right)$$

The problem is simplified by assuming spherical symmetry, in which case the Laplace operator becomes:

$$\Delta = \frac{1}{r}\frac{\partial^2(r\phi)}{\partial r^2} \tag{7.15}$$

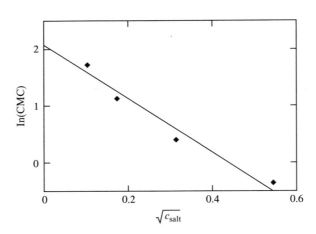

Figure 7.6 The variation of CMC with salinity. The straight line is a fit to the experimental data points. Debye–Hückel theory predicts a linear relationship between ln (CMC) and $\sqrt{c_{\text{salt}}}$ as seen from equation (7.20)

In order to proceed one has to find an expression for the change distribution i.e. how the ions distribute themselves in the electrostatic potential. By assuming that the ions are distributed according to Boltzmann's distribution law:

$$\rho(r) = Zec_0\exp\left[-Ze\phi(r)/(k_B T)\right] \tag{7.16}$$

around the central ion positioned at the origin, we arrive at the so-called Poisson–Boltzmann equation:

$$\varepsilon_0\varepsilon_r\Delta\phi(r) = -\Sigma Z_i ec_{0i}\exp\left[-Z_i e\phi(r)/(k_B T)\right] \tag{7.17}$$

where Z is the ion valency, k_B is the Boltzmann constant, T the temperature and c_{0i} the ionic concentration at infinity. This is a non-linear equation and usually difficult to solve, but by linearizing the exponential ($e^x = 1 + x + x^2/2 + \ldots$) one obtains a tractable equation:

$$\Delta\phi(r) = \kappa^2\phi(r); \quad \kappa^2 = \Sigma c_{0i} Z_i^2 e^2/(\varepsilon_0\varepsilon_r k_B T) \tag{7.18}$$

where the Debye–Hückel screening length, κ^{-1}, has been introduced. The final expression for the potential is given by the following:

$$\phi(r) = \left[\frac{Ze}{4\pi\varepsilon_0\varepsilon_r(1 + \kappa 2R)}\right]\left\{\frac{\exp\left[\kappa(2R - r)\right]}{r}\right\} \tag{7.19}$$

where R is the ionic radius. This is a simple, yet very important result. It shows that the potential around an ion in a salt solution does not decay as $1/r$ but as $\exp(-\kappa r)/r$. This means that the electrostatic interaction has a much shorter range, depending on κ^{-1}, which in turn depends on salt concentration and valency (see Figure 7.7). The physical interpretation is that the central ion attracts oppositely and repels equally charged ions, thereby creating the screening. Another important, yet simple equation is obtained for the excess chemical potential, which in the DH theory reads as follows:

$$\mu^{ex} = -\kappa Z^2 e^2/[(8\pi\varepsilon_0\varepsilon_r(1 + 2R\kappa)] \tag{7.20}$$

The screening length is an essential parameter for both qualitative and quantitative discussions. From equation (7.19), it can be seen that the role of electrostatic interactions is reduced at separations $> \kappa^{-1}$. For an aqueous 1M 1:1 salt solution at room temperature, the screening length is 3.0 Å. Since the latter is inversely proportional to $\sqrt{c}$, one finds that $\kappa^{-1} = 30$ Å in a 10 mM salt solution. The valency also plays a role and high-valency salts are very efficient in screening electrostatic interactions (see Table 7.7). A word of caution needs

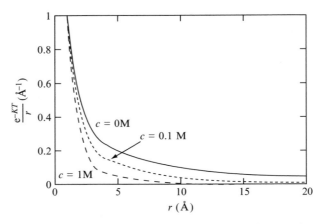

Figure 7.7   The decay of the electrostatic potential at three different salt concentrations

Table 7.7 The screening length for some simple salts in aqueous solution and at room temperature

Salt	κ^{-1} (Å)
1 M NaCl	3
10 mM NaCl	30
100 μM NaCl	300
1 mM NaCl	95
1 mM CaCl$_2$	55
1 mM AlCl$_3$	39

to be given: the DH theory works excellently for 1:1 salts in aqueous solution, but it strongly understimates the screening of charged aggregates in the presence of oppositely charged multivalent (counter)ions.

Bibliography

Israelachvili, J. N. *Intermolecular and Surface Forces*, Academic Press, London, 1991.
Tanford, C., *The Hydrophobic Effect. Formation of Micelles and Biological Membranes*, Wiley, New York, 1980.

8 COLLOIDAL FORCES

The interaction potential between large aggregates/particles in solution is nothing but an effective potential and could for that reason have been treated in the previous chapter. The interaction between colloidal particles, however, is of such importance that it deserves a chapter of its own.

Electric Double-Layer Forces are Important for Colloidal Stability

To stabilize or destabilize a colloidal solution is a central theme in many industrial processes. There are, of course, several mechanisms for achieving this, but the more general one is via electrostatic interactions, i.e. whenever a colloidal particle carries a net charge, special attention has to be paid to the direct electrostatic interactions. Charged colloids or particles appear almost everywhere—in clay and soils, membranes and proteins, paper, pulp, peat, paints, etc. The formation of the colloidal charge could be due to titrating groups or specific adsorption of (multivalent) ions on the particles. The geometric arrangements can vary—spherical sols, planar clay sheets or DNA cylinders. They all have a number of common features, which will be considered in some detail below.

The Poisson–Boltzmann Equation

For simplicity, the mathematical treatment will be restricted to two planar charged surfaces a distance $2a$ apart. The charge density, σ, is assumed to be uniformly distributed over the surfaces (Figures 8.1), which by means of simulations has been shown to be a very good approximation. Initially, one can also make the simplifying approximation that only the neutralizing counterions to the surface are present. The case with extra salt will be dealt with later on.

The primitive model was introduced in the previous chapter and will now be used in order to describe electric double layers, as depicted in Figure 8.1. Consider the potential $\phi(x)$ and the ion density outside the surface $\rho(x)$. By combining Poisson's and Boltzmann's equations, one obtains the so-called Poisson-Boltzmann (PB) equation, which describes the thermodynamics of a charged surface in contact with an aqueous solution:

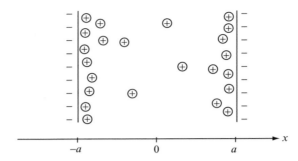

Figure 8.1 Schematic picture of two charged walls with an intervening solution containing counterions. The wall charges are supposed to create a uniform surface charge density

$$\mathrm{d}^2\phi/\mathrm{d}x^2 = [Ze\rho_0/(\varepsilon_r\varepsilon_0)]\exp([-Ze\phi/(k_BT)]) \tag{8.1}$$

where Z is the counterion valency and ρ_0 is a normalization constant with the dimension of density—note that equation (8.1) is a mean field approximation. The PB equation has for most cases no analytical solution and one has to resort to numerical solutions. The special case considered here has, however, an analytical solution:

$$\rho(x) = [2\varepsilon_r\varepsilon_0 k_B T s^2/(Zea^2)]/[\cos^2(sx/a)] \tag{8.2a}$$

where s is obtained from the following equation:

$$s\,\tan(s) = |\sigma|Zea/(2\varepsilon_r\varepsilon_0 k_B T) \tag{8.2b}$$

The solution to this equation can be obtained by simple iteration, starting by assuming that $s = \pi/4$ and solving for tan(s), which then gives a better value of s, and so on. Figure 8.2 shows a typical counterion concentration profile with a strong accumulation of ions at the charged walls. There is, however, one particularly simple solution, which is obtained when the right-hand side of equation (8.2b) becomes very large, in which case s must approach $\pi/2$. This solution has a number of interesting properties. Consider what happens when the system is diluted, i.e. when $a \to \infty$. The wall concentration will then go to a limiting value given by the following

$$\rho(a) = \sigma^2 Ze/(2\varepsilon_r\varepsilon_0 k_B T), \quad a \to \infty \tag{8.3}$$

i.e. the counterions cannot be diluted away. This effect is sometimes referred to as ion condensation. The condensed layer is, however, not a layer sitting directly bound to the surface, but rather a statement about the concentration in the neighbourhood of the surface.

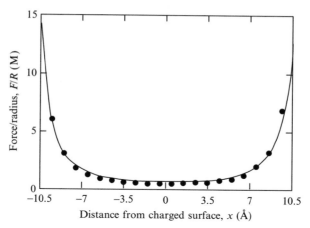

Figure 8.2 Monovalent counterion concentration profile between two charged surfaces 21 Å apart in water. The surface charge density is $0.224\,C/m^2$ and the smooth curve is obtained from the PB equation while the other is from a Monte Carlo simulation

The ion density at the mid-plane can also easily be obtained, as follows:

$$\rho(0) = 2\varepsilon_r\varepsilon_0 k_B T (\pi/2)^2/(Zea^2), \quad a \to \infty \tag{8.4}$$

Thus, the mid-plane concentration decays as $1/a^2$, independently of surface charge density, i.e. an ion far away from the surface can detect the sign of the surface charge density, but not the magnitude!

Contact Theorem and Osmotic Pressure

The interesting quantity for colloid stability is, of course, the force between the two charged walls. An expression for the force is easily derived from first principles and the resulting equations are sometimes referred to as the contact theorems. The osmotic pressure can be obtained from two independent contact relationships within the PB approximation:

$$p_{osm} = k_B T c(0), \quad c(x) = \text{ion concentration} = \rho(x)/(Ze) \tag{8.5a}$$

$$p_{osm} = k_B T c(a) - \sigma^2/(2\varepsilon_r\varepsilon_0) \tag{8.5b}$$

The first relationship is lacking a correlation term (which becomes important for divalent counterions) and is not an exact one, although the latter is. The curves in Figure 8.3 show how well the approximate PB results compare with Monte Carlo simulations.

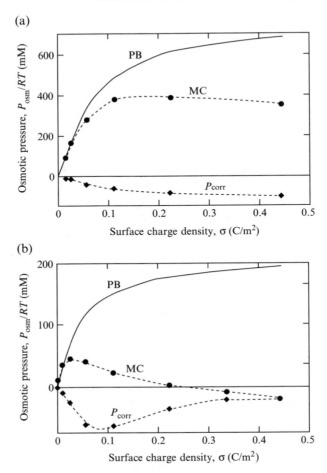

Figure 8.3 The osmotic pressure as a function of surface charge density in a salt-free double layer. The continuous lines are PB results and the exact Monte Carlo curves are shown as dashed. The distance between the walls is 21 Å: (a) monovalent; (b) divalent counterions. Reproduced by permission of the American Institute of Physics from L. Guldbrand *et al.*, *J. Chem. Phys.*, **80** (1984) 2221

The osmotic pressure is proportional to the ion concentration at the mid-plane, which clearly demonstrates that the double-layer repulsion is mainly an entropic force, i.e. when two charged surfaces are brought closer, the second surface will restrict the volume of the counterions belonging to the first surface and vice versa. It is this confinement of counterions that gives rise to the repulsion and it is somewhat misleading to look upon the repulsion between two charged colloids as a direct electrostatic interaction. The force is, of course, of electrostatic origin, since with neutral colloids no need for counterions exists, but it manifests itself as an entropic term. The correlation term shown in Figure 8.3 is always attractive and has the same mechanistic origin as the quantum mechanical dispersion force. It is also operating between spherical macroions surrounded by neutralizing counterions, in which case the correlation free energy decays like r^{-6}, with r being the separation between the macroions. Note that the correlation term is of approximately the same magnitude for monovalent as for divalent ions. It becomes, however, more significant in the latter case due to the relatively small entropic term. The general rule of thumb is that one should be careful in using the PB equation when divalent or multivalent counterions are present or when the salt concentrations is very high (>1 M).

So far we have only been treating a salt-free system, which is rarely seen in real systems. It is, however, important to realize that the salt concentration in many systems can often be much smaller than the concentration of counterions needed to neutralize the surface. Under such conditions, one finds the above expressions quite useful, because they are analytic.

Addition of Salt and the 'Weak Overlap Approximation'

In the more general and complex situation, when additional salt is present the PB equation requires numerical solutions. In the situation with an electrical double layer in equilibrium with a bulk salt reservoir we are normally only interested in the net osmotic pressure, i.e. the difference in osmotic pressure between two surfaces as compared to in the bulk solution. In other words:

$$P^{net} = P^{double\ layer} - P^{bulk} \qquad (8.6)$$

There is, however, one particularly simple asymptotic expression for the free energy and the force. This is based on what is called the Gouy–Chapman solution and is applicable for two weakly overlapping double layers. Under these circumstances, the PB equation can be linearized and solving the full non-linear equation is avoided. The free energy per unit area, G, is given by the following:

$$G = \frac{64 k_B T c_0 \gamma^2}{\kappa} \exp(-\kappa 2a) \qquad (8.7)$$

and the net osmotic pressure is:

$$P_{net} = -\frac{\partial G}{\partial(2a)} = 64 k_B T c_0 \gamma^2 \exp(-\kappa D) \qquad (8.8)$$

where c_0 is the bulk electrolyte concentration, κ the Debye–Hückel inverse screening length defined earlier in equation (7.18), D ($= 2a$) the surface separation and γ is related to the surface potential ϕ_0 via $\gamma = \tanh [Ze\phi_0/(k_B T)]$. Thus, the force between two polar surfaces decays exponentially with separation. This is a useful approximation, valid under many circumstances for describing the force between two planar charged walls immersed in an electrolyte solution. The asymptotic behaviour of the osmotic pressure as described by equation (8.8) has been verified in numerous experiments. The experiments (see Figure 8.4) not only give an exponentially decaying pressure, but the decay factor is within experimental errors equal to the screening length, κ^{-1}. This is a verification of the Debye-Hückel theory and the underlying dielectric continuum approximation. The force behaviour at short separation, $D < \kappa^{-1}$, is not as well described by theory, although it is possible to fit the full non-linear PB solution to the experimental curves using, among other factors, the surface charge density as a free parameter.

For two spherical charged colloids the free energy has a similar appearance and looks like the following:

$$G(\text{two spheres}) = \text{constant} \times \exp(-\kappa r)/r \qquad (8.9)$$

where r is their separation. Equation (8.9) is nothing but the Debye-Hückel expression already encountered above in equation (7.19). These two last expressions are approximate, but very useful for qualitative discussions.

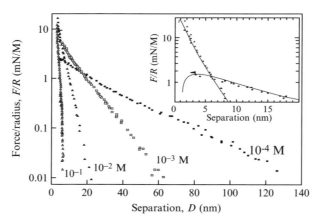

Figure 8.4 Forces between negatively charged mica surfaces in $LiNO_3$ solutions of varying concentration. Data obtained from a surface force apparatus experiment. Reproduced by permission of Academic Press from V. Shubin *et al.*, *J. Colloid Interface Sci.*, **155** (1993) 108

Other Types of Forces Exist

van der Waals Interaction and the Hamaker Constant

The famous theory of colloidal stability due to Derjaguin, Landau, Verwey and Overbeek (the DLVO theory) contains two ingredients: a repulsive double-layer force calculated via the PB equation and an attractive van der Waals force. The first has been treated in some detail in previous sections, including the attractive correlation component not accounted for in the PB approximation.

The van der Waals force contains several contributions. One is the quantum mechanical dispersion interactions (the London term) discussed in Chapter 7. A second term arises from the thermally averaged dipole–dipole interaction (the Keesom term) and a third contribution comes from dipole-induced–dipole interactions (Debye term). The van der Waals force operates between both apolar and polar molecules and varies rather little between different materials, i.e. compared to the double-layer force, which can change by orders of magnitude upon addition of small amounts of salt. The most straightforward way to calculate the van der Waals force is by assuming that the interaction is pairwise additive—this is usually called the Hamaker approach. For the interaction of two infinite planar walls, one obtains the following:

$$P_{vdW} = -A/(6\pi D^3) \tag{8.10}$$

where A is the Hamaker constant and D the separation between the walls. The Hamaker constant of most condensed phases is found in the range $0.4–4 \times 10^{-19}$ J.

The assumption of pairwise additivity is not entirely correct and a more rigorous theory has been presented by Lifshitz et al. The mathematical formalism of the Lifshitz theory is rather involved and in order to fully utilize the approach a knowledge of the frequency-dependent dielectric permittivity for all frequencies is needed. Fortunately, the simple Hamaker approximation turns out to be sufficient for most experimental situations (see Figure 8.5).

Equation (8.10) is also valid for the interaction of two particles across a medium. The only change is that the Hamaker constant is reduced and a useful rule of thumb for hydrocarbons interacting across water is that the interaction is reduced by approximately 90% compared to the interaction across air.

A deficiency in the Hamaker and Lifshitz theories is the assumption of a constant density, equal to its bulk value, of the confined liquid. It would be more correct to assume a constant chemical potential. Asymptotically, the choice is immaterial, but at intermediate separations the density of the confined liquid will differ from its bulk value. This density variation, increase or decrease, will cause a deviation from the simple expression of equation (8.10).

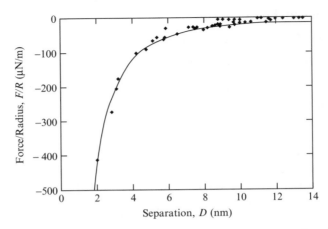

Figure 8.5 The attractive van der Waals force between two mica surfaces in air. From J. Israelachvili, *Intermolecular and Surface Forces*, Academic Press, London, 1992

Figure 8.6(a) demonstrates how a simple liquid tries to avoid the space between two inert surfaces, leading to a density lower than the corresponding bulk value. This density depression, which is not taken into account in the Hamaker and Lifshitz theories, will increase the attraction (see Figure 8.6b). A similarly increased attraction would be expected for water between two hydrophobic surfaces (see below).

Structural Forces

At separations of two surfaces comparable to the molecular size, packing effects will start to play a role in the interparticle force. This can, under certain circumstances, lead to an oscillatory density profile, as well as an oscillatory force as a function of the distance between the surfaces. The oscillations could be superimposed on both a net attractive and a net repulsive curve. The density profiles for such model systems have been investigated quite extensively by using computer simulations, but the net forces are more difficult to calculate. It is sometimes argued that an oscillatory density profile automatically leads to an oscillatory force profile, which of course might be true but has so far not been proven. One study of oscillatory forces between crossed mica cylinders used octamethylcyclotetrasiloxane (OMCTS) as the solvent. The period of the oscillations turned out to fit quite well with the molecular size (see Figure 8.7). The temperature dependence of these forces is weak, but small amounts of other solvents or contaminants completely change the behaviour and destroy the oscillatory pattern.

Linear alkanes give rise to oscillatory forces while branched alkanes do not (see Figure 8.8). Similar forces have also been seen in aqueous solution and in

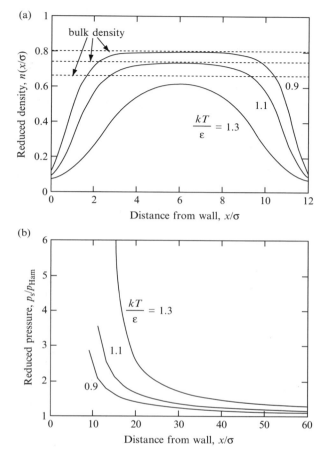

Figure 8.6   A Lennard–Jones liquid between two inert walls. (a) Reduced liquid density for three different reduced temperatures (kT/ε). (b) The increased attraction due to the density depression. In the Hamaker approach, P_{ham}, one assumes the density to be uniform in the slit and equal to the bulk density. Reprinted with permission from J. Forsman *et al.*, *J. Phys. Chem.*, **101** (1997) 4253. Copyright (1997) American Chemical Society

this case a shorter period is found when compared to that in OMCTS. This is, of course, explained by the difference in size between water and OMCTS molecules.

Repulsive Hydration Forces

It is easy to imagine that a charged or zwitterionic surface immersed in aqueous solution achieves one or several well-defined layers of water molecules hydrating the surface in the same way as a dissolved ion has a hydration shell.

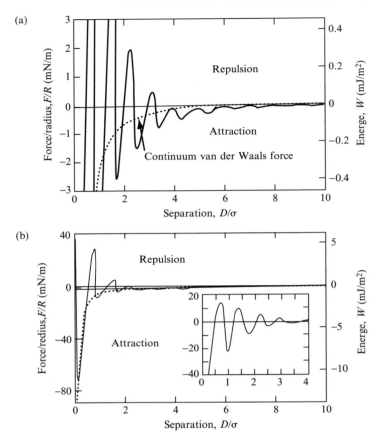

Figure 8.7 Oscillatory forces between mica surfaces in an inert silicone liquid, OMCTS, which has an approximate molecular diameter of 9 Å. From J. Israelachvili, *Intermolecular and Surface Forces*, Academic Press, London, 1992

Pushing two such surfaces together results in a dehydration and could be anticipated to be accompanied by a repulsive hydration force. This is the background to the notation of repulsive hydration forces.

Very strong short-ranged repulsive forces have been found to act between lipid bilayers. The range is somewhere between 10 and 30 Å and the repulsion is found to decay exponentially with separation. An osmotic stress technique was used to measure the force. Similarly, repulsive forces between solid mica surfaces are reported from experiments using surface apparatus. Repulsive hydration forces seem to exist both between neutral and charged surfaces. Despite the fact that the mica surfaces in the force apparatus are rigid and the bilayer systems studied have a certain flexibility, there is surprisingly good

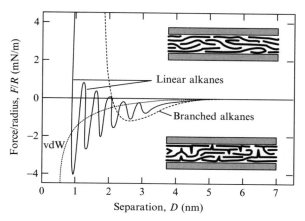

Figure 8.8 Forces between mica surfaces immersed in linear (*n*-tetradecane) and branched (2-methyloctadecane) alkane liquids. From J. Israelachvili, *Intermolecular and Surface Forces*, Academic Press, London, 1992

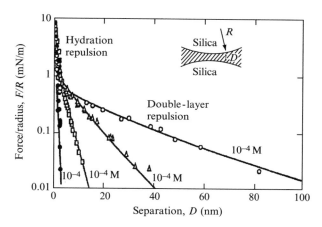

Figure 8.9 Repulsive hydration forces between mica surfaces in salt solution. Note that repulsion only appears at salt concentrations > 1 mM. From J. Israelachvili, *Intermolecular and Surface Forces*, Academic Press, London, 1992

agreement between the two techniques. The repulsive forces between mica surfaces have also been seen in other solvents.

These experiments have given rise to a surprisingly wild flora of theoretical explanations. Structural or H-bonding polarization at the surface has been suggested as the cause for the repulsion. Bilayer undulations and image charge interactions have also been put forward as possible mechanisms. Recently, it has been suggested that the lipids protrude out into the solvent and this protrusion is limited when two surfaces approach and a repulsive force appears. The mechanism is akin to the idea of undulation forces. The difference lies mainly in the range of the fluctuations. The original undulations model invoked long wavelength undulation, while in the protrusion model the 'undulations' are molecular in range.

Monte Carlo simulations show the existence of short-range repulsive forces even for perfectly smooth surfaces. As a matter of fact, both the repulsive hydration force and the attractive hydrophobic force described below can be modelled in a rather simple fashion by varying the strength of interaction between the solvent (water) and the surface. A strong attractive solvent–surface interaction leads automatically to a repulsive surface–surface force. If the surfaces are inert, i.e. no attractive surface–solvent interaction exists, then an attractive solvation force acts between the surfaces. The interaction is in both cases of limited range, i.e. < 100 Å (see Figure 8.10).

Attractive Hydrophobic Forces

There has been a steady accumulation of force measurements between hydrophobic surfaces, usually mica surfaces coated with monolayers exposing hydrocarbon or fluorocarbon groups to water. These studies have found a surprisingly long-ranged attractive force between such surfaces. The attraction ranges over several hundred Å. The attraction cannot be explained by van der Waals forces in a simple Hamaker approach and it seems to be rather insensitive to added salt. It is also difficult to explain the long range based on the same type of hydrophobic interaction as encountered between, for example, two neon atoms in water (cf. Figure 7.4). Note that although this 'traditional' hydrophobic interaction is supposed to be of short range, its magnitude could be increased by the density depression mechanism (see Figure 8.6).

It has been speculated that the attraction accounts for the rapid coagulation of hydrophobic particles in water and that it is also operational in protein folding. Like the repulsive hydration force, very little is known theoretically. One possible mechanism for the attraction can be cavity formation, i.e. small gas bubbles could form on the hydrophobized mica. Such a droplet formation is expected to give rise to both repulsive and attractive forces depending on the specific conditions (see Figure 8.11). Another possible source for the attractive interaction between hydrophobized surfaces is that they are locally non-neutral and that patches of negative and positive charges correlate and cause an attraction.

Depletion Forces

It is quite common to use poly(ethylene oxide) (PEO) when crystallizing proteins. In this case the polymer molecules most probably give rise to a depletion force between the macromolecules, i.e. PEO will not enter the gap between the

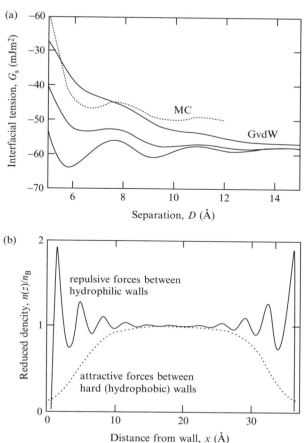

Figure 8.10 Repulsive hydration and attractive hydrophobic forces for a wetting and non-wetting wall, respectively. Theoretical data obtained from a generalized van der Waals theory. From J. Forsman *et al.*, *J. Colloid Interface Sci.*, **195** (1997) 264

macromolecules (too much restriction on its configurational freedom), but will stay in the bulk, where it will exert an osmotic pressure on the macromolecules. It is an interesting mechanism in the sense that the added polymer affects the colloid–colloid interaction by not being there! The range of the depletion attraction is of the same order of magnitude as the radius of gyration of the polymer and for an ideal polymer it scales as $r^{1/2}$, where r is the degree of polymerization.

Sometimes, a repulsive interaction appears at larger separation before the actual depletion attraction sets in (see Figure 8.12). This phenomenon is

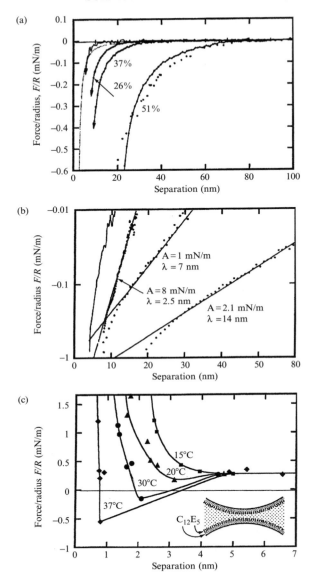

Figure 8.11 Attractive forces between two mica surfaces coated with a monolayer of a fluorocarbon surfactant measured (a) in ethylene glycol and (b) at various water contents. The dotted lines represent the expected van der Waals interactions. (c) Increasing attraction with temperature between surfaces coated with the non-ionic surfactant $C_{12}E_5$, which is believed to be more hydrophobic at higher temperatures. Reprinted with permission from J. Parker *et al.*, *Langmuir*, **8** (1992) 757. Copyright (1992) American Chemical Society

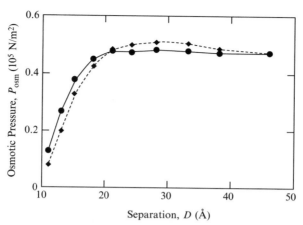

Figure 8.12   The interaction of two neutral surfaces enclosing a polymer solution as a function of separation. The slit is in equilibrium with a bulk solution of fixed concentration

somewhat ambiguously referred to as depletion repulsion. Both the attractive and repulsive parts have been observed experimentally and theoretically.

Colloidal Forces can be Measured Directly

Phase diagrams, coagulation kinetics, electrophoresis, etc., give indirect information on the forces acting between the aggregates. To directly measure the interaction between two charged colloids is more difficult. The osmotic stress technique is an excellent method for systems that spontaneously form, for example, lamellar structures. The system is in equilibrium with a bulk aqueous solution, whose chemical potential and osmotic pressure can be controlled. The aggregate spacing is measured by using X-ray diffraction and at equilibrium the bulk osmotic pressure should be equal to the pressure acting between the aggregates.

The surface force apparatus (SFA) is technically more involved, but also more versatile (Figure 8.13). It consists of two crossed cylinders covered by smooth negatively charged mica sheets. The separation between the cylinders is measured interferometrically with a resolution of only a few Å. The force is measured by a set of springs and a piezoelectric crystal connected to one of the cylinders. The force between curved surfaces, F^{cyl}, can be converted to a free energy for planar surfaces, G^{planes}, by the Derjaguin approximation:

$$G^{planes}(D) = F^{cyl}(D)/2\pi R \qquad (8.11)$$

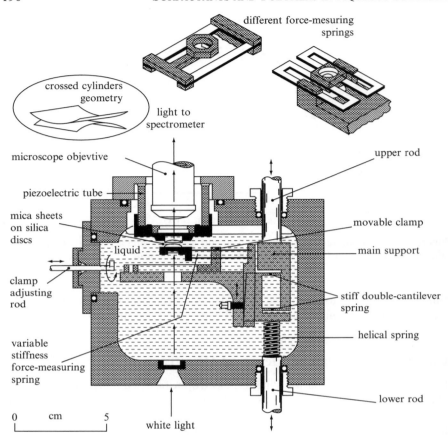

Figure 8.13   The surface force apparatus

where R is the cylinder radius. The SFA has the advantage that the mica surfaces can be modified by adsorption from solution and hence the interaction between, for example, polymer-covered surfaces can be measured. The surface can also be made hydrophobic by adsorption of a monolayer of cationic amphiphiles. Bilayers can be absorbed, thus giving rise to a huge spectrum of possible systems to study (see the previous figures in this chapter).

The atomic force microscope (AFM) and the scanning tunnelling microscope (STM) have also extended the possibility for direct measurements of colloidal forces.

Bibliography

Evans, D. F. and H. Wennerström, *The Colloidal Domain*, VCH, New York, 1994.

Israelachvili, J. N., *Intermolecular and Surface Forces*, 2nd Edn, Academic Press, London, 1991.

Jönsson, B., T. Åkesson and C. E. Woodward, Theory of interaction in charged colloids, in *Ordering and Phase Transitions in Charged Colloids*, A. K. Arora and B. V. R. Tata (Eds), VCH, New York, 1996.

9 POLYMERS IN SOLUTION

A polymer is a large molecule, a macromolecule, built up of a repetition of smaller chemical units, monomers. A natural division of the science of polymers is into biological and non-biological macromolecules. Biological macromolecules are, for example, proteins and polysaccharides while non-biological macromolecules include the common plastics and adhesives. This chapter mainly deals with non-biological water-soluble polymers but a few examples and principles from non-aqueous systems are included as well in order to present some basic principles.

Polymer Properties are Governed by the Choice of Monomers

Synthetic polymers are synthesized by the polymerization of monomers. The monomers in the polymerization constitute the repeat units, for example, acrylic acid is polymerized into poly(acrylic acid). A polymer can either be linear, branched or cross-linked, as depicted in Figure 9.1(a). It is important to characterize the category a certain polymer belongs to in order to understand its solution behaviour, since the latter differs considerably between the three categories.

If the polymer is synthesized with more than one kind of monomer, it is called a copolymer. The monomer units in a copolymer can either be (i) randomly distributed, (ii) distributed in blocks, or (iii) distributed such that one of the monomers is grafted in chains onto the backbone of the other monomer chain, as illustrated in Figure 9.1(b). Also here the solution properties and the surface chemical properties are very sensitive to which category a copolymer belongs. Some of the structures shown in Figure 9.1(b) are strikingly surface active. Polymers in this class are specially dealt with in Chapter 12.

The polarity of the monomer units is a convenient basis to categorise non-biological polymers: (i) non-polar polymers such as polystyrene and polyethylene; (ii) polar, but water-insoluble, polymers, such as poly(methyl methacrylate) and poly(vinyl acetate); (iii) water-soluble polymers, such as polyoxyethylene and poly(vinyl alcohol); (iv) ionizable polymers, or polyelectrolytes, such as poly(acrylic acid).

The configuration of a polymer in solution depends on the balance between the interaction of the segments with the solvent and the interaction of the

(a)

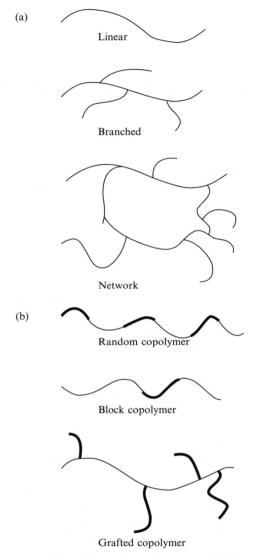

Linear

Branched

Network

(b)

Random copolymer

Block copolymer

Grafted copolymer

Figure 9.1 Polymers are differentiated with respect to (a) their structure, or (b) their chemical composition

segments with each other. In general terms, a polymer can form a random coil, an extended configuration or a helix. For synthetic polymers, the random coil is the most common configuration. Polyelectrolytes, where the monomer units are charged, can under certain circumstances form stiff rods.

The Molecular Weight is an Important Parameter

With the exception of proteins it is impossible to find a polymer batch where all macromolecules have exactly the same molecular weight. Synthetic polymers, as well as polysaccharides, have a molecular weight distribution that differs depending on the method of synthesis and on the fractionation procedure. A schematic of a molecular weight distribution is shown in Figure 9.2. It is important to realize that some physical properties are sensitive to the low molecular weight fraction while others are sensitive to the high molecular weight material. A molecular weight distribution such as that shown in Figure 9.2 is conveniently determined by size-exclusion chromatography. Another way to obtain a measure of the polydispersity in molecular weight is to determine the average molecular weight by two methods that give different averages. Conceptually the number average molecular weight, M_n, is the simplest and is defined as follows:

$$M_n = \frac{\Sigma N_i M_i}{\Sigma N_i} \qquad (9.1)$$

where N_i is the number of molecules with molecular weight M_i. M_n may be determined by freezing point depression, osmosis or chemical analysis of end groups.

Another molecular weight average is the weight average molecular weight, M_w, where the weight, w, is the weighting factor for each molecular weight species, and is thus defined as follows:

$$M_w = \frac{\Sigma w_i M_i}{\Sigma w_i} = \frac{\Sigma N_i M_i^2}{\Sigma N_i M_i} \qquad (9.2)$$

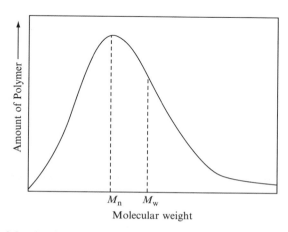

Figure 9.2   A schematic molecular weight distribution of a polymer

M_w can be determined by light scattering, either quasi-elastic or classic. This molecular weight is sensitive to high molecular weight species and is hence always larger than M_n. An illustrative example is the following calculation. If we have 100 molecules with a molecular weight of 100 and one molecule with a molecular weight of 10 000, the M_n is *ca.* 200 and M_w is *ca.* 5000. In light scattering experiments, it is therefore very important to control the conditions such that no high molecular weight contamination (dust for example) will affect the results. The ratio of the weight average to the number average molecular weights is a measure of the polydispersity of a sample. A sample is considered to be rather monodisperse if this ratio is less than 1.1.

Dissolving a Polymer can be a Problem

Dissolution of a polymer can sometimes be a problem, not only in the laboratory but also on an industrial scale. The dissolution process starts with the solvent, which is more mobile than the polymer chains, penetrating into, and hence swelling, the polymer. The polymer solution then becomes highly viscous and sticky with the result that all of the material will be found somewhere on the container walls. The next step in the dissolution process is that the polymer chains disentangle from the gel and diffuse into the solvent. This is a slow process since the polymer chain dynamics, which are dependent on the polymer molecular weight, are rate determining. The polymer, sticking to the container walls, will have a rather small exposure area to the solvent and hence the dissolution will indeed be time-consuming. Dissolving poly(vinyl pyridine) into water at neutral pH, for example, takes at least 24 h.

There are, however, some tricks that are handy. Technically, the problem of dissolving a polymer can be circumvented by handling polymers that are already dissolved at some high concentration in a solvent. Another way is used when the pure polymer is in the form of a powder. Let us take the poly(vinyl pyridine) example as an illustration. At high pH the polymer is not soluble in water and therefore the polymer powder can easily be dispersed into water at pHs say above 8 at high stirring rates. The pH is then quickly lowered to about 3–4 under vigorous stirring. The polymer particles will then not agglomerate but rather start to dissolve. Now the dissolution process will be orders of magnitude faster (*ca.* 30 min for the system above) since the total area of the polymer particles is much larger than that of a polymer adhering to the container walls.

Polymers in Solution can be Characterized by Viscosity Measurements

A convenient way to characterize a polymer in solution at low concentrations is to measure the viscosity. This can be achieved by measuring the efflux time, t, of a polymer solution for the flow through a capillary in an Ubbelohde viscometer and comparing with the corresponding time for the solvent, t_0. The viscosity is then proportional to the efflux time multiplied by the density of the liquid. Since dilute solutions can be approximated to have the same density as the pure solvent we then have the relative viscosity, η_{rel}, as $\eta_{rel} = t/t_0$. The following list gives the common nomenclature used in solution viscosity.

$$\text{Relative viscosity} = \eta_{rel} = \frac{\eta}{\eta_0} = \frac{t}{t_0} \tag{9.3}$$

$$\text{Specific viscosity} = \eta_{sp} = \frac{\eta - \eta_0}{\eta_0} = \eta_{rel} - 1 \tag{9.4}$$

$$\text{Reduced viscosity} = \eta_{red} = \frac{\eta_{sp}}{c} \tag{9.5}$$

$$\text{Inherent viscosity} = \eta_{inh} = \ln\left(\frac{\eta_{rel}}{c}\right) \tag{9.6}$$

$$\text{Intrinsic viscosity} = [\eta] = \left(\frac{\eta_{sp}}{c}\right)_{c=0} = \left(\ln\frac{\eta_{rel}}{c}\right)_{c=0} \tag{9.7}$$

The specific viscosity is a measure of the thickening effect of the polymer solution as compared to the solvent. This quantity is naturally very dependent on the polymer concentration and hence the reduced viscosity is a property telling us more about the specifics of a polymer system. If the reduced viscosity, or the inherent viscosity, is plotted versus the polymer concentration a straight line is normally obtained. The extrapolation of this line to zero polymer concentration gives us the intrinsic viscosity, $[\eta]$, which is also called the limiting viscosity number. The intrinsic viscosity is independent of polymer concentration but dependent on the solvent that is chosen. It is also dependent on the polymer molecular weight and can be used to obtain a viscosity average molecular weight, M_η, from the Mark–Houwink equation, as follows:

$$[\eta] = KM_\eta^\alpha \tag{9.8}$$

where K and α are constants. These constants can conveniently be found for common polymer/solvent combinations in the literature, e.g. the *Polymer Handbook* by Brandrup and Immergut (see the Bibliography at the end of this chapter). The viscosity average molecular weight is an average between the number average and the weight average molecular weights.

Polymer Solutions may Undergo Phase Separation

When dissolving two liquids the molecules can freely "wander around" in the container and hence the entropy is large for a liquid mixture. In a polymer, however, the segments are tied up to each other to form a 'string'. Therefore, in comparing a single segment with a free solvent molecule the entropy is far less for the segment. Polymer solutions thus have lower total entropy and are hence less stable and are more prone to phase separate when compared to mixtures of ordinary liquids. Flory and Huggins first quantified this in the late 1940s in what has later become known as the Flory–Huggins theory for polymer solutions. A derivation and description of the theory is given in Chapter 10. From what is said above it is obvious that a solution of a polymer with a high molecular weight is less stable towards phase separation when compared with a solution of the same polymer but of lower molecular weight. Hence, when a polymer phase-separates in solution, the high molecular weight species will separate out first, leaving the lower molecular weight species in solution. This phenomenon is used in fractionation of polymer samples with respect to molecular weight.

The temperature at which phase separation occurs for a 1% polymer solution is called the cloud point, due to the increased turbidity of the polymer solution as this temperature is reached. The highest, or lowest, temperature where a phase separation occurs is called the critical temperature and the corresponding polymer concentration is called the critical composition. This temperature/composition is called the critical point. Note the subtle difference between the critical point and the cloud point. The latter is the temperature at which phase separation occurs for a specific polymer concentration, normally 0.1 or 1%, while the former is the minimum, or maximum, temperature/concentration for phase separation in the phase diagram.

The question of whether or not a polymer will dissolve in a solvent is a matter of a balance between two free energy terms. The first is due to the *entropy of mixing*, which is low for polymers in solution for the reasons just mentioned. The entropy of mixing always acts in favour of mixing. The second is due to the *enthalpy of mixing*, which is a measure of the interaction energy between a polymer segment when compared to the interaction energy between segments and solvent molecules alone. This latter term is normally positive and hence opposes the mixing of the two components. Thus, in ordinary polymer/solvent systems the stability with respect to phase separation will decrease with a decrease in temperature. In these systems there will eventually be a phase separation when the temperature is lowered sufficiently. Here, a concentrated polymer phase will separate and be in equilibrium with a dilute polymer solution. Phase separation can also be achieved upon adding a non-solvent to the polymer–solvent system. To a good approximation, the solvent mixture can be considered as an average liquid giving an average of their interaction energies with the polymer segments.

Polymers Containing Oxyethylene Groups Phase-Separate Upon Heating in Aqueous Systems

Some aqueous systems display an extraordinary behaviour with respect to phase-separation. That is, they phase-separate when the temperature is *increased*. As mentioned in Chapter 4, almost all systems where polyoxyethylene is included in the polymer belong to this category. Figure 4.16 in Chapter 4 shows the phase diagram of pure polyoxyethylene (Note that this phase diagram has been determined at elevated pressures in order to reach high temperatures for aqueous systems.) Here, it is seen that at very high temperatures the two-phase region diminishes and the system is homogeneous again. This is due to the increased thermal energy that counteracts other forces at play. From this figure we see that the high molecular weight polyoxyethylene fractions phase-separate at *ca.* 98°C, while the lower molecular weight fractions separate at temperatures higher than this.

If the polymer is made more hydrophobic, for example, by attaching hydrocarbon moieties to the polymer, the phase separation temperature will be lowered. The phase separation temperature for ordinary non-ionic surfactants, which indeed can be considered as copolymers of polyoxyethylene and polyethylene, is normally called the cloud point and is determined at surfactant concentrations of 1% (by weight).

The phase separation temperature of the polyoxyethylene–water system is very dependent on the addition of salts. Both the salt concentration and the type of salt is important, as illustrated in Figure 9.3. Inspection of this figure reveals that the choice of cation of the salts plays a minor role while the type of anion matters considerably. Figure 9.3(b) shows the influence of halide ions on the cloud point. It is generally assumed that the difference in cloud point depression between the halide ions is due to adsorption of the more polarizable ions on the polyoxyethylene chain and hence a polyelectrolyte is created that counteracts the phase separation. Other polymers that contain polyoxyethylene also phase-separate upon heating. One example is ethylhydroxyethylcellulose (EHEC). Figure 9.4 shows the cloud points of aqueous EHEC solutions with various types of salts. We note that iodide and thiocyanate ions, which are both very polarizable, increase the cloud point. This is probably due to adsorption of these polarizable ions to the polymer. At higher salt concentrations, on the other hand, the cloud point decreases due to the higher ionic strength leading to a 'salting out' of the polyelectrolyte-type polyoxyethylene. Sulfate and phosphate salts belong to the class of salts which lowers the cloud point most efficiently.

Solvents and Surfactants have Large Effects on Polymer Solutions

Like the polarizable ions, higher alcohols adsorb on the EHEC polymer. These, however, decrease the cloud point of the polymer since they give a

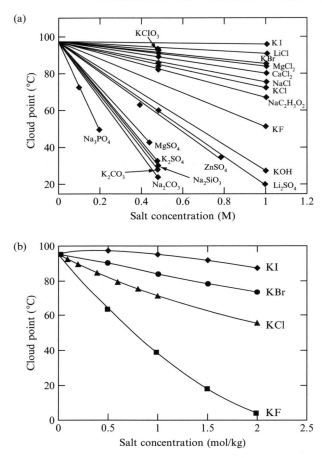

Figure 9.3   Cloud point depression of polyoxyethylene in various salt solutions. Reproduced by permission of Academic Press from O. Bailey and O. Koleske, *Polyethylene Oxide*, Academic Press, London, 1976, p. 98

higher hydrophobicity of the polymer /alcohol complex. This is illustrated in Figure 9.5.

In addition, surfactants adsorb on EHEC polymers. This is illustrated later in Figure 13.16, which shows that the cloud point decreases on the addition of sodium dodecyl sulfate (SDS). This decrease is more pronounced with increasing ionic strength, which is due to shielding of the charges created by the adsorption of the surfactant. At higher concentrations, the cloud point increases again, which is due to the formation of micelles at the polymers.

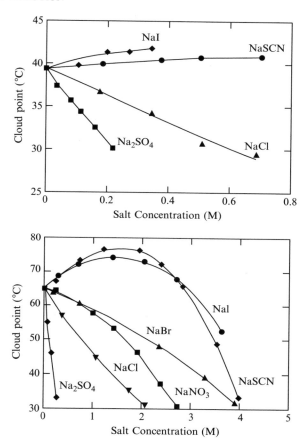

Figure 9.4 Cloud point depression of EHEC in various salt solutions. Reprinted with permission from M. Malmsten and B. Lindmnan, *Langmuir*, **6** (1990) 357, and G. Karlström, A. Carlsson and B. Lindman, *J. Phys. Chem.*, **94** (1990) 5005. Copyright (1990) American Chemical Society

The Solubility Parameter Concept is a Useful Tool for Finding the Right Solvent for a Polymer

In the search of a solvent for a certain polymer one can conveniently use the solubility parameter concept. The solubility parameter is based on the assumption that 'like dissolves like'. As was stated previously, a polymer is not soluble in certain liquids due to a large difference in the interaction energy between segments of the polymer and solvent molecules when compared to the interaction energy between segment–segment and solvent–solvent molecules. Thus,

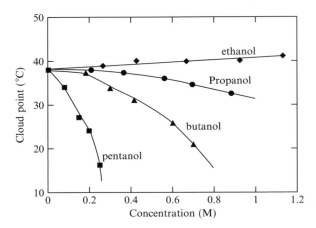

Figure 9.5 The cloud point of a salt-free 0.1% EHEC aqueous solution is very sensitive to the addition of alcohols. Reprinted with permission from M. Malmsten and B. Lindman, *Langmuir*, **6** (1990) 357. Copyright (1990) American Chemical Society

in order to achieve some solubility the segment–solvent interaction energy should be as close as possible to the interaction energy between the segment–segment and solvent–solvent molecules. The enthalpy of vaporisation, which is a reflection of the cohesive forces in the liquid, is used as a measure of these interaction energies. However, since the enthalpy of vaporization, ΔH_{vap}, depends on the molar volume, V, it should be normalized, thus giving the cohesive energy density as follows:

$$\delta^2 = \frac{\Delta H_{vap}}{V} \tag{9.9}$$

where δ is called the solubility parameter. The units for the solubility parameter are $cal^{1/2}cm^{-3/2}$, or in SI units, $J^{1/2}cm^{-3/2}$ ($= MPa^{1/2}$). These parameters can be found from tables for solvents as well as those for polymers, for example, the *Polymer Handbook* (Brandrup and Immergut), or the *Handbook of Solubility Parameters and Other Cohesion Parameters* (Barton) (see the Bibliography at the end of this chapter). In order to find a suitable solvent for a polymer one should first find the solubility parameter for the polymer and then select solvents that have solubility parameters that are close to those of the polymer. Polystyrene, for example, has a solubility parameter of $9.1\ cal^{1/2}cm^{-3/2}$ and suitable solvents are cyclohexane ($\delta = 8.2$), benzene ($\delta = 9.2$) and methyl ethyl ketone ($\delta = 9.3$), while *n*-hexane ($\delta = 7.3$) and ethanol ($\delta = 12.7$) are non-solvents, i.e. they do not dissolve the polymer.

In a more elaborate analysis, the solubility parameter is split up into contributions originating from dispersion forces, polar forces and hydrogen bonding. Hence the total solubility parameter can be written as follows:

$$\delta^2 = \delta_d^2 + \delta_p^2 + \delta_h^2 \tag{9.10}$$

where the subscripts d, p and h represent the dispersion, polar and hydrogen bonding contributions to the total solubility parameter. Another approach is to split the solubility parameter into van der Waals and acid–base contributions.

The Theta Temperature is of Fundamental Importance

The theta (θ) temperature is one particularly important parameter which describes a polymer–solvent system. At this temperature a polymer segment will not be able to tell whether it is in contact with another segment or a solvent molecule. The polymer will have the same configuration as it would have in its own liquid. The polymer is said to be in its 'unperturbed dimension'. In normal solvents the solvent quality increases as the temperature is raised (due to the larger thermal energy) and hence each polymer segment will have a tendency to prefer to be in contact with the solvent molecules rather than in contact with the polymer's own segments. Thus, the polymer will expand its configuration. On the other hand, at temperatures below the theta temperature the polymer segments will prefer to be in contact with other polymer segments rather than with the solvent molecules. Therefore, the polymer will contract. The theta temperature is also called the Flory temperature and the solvent, or solvent mixture, used at this temperature is called a theta solvent.

The dimension of a polymer coil is normally given by its radius of gyration, R_g, and the expansion, or contraction, when compared to its unperturbed dimensions at theta conditions, is given by the following:

$$\alpha = \frac{R_g}{R_g^0} \tag{9.11}$$

where R_g^0 is the radius of gyration at the theta temperature. In passing, we will just mention that polymers are used in lubricating oils in order to counteract the decrease in viscosity with increasing temperature that is normally found in oils. Thus, at higher temperatures the polymer expands and the normal decrease in viscosity is counteracted by this polymer expansion.

Returning to the phase behaviour, it is found that the critical temperature for a system asymptotically approaches a certain value as the molecular weight of the polymer is increased. The limiting value when the polymer molecular weight is infinitely large is identified as the theta temperature. The latter can be found from a plot where the reciprocal of the critical temperature is plotted as a function of the sum of the reciprocal of the molecular weight and the reciprocal of the square root of the molecular weight. An example of such a plot is given in Figure 9.6 for two polymer–solvent systems.

In colloidal chemistry, a particle, or emulsion droplet, can be stabilized against flocculation by the adsorption of polymers on the particle surface, as illustrated in Figure 9.7. If the particle is sufficiently covered with the polymer and if the polymer segments extend into the solution, as is shown in this figure, *the particle will behave as if it is a polymer with an infinitely high molecular weight.* This is of fundamental importance in the understanding of colloidal stability. Thus, a colloidal system is expected to flocculate at the same conditions where a polymer–solvent system reaches the theta condition, i.e. when a polymer with infinitely high molecular weight phase-separates. This has been proven true, without any exceptions, for numerous systems. A knowledge of the theta conditions is therefore of fundamental importance when designing a dispersed system stabilized with polymers.

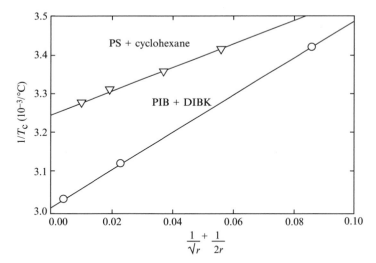

Figure 9.6 Determination of the theta temperatures for the systems polystyrene (PS) in cyclohexane and polyisobutylene (PIB) in diisobutyl ketone (DIBK); see also Chapter 10. Reproduced by permission of John Wiley & Sons, Inc. from O. Billmeyer, *Textbook of Polymer Science*, Wiley, New York, 1982

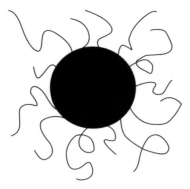

Figure 9.7 Adsorption of polymers on particles results in systems which behave as polymers, with infinitely high molecular weights, in solution

There are Various Classes of Water-Soluble Polymers

We will give here some example of water-soluble polymers, together with their properties and uses. We first consider non-ionic water-soluble polymers with an oxygen or nitrogen in the backbone of the polymer. Of the polyoxyalkylenes only polyoxyethylene (POE) is water-soluble. Polyoxymethylene is not water soluble despite the fact that it contains a higher portion of oxygen than poly-oxyethylene. The reason for the solubility of POE is discussed in Chapter 4. This polymers can be synthesized with molecular weights up to the millions. POEs are used in various applications such as cosmetic and pharmaceutical formulations, ceramic binders, etc. Polyoxypropylene (POP) is a polymer where only the oligomers are soluble in water and longer polymer chains sometimes serve as hydrophobic entities in surfactants. If one changes the oxygen atom in POE for nitrogen, we then obtain polyethyleneimine (PEI). Commercial samples of this polymers are normally branched and the ratio between the secondary, tertiary and quaternary amine groups is normally 1:2:1.

Secondly, there are water-soluble polymers containing an acrylic group. The first examples are poly(acylic acid) (PAA) and poly(methacrylic acid) (PMA). Strangely enough, the latter polymer shows a higher water solubility than the former. The reason is that the latter material forms a helical configuration, thus burying the hydrophobic backbone inside the helix. PAA and POE form a complex in aqueous solution where the hydrogens in the PAA are attracted to the oxygens in the POE. The second example of a water-soluble polymer with an acrylic group is polyacrylamide (PAAm). This is a very hydrophilic polymer which is insensitive to the addition of salts. This polymer is used as a flocculent since it has a high affinity to surfaces due to its cationic nature at lower pH values.

Thirdly, we consider water-soluble non-ionic polymers containing a vinyl group. The first example is poly(vinyl alcohol) (PVAl), which is synthesized by the hydrolysis of poly(vinyl acetate). Polymers with a degree of hydrolysis over *ca.* 86% are soluble in water. If the degree of hydrolysis is higher than 90%, the system needs to be heated in order to obtain a complete solution. Once dissolved in hot water, the polymer remains in solution when cooled. This apparent irreversibility is due to the formation of internal hydrogen bonds in the solid polymer. Another example is poly(vinyl pyrrolidone) (PVP), which is highly soluble in water. This polymer has a weak basic character and associates with anionic surfactants, such as SDS, in aqueous solution. Aqueous solutions of PVP are used in pharmacy, cosmetics and medicine due to its low toxicity and high water solubility. PVP is also used in detergent formulations where its role is to prevent re-deposition of soil on fibres.

In the fourth and final group, the polymers have a natural origin. The first example is the class of cellulose derivatives. Cellulose can be made water-soluble by chemical derivatization. Most commonly, the three hydroxyl groups on the β-anhydroglycose unit, which constitutes the cellulose chain, are used as starting points in the derivatization. The extent of the reaction of these cellulose hydroxyls is called the degree of substitution (DS) and is defined as the average number of hydroxyls that have reacted; the DS is thus a number between 0 and 3. Carboxy methyl cellulose (CMC) is manufactured by reacting the cellulose hydroxyls with monochloroacetate, giving a sodium salt of the carboxylic acid with a DS varying between 0.4 and 1.4. CMC is normally sold in the salt form. The pK_a of the polymer is normally around 4.4, and depends slightly on the DS. Thus, at neutral pH values most of the carboxylic acid groups are in the dissociated form and therefore it displays almost no surface activity (see Chapter 16). The largest use of CMC is in detergents where the role is to prevent soil re-deposition after such soil has been removed from the fabrics. CMC is also used as a dispersant in water-borne paints and paper coatings.

Hydroxy ethyl cellulose is manufactured by the reaction of alkali-swollen cellulose with ethylene oxide. The product is a versatile water-soluble polymer which is used in numerous of applications as thickeners, protective colloids, binders, etc. The molar substitution (MS) is the molar ratio of ethylene oxide to cellulose hydroxyl groups. The aqueous solution properties depend both on the DS and the MS in that there is required both a minimum DS (*ca.* 0.65) and MS (*ca.* 1.0) for water solubility.

Ethyl hydroxy ethyl cellulose (EHEC) is produced by first reacting cellulose with ethylene oxide and then adding ethylene chloride. The polymer structure is shown below in Figure 13.8. The solution properties of this polymer are thoroughly described in Chapter 13.

Polysaccharides are linear or branched polymers made up of sugar-based units. The solution properties are highly dependent on substitution, degree of branching and molecular weight. Polysaccharides are widely used in the food industry as gelants. Examples of polysaccharides include gum arabic, guar gum, agar, carrageenan and dextran.

Polyelectrolytes are Charged Polymers

Polyelectrolytes in solution have many applications and are used technically as thickeners, dispersants, flocculation aids, etc. The word 'polyelectrolyte' is sometimes used for all types of aggregates which carry a high net charge. In some of the literature it is reserved for charged polymers and it is the latter type we will consider now. A flexible polymer often achieves a net charge from carboxylate or sulfate groups ($-COO^-$, $-SO_4^-$) or from ammonium or protonated amines. Polyelectrolytes are often classified as strong (quenched) or weak; the net charge of the latter varies with pH. If a polyelectrolyte carries only one type of ionizable monomer unit, for example, acrylic acid, one can easily define the degree of ionization, α, as the fraction of ionizable groups that are ionized. This property is, of course, dependent on the pH and the relationship can be written as follows:

$$pH = pK + \log\frac{\alpha}{1 - \alpha} \tag{9.12}$$

where pK is the acidity (or basicity) constant of the monomer unit. This 'constant' is not a constant in a strict sense since it depends on the degree of ionization. This is accounted for in more elaborate descriptions of polyelectrolyte behaviour in aqueous solution.

The degree of expansion of the polyelectrolyte increases with the degree of ionization, α, due to repulsion between the ionized groups. There are limits, however. Figure 9.8 shows the radius of gyration of a polyelectrolyte, poly (methacrylic acid), as a function of the degree of ionization. This figure reveals that, to a first approximation, the radius of gyration reaches a saturation value at a degree of ionization of about 0.3. Thus, if elongated polyelectrolyte molecules are required there is no point in charging the polyelectrolyte beyond this value since the dimensions will not change much once *ca.* 30% of the ionizable groups have become charged. This correspond to *ca.* half a unit from the pK value of the polymer. Below, the dimensions of polymers in solution will be described by using scaling theories.

Polymer Configurations Depend on Solvent Conditions

In the preceding sections it is stressed that the low solubility of polymers and related phenomena are due to the low entropy of the polymer when compared

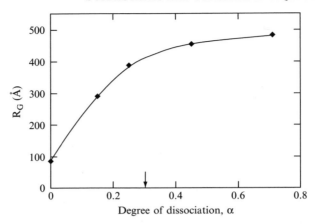

Figure 9.8 The radious of gyration of poly(methacylic acid) as a function of the degree of ionization

to the free monomers. It is also clear that the different solution behaviour of neutral polymer and polyelectrolyte systems is largely due to the counterion entropy and only indirectly to electrostatic interactions.

Summary of Scaling Behaviour

With these results in mind one may ask if the polymer entropy is always negligible? This is definitely not the case and once in solution the configurational entropy plays an important role in determining the physical properties of a polymer solution. One conceptually simple property is the end-to-end separation, R_{ee}, usually defined as the average distance between the first and the last monomer of the chain. For a chain without any interaction (Gaussian chain) consisting of r monomers, the end-to-end separation scales as follows:

$$R_{ee} \propto r^{1/2} \qquad (9.13)$$

The Gaussian chain can be seen as a random walk, where each bond can take any direction in space. Equation (9.13) is analogous to the result obtained from the diffusion equation where one would interpret the number of monomers, r, as the time. The charged monomers in a polyelectrolyte experience a long-range repulsion and a more extended chain results in this case, namely:

$$R_{ee} \propto r \qquad (9.14)$$

Equation (9.14) is true for a polyelectrolyte at infinite dilution and without added salt, that is, in the limit when the screening length $\kappa^{-1} \gg R_{ee}$. At finite polymer concentrations and/or with added salt, screening will take place and if $\kappa^{-1} < R_{ee}$, then the end-to-end separation will scale as follows:

$$R_{ee} \propto r^{3/5} \tag{9.15}$$

This is the typical behaviour for a chain with short-ranged repulsive interactions. For a chain with attractive interactions, the limiting case is a closed packed globule, in which case:

$$R_{ee} \propto r^{1/3} \tag{9.16}$$

For a neutral polymer one can modify R_{ee} by changing the solvent. A good solvent corresponds to an effectively repulsive interaction between monomers, while a poor solvent would cause the polymer to contract due to effectively attractive interactions. The end-to-end separation for a polyelectrolyte can be modified in several other ways. For example, increasing the polymer concentration will increase the screening and lead to a shorter R_{ee}. The same is true for the addition of salt. Another way is to change the pH, thereby neutralizing or charging up ionizable monomers. In this case, the end-to-end separation increases with the degree of ionization, α. For a weakly charged polyelectrolyte, (even) in the presence of added salt, it will vary approximately as:

$$R_{ee} \propto \alpha^{2/5} \tag{9.17}$$

Simple Scaling Theory—Without Salt

In this section, we derive some of the relationships for the end-to-end separations of polyelectrolyte chains presented previously. This can be done in a simplified fashion by considering how the different (free) energy contributions depend on the polymer configuration. This approach has certain limitations, but it still captures several important physical features. We first construct a polyelectrolyte chain as a collection of charges connected with rigid bonds, as shown in Figure 9.9.

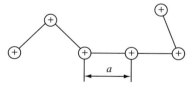

Figure 9.9 Schematic picture of the polyelectrolyte model

When calculating the electrostatic energy we make the assumption that the charges are lined up on a straight line of length $R = ra$, where a is the average monomer–monomer bond length. It is then straightforward to calculate the electrostatic energy by summing up all pair interactions and we find for a chain with r monomers;

$$E_c = kT \cdot \frac{r^2 l_B \ln R}{R} \qquad (9.18)$$

where l_B ($= e^2/(4\pi\varepsilon_0\varepsilon_r kT)$) is known as the Bjerrum length and a negligible constant term has been omitted. The entropy due to stretching of the chain is assumed to be the same as that of the ideal chain:

$$-TS \propto R^2/r \qquad (9.19)$$

It is convenient to omit all trivial constants, which allows us to write the total chain free energy as follows:

$$F = E_c - TS \propto \frac{r^2 \ln R}{R} + R^2/r \qquad (9.20)$$

The optimal free energy is found by optimizing equation (9.20) with respect to $R = R_{ee}$, neglecting the variation of the logarithmic term:

$$\frac{dF}{dR_{ee}} = 0 \rightarrow -\frac{r^2}{R_{ee}}\ln r + R_{ee}/r = 0$$

from which one finds that Ree varies as:

$$R_{ee} \propto r(\ln r)^{1/3} \qquad (9.21)$$

This expression is valid for a polyelectrolyte chain at infinite dilution and in the absence of salt. (Strictly speaking, the logarithmic term should disappear when r approaches infinity.)

The extension of equation (9.21) to a weak poly electrolyte is trivial. Here, the polyelectrolyte consists of a titrating chain, where the charge varies with pH. We note that with the assumption that all monomers carry a fractional charge equal to the degree of dissociation, α, one may rewrite the electrostatic energy as follows:

$$E_c = kT \cdot \frac{\alpha^2 l_B r^2 \ln r}{R} \qquad (9.22)$$

Using the same procedure, one finds that the end-to-end separation scales with α to the power of 2/3 (see Figure 9.10):

$$R_{ee} \propto \alpha^{2/3} r(\ln r)^{1/3} \qquad (9.23)$$

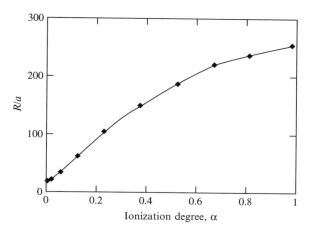

Figure 9.10   The end-to-end separation, R_{ee}, as a function of the degree of ionization, α, using data taken from a Monte Carlo simulation; $r = 320$ (cf. Figure 9.8)

Simple Scaling Theory—With Salt

In real systems, salt will always be present and screen the electrostatic interactions between the charged monomers. The interaction will also be screened by other chains and counterions, which will reduce the range of the Coulomb interaction. As a first approximation, the Debye–Hückel expression, derived in Chapter 7, may be used to describe the interaction between two monomers a distance r apart:

$$e^2 \exp(-\kappa r)/(4\pi\varepsilon_0\varepsilon_r r) \tag{9.24}$$

The interaction is now of short range and it does not seem appropriate to use a straight line of charge as a model for the chain in order to find the electrostatic energy. Instead, one makes the assumption that the chain on average is spherical with a radius equal to R. The Debye–Hückel potential is awkward to evaluate, but can be simplified by assuming that the monomers interact with a pure Coulombic potential, but only up to a distance κ^{-1}. Using the same ideal chain entropy as above, this leads to the following expression for the free energy:

$$F \approx r^2/(R^3\kappa^2) + R^2/r \tag{9.25}$$

Optimizing the free energy with respect to R leads to the following scaling relations for the non-titrating chain (Figure 9.11):

$$R_{ee} \propto r^{3/5}\kappa^{-2/5} \propto r^{3/5}C^{-1/5} \tag{9.26}$$

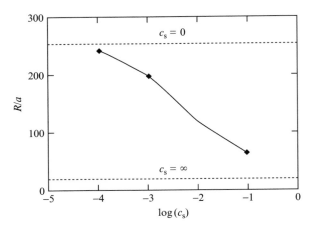

Figure 9.11 The end-to-end separation, R_{ee}, as a function of the salt concentration, using data taken from a Monte Carlo simulation; $r = 320$

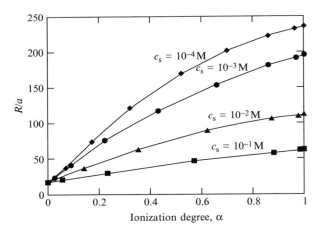

Figure 9.12 The end-to-end separation, R_{ee}, as a function of the degree of ionization, α, for chains at various salt concentrations. Data taken from a Monte Carlo simulation; $r = 320$

It is straightforward to generalize equation (9.26) to a titrating chain in which case the chain expands as $\alpha^{2/5}$ as the degree of ionization increases (Figure 9.12). This describes the initial expansion seen experimentally, but it lacks the plateau value at high α (see Figure 9.8). This is due to non-linear screening effects not incorporated into the Debye–Hückel approximation. A more sophisticated theory or numerical simulations will also predict a weaker dependency on α. With multivalent counterions, one finds a much more complex

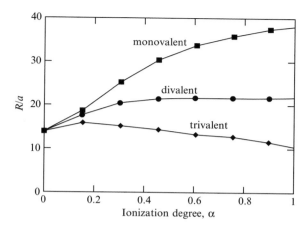

Figure 9.13 The end-to-end separation, R_{ee}, as a function of the degree of ionization, α, for various counterion valencies. Note that a is the monomer–monomer separation distance. Figure provided by courtesy of M. Khan

behaviour than is described by equation (9.26). The correlation between the highly charged counterions can then give rise to effectively attractive interactions between the monomers, thus leading to a reduced end-to-end separation when the degree of ionization increases (see Figure 9.13) This is analogous to the attraction seen between electric double-layers, e.g. in the presence of divalent counterions.

The Electrostatic Persistence Length

The configuration of polymers is sometimes described in terms of the persistence length, which can be seen as a measure of the chain stiffness. This is intimately coupled to the end-to-end separation and can be written as follows:

$$l_p = \frac{R_{ee}^2}{2ra} \tag{9.27}$$

where ra is the contour length of the chain. For polyelectrolytes, it is sometimes useful to split up the persistence length into one electrostatic and one non-electrostatic part:

$$l_{p,tot} = l_{p,0} + l_{p,e} \tag{9.28}$$

where $l_{p,0}$ is the persistence length for the neutral chain. For a freely jointed chain, as in the scaling analysis above, from equation (9.26) we find that the electrostatic persistence length scales as $\kappa^{-4/5}$. For a chain where $l_{p,e} \gg l_{p,0}$,

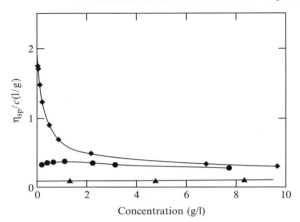

Figure 9.14 The reduced viscosity for polyvinylpyridine with bromide as the counterion as a function of the polymer concentration: (♦) pure water; (●) 0.001 M KBr; (▲) 0.0335 M KBr

exact Monte Carlo simulations indicate a κ^{-1} dependence. For a chain with an intrinsic stiffness, such as DNA, one usually finds that $l_{p,e} \ll l_{p,0}$. Under these conditions, the electrostatic persistence length scales according to κ^{-2}.

The viscosity in a polymer solution is usually a good measure of the geometrical extensions of the chain. Figure 9.14 shows a classical experiment on polyvinylpyridine H with bromide as the counterion (PVPBr) as a function of the polymer and salt concentrations. We can note two things. First, the polyelectrolyte itself and its counterions contribute to the screening, which is most pronounced in the 'salt-free' curve—the viscosity decreases as more polymer is added. Secondly, once a certain amount of salt is added, the viscosity variation is rather modest (see the lower curve in Figure 9.14). This is in agreement with equation (9.26), where it is found that R_{ee} varies as $c^{-1/5}$.

Bibliography

Barton, A. F. M., *Handbook of Solubility Parameters and Other Cohesion Parameters*, 2nd Edn, CRC Press, Boca Raton, FL, 1991.

Billmeyer Jr, F. W., *Textbook of Polymer Science*, 3rd Edn, Wiley, New York, 1984.

Brandrup, J. and E. H. Immergut, *Polymer Handbook*, Wiley, New York, 1975.

Flory, P. J., *Principles of Polymer Chemistry*, Cornell University Press, Ithaca, NY, 1953.

Flory, P. J., *Statistical Mechanics of Chain Molecules*, Wiley, New York, 1969.

McClelland, B. J., *Statistical Thermodynamics*, Chapman & Hall, London, 1973.

Napper, D. H., *Polymeric Stabilization of Colloidal Dispersions*, Academic Press, London, 1983.

10 REGULAR SOLUTION THEORY

Bragg–Williams Theory Describes Non-ideal Mixtures

The Free Energy of Mixing

The Bragg–Williams (BW) model is sometimes also called the theory of regular solutions and is one of the simplest statistical mechanical approaches to liquid mixtures. It does not involve any sophisticated mathematical treatments beyond simple combinatorics. Despite its simplicity it gives a surprisingly good qualitative description of a number of very complex processes in liquid mixtures. It is also the basis for the Flory–Huggins theory of polymer solutions. Several of the concepts, i.e. the χ-parameter, in the BW model appear in many different situations and it is of some value to know their origins.

The BW model is based on a lattice model, where each site can accommodate one molecule irrespective of type and size (see Figure 10.1). This means that the number of neighbours is always constant, assuming that all lattice positions are occupied and that the volume does not change upon mixing. The basic assumptions in the model are as follows:

1. The mixture is random.

2. The number of nearest neighbours is constant.

3. The interaction is limited to nearest neighbours.

This means that we allow the mixing energy to be non-zero, $\Delta E \neq 0$, while we assume that the mixing entropy is ideal, $\Delta S = \Delta S_{ideal}$. This type of mean

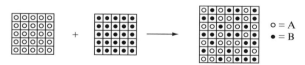

Figure 10.1 Representation of the lattice model for the random mixing of two liquids (Filled and open circles)

field approximation is clearly inconsistent and will be discussed further below.

Consider the mixing of two substances A and B.

$$\Delta S = -k(N_A + N_B)[x_A \ln x_A + x_B \ln x_B] \tag{10.5}$$

The enthalpy of mixing, ΔH, is calculated from the difference in contact energy between the two kind of molecules. Denote these contact energies by w_{AA}, w_{BB} and w_{AB}, respectively. The total energy of pure A and B then becomes:

$$E_{AA} = zN_A w_{AA}/2 \tag{10.6a}$$

$$E_{BB} = zN_B w_{BB}/2 \tag{10.6b}$$

where z is the number of neighbours. The factor '2' appears from calculating the number of contacts twice.

The total energy for the mixture is written as follows:

$$E_{AB} = N_{AA} w_{AA} + N_{BB} w_{BB} + N_{AB} w_{AB} \tag{10.7}$$

where N_{ij} is the number of ij pairs in the mixture. Note that N_{ij} is the number of ij contacts while N_i is the number of 'i' molecules. The number of i–j contacts = (total number of i–x pairs) $\times$ (the probability that the pair is an i–j contact), namely:

$$\frac{1}{2}(N_i z)\left(\frac{N_j}{N_i + N_j}\right) \tag{10.8}$$

Hence, we have:

$$N_{AA} = N_A \frac{z}{2} N_A/(N_A + N_B) \tag{10.9a}$$

$$N_{BB} = N_B \frac{z}{2} N_B/(N_A + N_B) \tag{10.9b}$$

$$N_{AB} = N_A z N_B/(N_A + N_B) \tag{10.9c}$$

The change in internal energy upon mixing is equal to the energy of the mixture minus the energy of the two single liquids. Hence:

$$\Delta E = E_{AB} - E_{AA} - E_{BB}$$

$$= N_{AA}w_{AA} + N_{BB}w_{BB} + N_{AB}w_{AB} - zN_A w_{AA}/2 - zN_B w_{BB}/2$$

$$= N_A \frac{z}{2} \frac{N_A}{(N_A + N_B)} w_{AA} + N_B \frac{z}{2} \frac{N_B}{(N_A + N_B)} w_{BB} + N_A z \frac{N_B}{(N_A + N_B)} w_{AB}$$

$$- \frac{z}{2} N_A w_{AA} - \frac{z}{2} N_B w_{BB}$$

$$= z \frac{N_A N_B}{N_A + N_B} \left(w_{AB} - \frac{w_{AA}}{2} - \frac{w_{BB}}{2} \right) = z \frac{N_A N_B}{N_A + N_B} \Delta w \qquad (10.10)$$

where we have introduced Δw as:

$$\Delta w = w_{AB} - \frac{w_{AA}}{2} - \frac{w_{BB}}{2} = w_{AB} - \frac{1}{2}(w_{AA} + w_{BB}) \qquad (10.11)$$

One should note that equation (10.10) only involves Δw and that the result is implicitly dependent upon the individual interaction parameters w_{AA}, w_{BB} and w_{AB}. The change in interaction upon mixing Δw can obviously be both positive and negative, with the sign depending on whether the AB interaction is more attractive than the average of AA and BB.

In a lattice model there is no pressure–volume term and one can safely write the enthalpy change upon mixing as follows:

$$\frac{\Delta H}{N_A + N_B} = \frac{\Delta E}{N_A + N_B} = z\Delta w x_A x_B \qquad (10.12)$$

Finally, by introducing the interaction parameter χ, defined as:

$$\chi = \frac{z\Delta w}{kT} \qquad (10.13)$$

we have the enthalpy of mixing, expressed per mole of substance, as follows:

$$\frac{\Delta H}{n_A + n_B} = RT\chi x_A x_B \qquad (10.14)$$

where n_i is the number of moles of component i. The important quantity to calculate is the free energy of mixing per mole of substance, namely:

$$\frac{\Delta G}{n_A + n_B} = \frac{\Delta H}{n_A + n_B} - \frac{T \Delta S}{n_A + n_B}$$
$$= RT(\chi x_A x_B + x_A \ln x_A + x_B \ln x_B) \qquad (10.15)$$

With an explicit expression for the free energy a number of interesting quantities can be calculated. For example, the chemical potential of component A in the mixture is given by:

$$\mu_A = \frac{d}{dn_A}(G_A + G_B + \Delta G) = \mu_A^0 + \frac{d\Delta G}{dn_A} \qquad (10.16)$$

Note that in equation (10.16) one obtains, by rather simple means, a non-ideal expression for the chemical potential, where the last term is the interesting excess quantity:

$$\frac{d\Delta G}{dn_A} = RT(\chi x_B^2 + \ln x_A) \qquad (10.17)$$

Thus, the chemical potentials for A in the mixture can be written as follows:

$$\mu_A = \mu_A^0 + RT \ln a_A = \mu_A^0 + RT(\chi x_B^2 + \ln x_A) \qquad (10.18)$$

where a_A is the activity, and hence the activity coefficient of component A can be identified as follows:

$$\frac{a_A}{x_A} = \exp\,(\chi x_B^2) \qquad (10.19)$$

Historically, the χ-parameter was first treated as an enthalpic quantity, as in equation (10.14), but has later been identified as a free energy quantity, as suggested from equation (10.19).

Raoult's Law

The vapour pressure above an ideal mixture is given by Raoult's law, viz. $p_A = x_A p_A^0$. For a non-ideal mixture, $p_A = a_A p_A^0$ and, following the Bragg–Williams model using equation (10.19), we then obtain:

$$p_A = x_A \exp\,(\chi x_B^2) p_A^0 \qquad (10.20)$$

The interaction parameter χ can be both positive and negative as one in reality obtains both positive and negative deviations from Raoult's law, as can be seen in Figure 10.2.

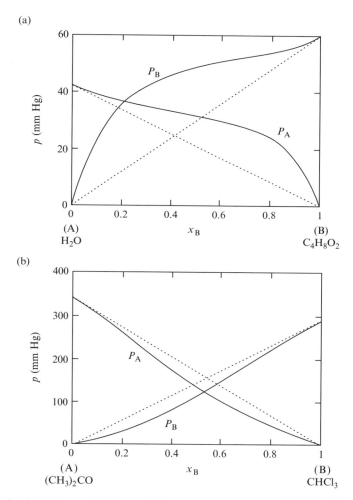

Figure 10.2 (a) Water and dioxane behave as two non-compatible liquids, while (b) acetone and chloroform demonstrate the opposite situation. From W. J. Moore, *Physical Chemistry*, Longmans, London, 1962. © 1962 by Prentice-Hall Inc., Englewood Cliffs, NJ, reprinted by permission of Pearson Education Limited

Phase Separation

The requirement for two liquids to be miscible is that $\Delta G < 0$. This condition is, however, not sufficient for complete miscibility at all concentrations. Consider the free energy change on mixing for one mole of a mixture:

$$\frac{\Delta G}{RT(n_A + n_B)} = (\chi x_A x_B + x_A \ln x_A + x_B \ln x_B) \tag{10.21}$$

Figure 10.3 shows the left-hand side of equation (10.21) for different values of χ. The upper curve will lead to two phases if the initial mole fraction is 0.5. We will then obtain two phases with concentrations given by points c and d.

Equation (10.21) predicts that phase separation will always first occur at $x_A = x_B$. Phase separation will occur when $d^2(\Delta G/RT)/dx_A^2 < 0$:

$$d(\Delta G/RT)/dx_A = \chi(1 - 2x_A) + 1 + \ln x_A - 1 - \ln(1 - x_A) \tag{10.22a}$$

$$\frac{d^2(\Delta G/RT)}{dx_A^2} = -2\chi + \frac{1}{x_A} + \frac{1}{1 - x_A} = 0 \tag{10.22b}$$

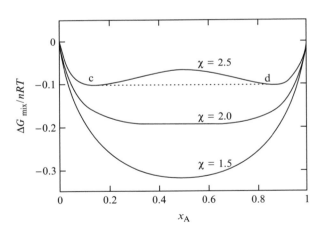

Figure 10.3 The free energy of mixing per mole of solution as a function of composition. The two lower curves describe complete miscibility, while the upper curves display a phase separation

The latter equation gives that $\chi = 2$ when $x_A = x_B = 0.5$. Hence, there will be a phase separation if $\chi \geq 2$. The temperature when phase separation first occurs is called the critical temperature:

$$\chi_c = 2 = \frac{z\Delta w}{kT} \Rightarrow T_c = \frac{z\Delta w}{2k} \tag{10.23}$$

The BW theory captures the main physics of the phase separation in a simple two-component system. In real systems, however, one may sometimes find a mixture of two liquids, which are completely miscible at low temperature but phase-separate with increasing temperature (see, for example, the polyoxyethylene–water system in Chapter 3). Figure 10.4 shows the different possibilities that could occur—eventually at sufficiently elevated temperature all systems will show an upper consulate point, i.e. a closed miscibility gap, as shown in Figure 10.4(c).

The lower consulate point is sometimes described as being due to a temperature-dependent interaction difference, Δw. This goes outside the BW model and does not really improve our physical understanding of the phenomenon. In the polyoxyethylene–water system, the lower consulate point is probably due to a temperature-induced structural change of the EO groups; i.e. internal degrees of freedom that are absent from the BW model play an important role.

It is also instructive to consider the mixing of water and a hydrophobic molecule, such as a hydrocarbon. The water–water interaction (w_{ww}) is strongly negative, while the water–hydrocarbon (w_{wh}) and hydrocarbon–hydrocarbon (w_{hh}) interactions are relatively small, in comparison to w_{ww}, and can be put to zero. The effective pair interaction then becomes:

$$\Delta w = w_{wh} - \frac{1}{2}(w_{hh} + w_{ww}) \approx -\frac{w_{ww}}{2} > 0 \tag{10.24}$$

The χ-parameter for such a system at room temperature would be of the order of 5 (cf. Chapter 7). Hence, the BW theory correctly predicts that oil and water do not mix.

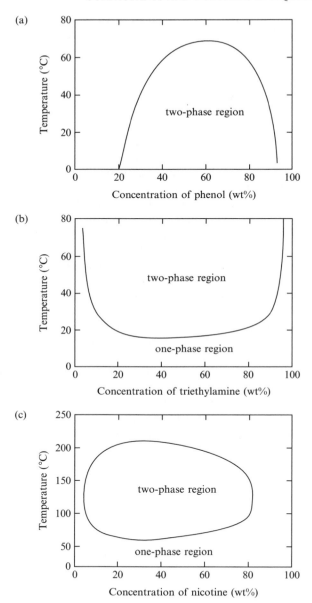

Figure 10.4 Phase diagrams showing both upper and lower consulate points for differ-
ent systems (a) phenol–water; (b) triethylamine–water; (c) nicotine–water system. From
W. J. Moore, *Physical Chemistry*, Longmans, London, 1962. © 1962 by Prentice-
Hall Inc., Englewood Cliffs, NJ, reprinted by permission of Pearson Education
Limited

Flory–Huggins Theory Describes the Phase Behaviour of Polymer Solutions

The Entropy of Mixing for Polymer + Solvent

Consider the mixture of two liquids, one of which is a polymer (see Figure 10.5). One can use the same type of approach as above, but the mixing entropy will be different. It is obvious that the entropy change will be smaller, since the monomers in the polymer will not be fully capable of exploiting the volume increase upon mixing—the connectivity of the polymer prevents this. The mixing energy, however, will be of the same form as in the mixing of two monomeric liquids.

Consider a solution consisting of N_1 solvent molecules and N_2 polymers, each with r monomers, i.e. we have altogether $N = N_1 + N_2 r$ solvent molecules and monomers. The mixing entropy will then be given by the following:

$$\Delta S = -k(N_1 + N_2)\left[\left(\frac{N_1}{N_1 + N_2}\right)\ln\left(\frac{N_1}{N_1 + rN_2}\right)\right.$$
$$\left. + \left(\frac{N_2}{N_1 + N_2}\right)\ln\left(\frac{rN_2}{N_1 + rN_2}\right)\right] \qquad (10.25)$$

where:

$$\frac{N_1}{N_1 + rN_2} = \phi_1 = \text{volume fraction of solvent molecules}$$

$$\frac{rN_2}{N_1 + rN_2} = \phi_2 = \text{volume fraction of polymer molecules}$$

We can now write the mixing entropy in the usual form:

$$\Delta S = -k(N_1 + rN_2)(x_1\ln\phi_1 + x_2\ln\phi_2) \qquad (10.26)$$

The quantity base-mole is defined as the total number of moles of solvent molecules and polymer segments in the system and equals $n_1 + rn_2$. The entropy of mixing, expressed per base-mole of substance, is then given by:

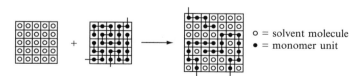

$\circ$ = solvent molecule
$\bullet$ = monomer unit

Figure 10.5 Lattice model for the random mixing of a polymer (filled circles) and a liquid (open circles)

$$\frac{\Delta S}{n_1 + rn_2} = -R(x_1 \ln \phi_1 + x_2 \ln \phi_2) \tag{10.27}$$

The mixing energy will be of the same form as for a monomeric mixture and from equation (10.5) we have:

$$\frac{\Delta H}{n_2 + rn_2} = \frac{N_1 rN_2}{(N_1 + rN_2)^2} \chi RT = \phi_1 \phi_2 \chi RT \tag{10.28}$$

It is interesting to see the difference in entropy change when mixing two simple liquids (ΔS_{mon}) and when mixing a simple liquid and a polymer (ΔS_{pol}). Let us denote this difference by $\Delta\Delta S$:

$$\Delta\Delta S = \Delta S_{mon} - \Delta S_{pol} = -k\left[N_2(r-1)\ln\left(\frac{rN_2}{N_1 + rN_2}\right)\right] \tag{10.29}$$

That is, the difference increases with the length of the polymer and as a consequence one expects to see phase separation in polymer mixtures at an earlier stage, i.e. at lower temperatures, than in ordinary monomeric liquid mixtures.

Phase Equilibrium in the Flory–Huggins Theory

From equations (10.27) and (10.28) one may write the free energy of mixing, expressed per base-mole, as follows:

$$\frac{\Delta G}{RT(n_1 + rn_2)} = \phi_1 \phi_2 \chi + (x_1 \ln \phi_1 + x_2 \ln \phi_2) \tag{10.30}$$

where the first term represents the energy and the second the entropy change upon mixing. From the derivative with respect to component 1, the solvent, one obtains the chemical potential, as follows (Figure 10.6):

$$\Delta\mu_1 = \mu_1 - \mu_1^0 = \frac{d\Delta G}{dn_1} = RT\left[\chi\phi_2^2 + \ln(1 - \phi_2) + \left(1 - \frac{1}{r}\right)\phi_2\right] \tag{10.31}$$

A non-monotonic behaviour signals a phase separation. One could ask at which value of ϕ this first happens? After some algebra, one finds the following expression for the critical point:

$$\chi_c \approx \frac{1}{2} + \frac{1}{\sqrt{r}} \tag{10.32}$$

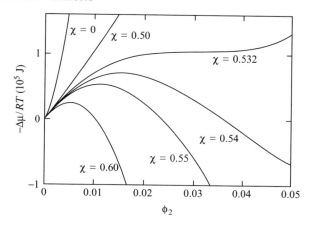

Figure 10.6 The excess chemical potential of the solvent in a binary solution containing a polymer with volume fraction ϕ_2. The degree of polymerization is 1000 and the χ-values are indicated on the diagram. Reprinted from Paul J. Flory, *Principles of Polymer Chemistry*. Copyright © 1953 Cornell University Press and Copyright © 1981 Paul J. Flory. Used by permission of the publisher, Cornell University Press

and:

$$\phi_{2c} \approx \frac{1}{\sqrt{r}} \tag{10.33}$$

These values should be compared with the regular solution for which $\chi_c = 2$ and $\phi_c = 0.5$. Thus, we see that polymer solutions become more easily incompatible and phase-separate more easily, as discussed in connection with equation (10.29).

The Θ Temperature (Good and Bad Solvents)

A common concept in polymer literature is the Θ temperature and the notion of good and bad solvents. In order to introduce these concepts, let us return to equation (10.31) and the excess chemical potential, $\Delta\mu$, which for small volume fractions of solute may be expanded and rewritten as

$$\Delta\mu_1 = -RT\left[\left(\frac{1}{2} - \chi\right)\phi_2^2 + \ldots\right] = -RT\left[\frac{1}{2}\left(1 - \frac{\Theta}{T}\right)\phi_2^2 + \ldots\right] \tag{10.34}$$

where the theta temperature is defined as follows:

$$\Theta = \frac{2z\Delta w}{k} = 2\chi T \tag{10.35}$$

Equation (10.34) shows that when the physical temperature is equal to the theta temperature, then the solution behaves as if it were ideal (perfect), i.e. $\Delta\mu_1 = 0$. If $T > \Theta$, we describe the solvent as a good solvent for the polymer and if $T < \Theta$ it is a poor solvent. One can also interpret the Θ temperature in a slightly different manner by rewriting the critical temperature at which phase separation first occurs as follows:

$$\Theta = T_c \left(1 + \frac{2}{\sqrt{r}}\right) \tag{10.36}$$

Thus, Θ is equal to the critical temperature for an infinitely long polymer, i.e. when $r = \infty$ (see Figure 10.7).

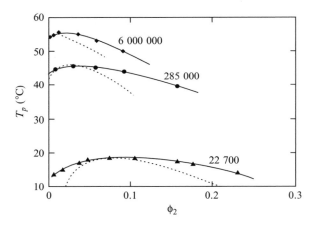

Figure 10.7 Phase diagrams for three polyisobutylene fractions in diisobutyl ketone. The continuous curves are drawn through experimental points, while the dashed curves represent theoretical results. Reprinted from Paul J. Flory, *Principles of Polymer Chemistry*. Copyright © 1953 Cornell University Press and Copyright © 1981 Paul J. Flory. Used by permission of the publisher, Cornell University Press

Bibliography

McClelland, B. J., *Statistical Thermodynamics*, Chapman & Hall, London, 1973.
Flory, P. J., *Principles of Polymer Chemistry*, Cornell University Press, Ithaca, NY, 1953.

11 NOVEL SURFACTANTS

Gemini Surfactants have an Unusual Structure

Introduction

A gemini surfactant may be viewed as a surfactant dimer, i.e. two amphiphilic molecules connected by a spacer. Figure 11.1 shows the general structure. Gemini surfactants are also referred to as twin surfactants, dimeric surfactants or bis-surfactants.

The spacer chain, which can be hydrophilic or hydrophobic, rigid or flexible, should bind the two moieties together at, or in close proximity to, the head groups. Connecting two surfactant moieties towards the end of their hydrophobic tails results in a so-called 'bolaform' surfactant and the physicochemical properties of such species are very different from those of gemini surfactants. Most geminis are composed of two identical halves, but unsymmetrical gemini surfactants have also been synthesized, either having different hydrophobic tail lengths, or different types of polar groups (heterogemini surfactants), or both.

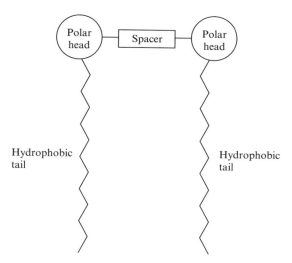

Figure 11.1   Schematic structure of gemini surfactants

In recent years, 'higher oligomers' of single surfactants, i.e. tris-surfactants, tetra-surfactants, etc., have also been synthesized. Little is known at the present time, however, about the properties and usefulness of these species.

Gemini surfactants have not yet reached the market in large-scale quantities. They are, however, attracting considerable attention, both in industry and in academia. Some gemini surfactants, in particular, symmetrical cationic ones, are made from readily available raw materials by a straightforward synthesis, as will be discussed below.

Figure 11.2 shows some examples of gemini surfactants. Compounds 1–3 are cationics differing in the type of spacer unit connecting the two ionic moieties. The spacer of Compound 1 is hydrophobic and flexible, that of Compound 2 is hydrophilic and flexible, while that of Compound 3 is hydrophobic and rigid. Compound 4 is a typical nonionic gemini and Compound 5 is an anionic one, based on the same backbone structure. Compounds 6 and 7, finally, are examples of heterogemini surfactants.

Synthesis

Cationic geminis such as Compounds 1–3 of Figure 11.2 are conveniently prepared by reacting an alkyldimethylamine with an α, ω-dihalo compound. Dibromo reagents are more reactive and are usually employed in laboratory synthesis but the corresponding dichloro compounds may be preferred in large scale preparation:

$$2R-N(CH_3)_3 + Br-X-Br \longrightarrow R-N^+(CH_3)_2-X-N^+(CH_3)_2-R2Br^-$$

where R is an alkyl of normal surfactant chain length, such as $C_{12}H_{25}$. X can be alkylene to give a hydrophobic, flexible spacer, $CH_2CH(OH)CH_2$ or CH_2 $(CH_2OCH_2)_nCH_2$ to give a hydrophilic flexible spacer, or $CH_2 - \phi - CH_2$ to give a hydrophobic, rigid spacer (ϕ denotes a phenyl ring, i.e. in this case, C_6H_4).

For the specific (but important) case when X equals CH_2CH_2 in the formula above, the dihalo compound is not very reactive. The preferred synthesis route is then as follows:

$$2RBr + (CH_3)_2N-CH_2CH_2-N(CH_3)_2 \longrightarrow$$

$$R-N^+(CH_3)_2-CH_2CH_2-N^+(CH_3)_2-R + 2Br^-$$

Anionic and nonionic geminis are often prepared by the ring opening of a bisepoxide, giving a bishydroxyether intermediate. Below is shown a synthesis of a gemini surfactant with sulfate as the polar head group (Compound 5 of Figure 11.2):

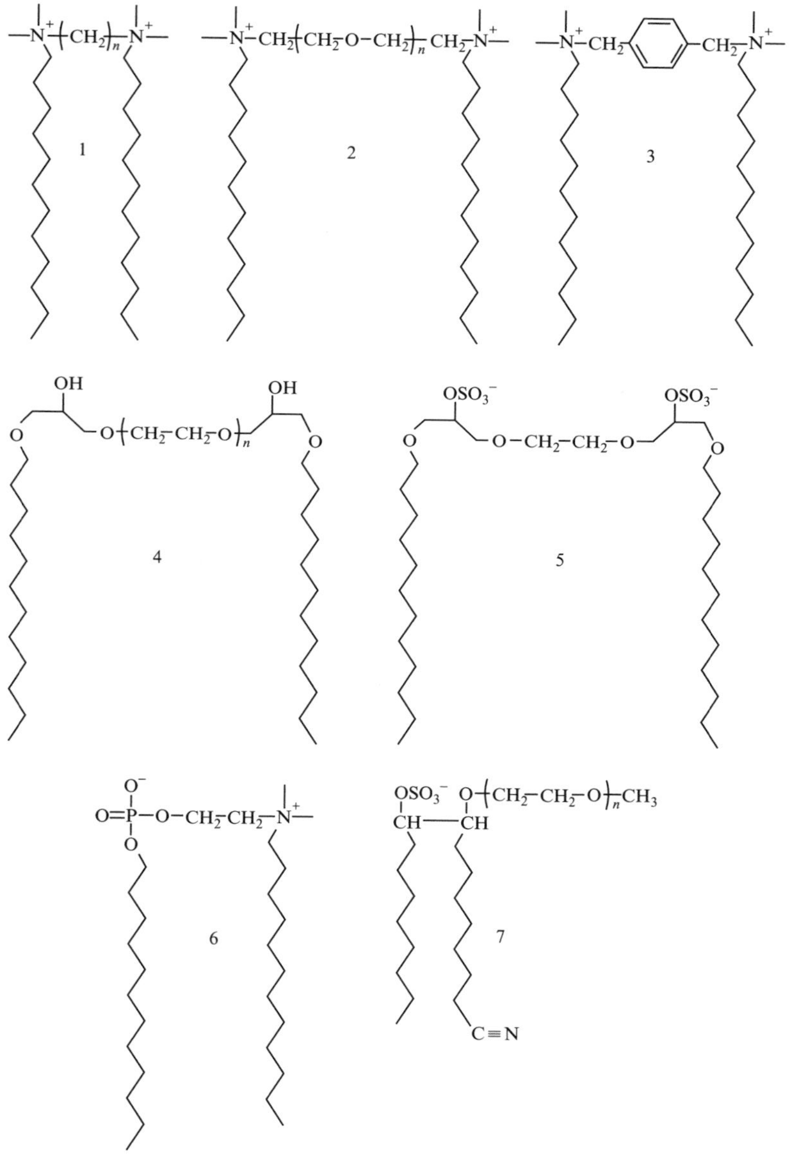

Figure 11.2 Structures of gemini surfactants: Compounds 1–3 are cationics differing in the type of spacer unit, Compounds 4 and 5 are non-ionic and anionic surfactants, respectively, based on the same backbone structure, and Compounds 6 and 7 are heterogemini surfactants, having non-equal polar head groups

$$2 \text{ ROH} + \underset{O}{CH_2\text{-CH-Y-CH-CH}_2} \longrightarrow R\text{-O-CH}_2\text{-}\underset{OH}{CH}\text{-Y-}\underset{OH}{CH}\text{-CH}_2\text{-O-R}$$

$$\xrightarrow{SO_3} R\text{-O-CH}_2\text{-}\underset{OSO_3^-}{CH}\text{-Y-}\underset{OSO_3^-}{CH}\text{-CH}_2\text{-O-R}$$

where R is an alkyl of normal surfactant chain length and Y is OCH_2CH_2O.

Below is shown the synthesis for the heterogemini surfactant Compound 7 of Figure 11.2:

$$CH_3(CH_2)_7CH=CH(CH_2)_7CN \xrightarrow{H_2O_2} CH_3(CH_2)_7\underset{O}{CH\text{-}CH}(CH_2)_7CN$$

$$\xrightarrow{HO-(CH_2CH_2O)_n\text{-}CH_3} CH_3(CH_2)_7\underset{OH}{CH}\text{-}\underset{O-(CH_2CH_2O)_n\text{-}CH_3}{CH}(CH_2)_7CN$$

$$\xrightarrow[NaOH]{SO_3} CH_3(CH_2)_7\underset{^-O_3SO}{CH}\text{-}\underset{O-(CH_2CH_2O)_n\text{-}CH_3}{CH}(CH_2)_7CN \qquad Na^+$$

Micellization and Behaviour at the Air–Water Interface

A very striking feature of gemini surfactants is that they start to form micelles at a concentration more than one order of magnitude lower than that of the corresponding 'monomer' surfactant. The low value of the critical micelle concentration (CMC) is an important property, implying, for instance, that geminis are very effective in solubilizing oily components. The efficiency of geminis, expressed as the C_{20} value, i.e. the surfactant concentration at which the surface tension (γ) is lowered by $20\,mN/m$, is also very high, again when compared with the monomeric species. The effectiveness of geminis, which is given by the value of surface tension at the CMC (γ_{CMC}) is usually also somewhat better than for the corresponding monomeric surfactants. Figure 11.3 shows surface tension plots for a cationic gemini surfactant and for the corresponding monomeric amphiphile. It is also interesting to compare the values for the cationic gemini with two dodecyl tails and two hydrophilic head groups with those of normal double-tailed cationic surfactants which only contain one polar head. Didecyldimethylammonium bromide has a CMC value of 1.8×10^{-3}, while the longer homologue didodecyldimethylammonium bromide has a CMC value of 1.7×10^{-4}. In addition, by using this comparison the CMC value of the gemini surfactant of Figure 11.3 is very low.

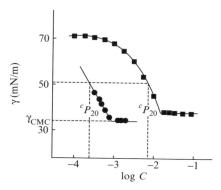

Figure 11.3 Surface tension versus log concentration plots of the gemini surfactant $C_{12}H_{25}N(CH_3)_2-(CH_2)_3-N(CH_3)_2C_{12}H_{25}^{2+}$ $2Br^-$ ($\bullet$), and the monomeric surfactant $C_{12}H_{25}N(CH_3)_3^+$ Br^- ($\blacksquare$). From R. Zana, Dimeric (gemini) surfactants, in *Novel Surfactants*, K. Holmberg (Ed.), Surfactant Science Series, 74, Marcel Dekker, New York, 1998

The dynamic surface tension is an important property, relevant to many practical, non-equilibrium processes such as emulsification and foaming. Dynamic surface tension is a measure of how fast, in the millisecond range, a surfactant reduces the surface tension from the value of pure water (around 70 mN/m) to values in the range of 30 mN/m. It has been found that the type of spacer governs the dynamic surface tension of geminis to a considerable degree, with the longer and the more flexible the spacer, then the better, i.e. the shorter, the time before onset of surface tension reduction.

Geminis with flexible spacers, both hydrophilic and hydrophobic, generally show lower γ_{CMC} values than the corresponding surfactants with rigid spacers. This is probably due to better packing of the former at the air–water interface (see below).

As mentioned above, geminis have very low CMC values when compared with conventional surfactants of the same hydrocarbon chain length. CMC ratios as high as 80 between monomeric and dimeric species have been reported. The CMC values are not very dependent on the polarity of the spacer. The values do change with the length of the spacer, however, as is illustrated in Figure 11.4 for three series of cationic gemini surfactants, all having hydrophobic spacers. For all three series, there is a CMC maximum at a spacer length of 5–6 carbon atoms. This maximum has been attributed to changes of spacer conformation and its resulting effect on head group hydration and alkyl chain orientation. When the spacer becomes long enough, it may twist in order to allow its middle portion to reside in the micelle interior, thus contributing to the hydrophobicity of the surfactant. For geminis with a hydrophilic spacer such as oligo(ethylene glycol) the CMC values continue to increase with increasing spacer chain length.

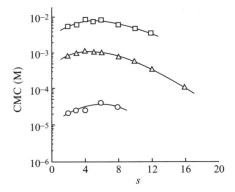

Figure 11.4 Variation of CMC against the number of methylene groups, s, in the spacer for three series of cationic gemini surfactants: ($\bullet$)$C_{10}H_{21}N(CH_3)_2-(CH_2)_s-N(CH_3)_2$ $C_{10}H_{21}^{2+}$ $2Br^-$, ($\triangledown$)$C_{12}H_{25}N(CH_3)_2-(CH_2)_s-N(CH_3)_2C_{12}H_{25}^{2+}$ $2Br^-$;($\circ$) $C_{16}H_{33}N(CH_3)_2-(CH_2)_s-N(CH_3)_2C_{16}H_{33}^{2+}$ $2Br^-$. From R. Zana, Dimeric (gemini) surfactants, in *Novel Surfactants*, K. Holmberg (Ed.), Surfactant Science Series, 74, Marcel Dekker, New York, 1998

The area occupied by a surfactant at the air–water interface can be obtained from the slope of the curve of surface tension, γ, versus the logarithm of surfactant concentration, $\ln C$, by using the Gibbs equation, which is discussed later in Chapter 15. The Gibbs equation can be used to study the packing at the air–water interface, and Figure 11.5 shows the variation of area per surfactant

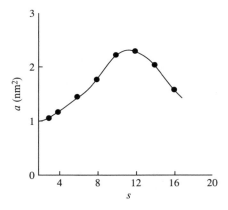

Figure 11.5 Variation of the area, a, per surfactant molecule at the air–water interface with the number of methylene groups, s, in the spacer for the series of gemini surfactants $C_{12}H_{25}N(CH_3)_2 - (CH_2)_s - N(CH_3)_2C_{12}H_{25}^{2+}$ $2Br^-$. From R. Zana, Dimeric (gemini) surfactants, in *Novel Surfactants*, K. Holmberg (Ed.), Surfactant Science Series, 74, Marcel Dekker, New York, 1998

molecule at the surface with the spacer length for a series of cationic geminis. As can be seen from this figure, there is a pronounced maximum at 10–12 carbon atoms in the spacer. This maximum has been explained in terms of a change of location of the hydrophobic spacer. At chain lengths below 10, the spacer lies more or less flat at the interface, thus occupying a larger and larger area as the number of methylene groups increases. Above around 12 carbon atoms, the spacer chain starts to fold, forming a loop which projects out into the air. This is analogous to the explanation of the relationship between the CMC and the chain lengths of hydrophobic spacers given above.

Micelle Shape and Effect on the Rheology of Solutions of Gemini Surfactants

At low concentrations, i.e. just above the CMC, cationic geminis form spherical micelles, just as their monomeric counterparts do. The aggregation number, i.e. the number of molecules that make up a micelle, for surfactants with 12 carbon atoms in the hydrophobic tails goes from around 40 for species with two methylene groups as a spacer to around 25 for surfactants with a ten-methylene spacer unit. The surfactants with short spacers (2, 3 and 4 methylene groups) show a very steep growth with concentration, indicating a transition from spherical to elongated micelles already at very low surfactant concentrations. Surfactants with longer spacer units show a much less pronounced micelle growth with increasing concentration, in this respect resembling the behaviour of the corresponding monomeric surfactants. The difference in aggregation numbers between geminis with short spacers and those with long spacers, schematically illustrated in Figure 11.6, can be explained as follows. When the

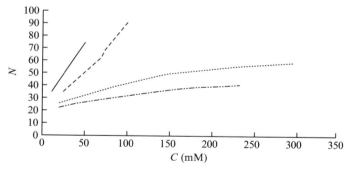

Figure 11.6 Variation of micelle aggregation number, N, with surfactant concentration, C, for cationic geminis with $C_{12}H_{25}$ hydrophobic tails and with different lengths of the spacer. Spacer lengths: (——) three methylene groups (_ _ _) four methylene groups; (......) five methylene groups. The monomeric analogue, didodecyldimethylammonium bromide, is shown as a reference (_.._.._). Data taken from K. Esumi *et al., Langmuir,* **13** (1997) 2585 and 2803

spacer consists of four carbon atoms or less, the distance between the charged head groups becomes shorter than the inter-head-group distance in conventional micelles. The surfactant packing in the micelles will therefore be different from the packing of the corresponding normal surfactant having the same hydrophobic tail. The gemini surfactant can be said to pack as if the polar head group were smaller than it really is. This is the reason why the transition from spherical to rod-like aggregates occurs so readily; for geometrical reasons, surfactants with small head group areas prefer to arrange themselves in elongated structures. When the spacer length becomes equivalent or larger than the normal inter-head-group distance in micelles of cationic surfactants, the growth of the micelle with surfactant concentration is similar for geminis to that of monomeric surfactants.

The shape of the micelles governs the solution viscosity. The cationic gemini surfactant with $C_{12}H_{25}$ hydrophobic tails and a two-methylene spacer unit exhibits a dramatic viscosity increase at about $2\,wt\%$ surfactant. As can be seen from Figure 11.7, the viscosity increases between 6 and 7 orders of magnitude within a rather narrow concentration range. This is a much more pronounced viscosity increase with concentration than what one normally encounters with cationic surfactants and can be accounted for by the sharp transition from spherical to thread-like micelles characteristic of geminis with short spacer units. It can also be noted that thread-like micelles, and thus an increased vicosity, can be induced by the use of an applied shear, already at a concentration below that where they form under static conditions. The unusual rheological behaviour of gemini surfactants, which takes place at relatively low surfactant concentrations, can have important practical consequences.

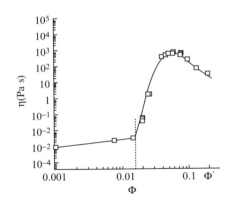

Figure 11.7 Variation of the zero shear viscosity, η, with the surfactant volume fraction, Φ, of solutions of the cationic gemini surfactant $C_{12}H_{25}N(CH_3)_2 - (CH_2)_2 -N(CH_3)_2C_{12}H_{25}^{2+}$. $2Br^-$ at $20°C$. Reprinted with permission from F. Kern *et al.*, *Langmuir*, **10** (1994) 1714. Copyright (1994) American Chemical Society

Concluding Remarks

Gemini surfactants are subject to much evaluation work at the present time and one may foresee that the attractive features of these surfactants, such as high efficiency, low CMC and γ_{CMC}, values, and a very steep rise in viscosity with concentration, will find practical use. The high surfactant efficiency and the low CMC values have triggered much effort into the potential use of gemini surfactants as solubilizers of various kinds. In model experiments, using hydrocarbons as the compounds to be solubilized, geminis have been found to be better than conventional surfactants, both on a molar and a weight basis. Geminis are also of interest as lubricating agents as a result of their tight packing at surfaces.

Much effort is also being devoted to exploit the specific geometry of gemini surfactants to create structures of well-defined geometry. Gemini surfactants form vesicles and liquid crystalline phases over broad concentration ranges, a property that can be taken advantage of for a variety of applications. One example of such work is the preparation of mesoporous molecular sieves. Using gemini surfactants as templates, materials with a cubic geometry and with very well-defined pore dimensions have been obtained. Mesoporous structures made from surfactant liquid crystals, including liquid crystals from gemini surfactants, are further discussed later in Chapter 23.

Cleavable Surfactants are Environmentally Attractive but are of Interest for other Reasons as well

Introduction

By tradition, surfactants are stable species. Among the surfactant work-horses—anionics such as alkylbenzene sulfonates and alkyl sulfates, non-ionics such as alcohol ethoxylates and alkylphenol ethoxylates, and cationics such as alkyl quats and dialkyl quats—only alkyl sulfates are not chemically stable under normal conditions. Throughout the years the susceptibility of alkyl sulfates to acid-catalysed hydrolysis has been seen as a considerable problem, and is particularly well-known for the most prominent member of the class, i.e. sodium dodecyl sulfate (SDS). The general attitude has been that weak bonds in a surfactant may cause handling and storage problems and should therefore be avoided.

In recent years, the attitude towards easily cleavable surfactants has changed. Environmental concern has become one of the main driving forces for the development of new surfactants and the rate of biodegradation has become a major issue. One of the main approaches taken to produce readily biodegradable surfactants is to build into the structure a bond with limited stability. For practical reasons, the weak bond is usually the bridging unit between the polar head group and the hydrophobic tail of the surfactant which means that degradation immediately leads to destruction of the surface activity of the molecule, an event usually referred to as the primary degradation of the surfactant. Biodegradation then proceeds along various routes depending on the type of primary degradation product. The ultimate decomposition of the surfactant, often expressed as the amount of carbon dioxide evolved during four weeks exposure to appropriate microorganisms (counted as a percentage of the amount of carbon dioxide that could theoretically be produced) is the most important measure of biodegradation (see page 000). In addition, it seems that for most surfactants containing easily cleavable bonds the values for ultimate decomposition are higher than those of the corresponding surfactants lacking the weak bond. Thus, the strong trend towards more environmentally benign products favours the cleavable surfactant approach on two accounts.

A second incentive for the development of cleavable surfactants is to avoid complications such as foaming or the formation of unwanted, stable emulsions after the use of a surfactant formulation. Cleavable surfactants present the potential for elimination of some of these problems. If the weak bond is present between the polar and the non-polar part of the molecule, cleavage will lead to one water-soluble and one water-insoluble product. Both moieties can usually be removed by standard work-up procedures. This approach has been of particular interest for surfactants used in preparative organic chemistry and in various biochemical applications.

A third use of surfactants with limited stability is to have the cleavage product impart a new function. For instance, a surfactant used in personal care formulations may decompose on application to form products beneficial to the skin. Surfactants that after cleavage impart a new function are sometimes referred to as 'functional surfactants'.

Finally, surfactants that in a controlled way break down into non-surfactant products may find use in specialized applications, e.g. in the biomedical field. For instance, cleavable surfactants that form vesicles or microemulsions can be of interest for drug delivery, provided that the metabolites are non-toxic.

Most cleavable surfactants contain a hydrolysable bond. Chemical hydrolysis can be either acid- or alkali-catalysed and many papers discuss the surfactant breakdown in terms of either of these mechanisms. In the environment, bonds susceptible to hydrolysis are often degraded by enzymatic catalysis, although few papers dealing with cleavable surfactants have investigated such processes *in vitro*. Other approaches that have been taken include incorporation of a bond that can be destroyed by UV-irradiation or the use of an ozone-cleavable bond. This presentation is subdivided according to the type of weak linkage present in the surfactant.

Acid-Labile Surfactants

Cyclic acetals

Cyclic 1,3-dioxolane (5-membered ring) and 1,3-dioxane (6-membered ring) compounds, illustrated in Figure 11.8, are early examples of acid-labile

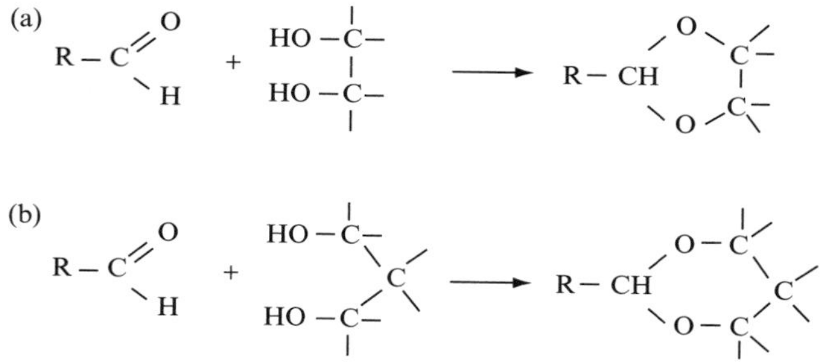

Figure 11.8 Preparations of (a) 1,3-dioxolane, and (b) 1,3-dioxane surfactants from a long-chain aldehyde and a 1,2- and a 1,3-diol, respectively. From K. Holmberg, Cleavable surfactants, in *Novel Surfactants*, K. Holmberg (Ed.), Surfactant Science Series, 74, Marcel Dekker, New York, 1998

surfactants. They are typically synthesized from a long-chain aldehyde by reaction with a diol or a higher polyol. Reaction with a vicinal diol gives the dioxolane, while 1,3-diols yield dioxanes.

If the diol contains an extra hydroxyl group, such as in glycerol, a hydroxy acetal is formed and the remaining hydroxyl group can subsequently be derivatized to give anionic or cationic surfactants, as illustrated in Figure 11.9. It is claimed that glycerol gives ring closure to dioxolane, yielding a free, primary hydroxyl group but it is likely that some dioxane with a free, secondary hydroxyl group is formed as well. The free hydroxyl group can be treated with SO_3 and then neutralized to give the sulfate, it can be reacted with propane sultone to give the sulfonate, or it can be substituted by bromide or chloride and then reacted with dimethylamine to give a tertiary amine as the polar group. Quaternization of the amine can be made in the usual manner, e.g. with methyl bromide. The remaining hydroxyl group may also be ethoxylated to give non-ionic chemodegradable surfactants. The rate of decomposition in

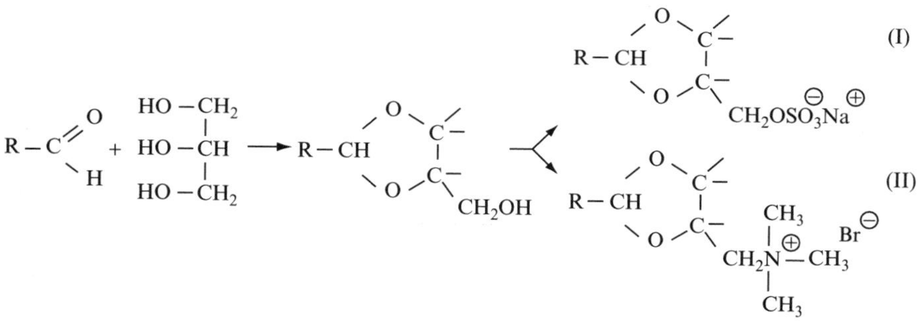

Figure 11.9 Examples of (I) anionic, and (II) cationic 1,3-dioxolane surfactants

sewage plants of this class of non-ionic surfactants is much higher than for normal ethoxylates.

Hydrolysis splits acetals into aldehydes, which are intermediates in the biochemical β-oxidation of hydrocarbon chains. Acid-catalysed hydrolysis of unsubstituted acetals is generally facile and occurs at a reasonable rate, at pH 4–5, at room temperature. Electron-withdrawing substituents, such as hydroxyl, ether oxygen and halogens, reduce the hydrolysis rate, however. Anionic acetal surfactants are more labile than cationic materials, a fact that can be ascribed to the locally high oxonium ion activity around such micelles. The same effect can also be seen for surfactants forming vesicular aggregates, again undoubtly due to differences in the oxonium ion activity in the pseudo-phase surrounding the vesicle. Acetal surfactants are stable at neutral and high pH levels.

The 1,3-dioxolane ring has been found to correspond to approximately two oxyethylene units with regard to the effects on both the CMC and adsorption characteristics. Thus, the surfactant type I shown in Figure 11.9 should resemble ether sulfates of the general formula $R - (OCH_2CH_2)_2OSO_3Na$. This is interesting since the commercial alkyl ether sulfates contain two – three oxyethylene units.

Acyclic acetals

Alkyl glucosides, often somewhat erroneously referred to as alkyl polyglucosides or APGs, are cyclic compounds but since the ring does not involve the two geminal hydroxyl groups of the aldehyde hydrate, it is here included in the category of acyclic acetals. Alkyl glucosides are by far the most important type of acetal surfactant. The most common routes of preparation of this surfactant class were given in Chapter 1 (see Figure 1.17).

Alkyl glucoside surfactants break down into glucose and long-chain alcohol under acidic conditions. On the alkaline side, even at very high pH, they are stable to hydrolysis. Their cleavage profile, along with their relatively straightforward synthesis route makes these surfactants interesting candidates for various types of cleaning formulations.

Polyoxyethylene-based cleavable surfactants have been synthesized by reacting end-capped poly(ethylene glycol) (PEG) with a long-chain aldehyde. During acid hydrolysis, these compounds will revert to the original fatty aldehyde and end-capped PEG. Studies of the relationship between structure and hydrolytic reactivity have shown that the hydrolysis rate increases as the hydrophobe chain length decreases when the hydrophilic part was kept the same. This has been attributed to a decreased hydrophobic shielding of the acetal linkage from the oxonium ions. No effect on the hydrolysis rate was seen when the hydrophilic part was varied while the hydrophobic part was kept constant, or when the structure of the hydrophobe was varied from linear to branched. Furthermore, the hydrolytic reactivity is higher for non-aggregated than for micellized surfactants.

Acetal surfactants have been found to resemble traditional surfactants in terms of their physicochemical properties. However, the CMC values for acetal-containing surfactants are somewhat lower than for the corresponding conventional surfactants. Furthermore, the efficiency of the surfactants, expressed as the concentration required to produce a 20 mN/m reduction in the surface tension, was higher for the cleavable surfactants. Evidently, the acetal linkage connecting the hydrophobic tail and the polar head group gives a contribution to the surfactant hydrophobicity, resulting in higher adsorption efficiency at the air–water interface and an increased tendency to aggregate into micelles.

Ketals

Surfactants containing ketal bonds can be prepared from a long-chain ketone and a diol in analogy with the reaction schemes given in Figures 11.8 and 11.9 for the preparation of acetal surfactants. Non-ionic cleavable surfactants based on a long-chain carbonyl compound, glycerol, and a polyoxyethylene chain have been commercialized. Both long-chain ketones and aldehydes can be used and they form cyclic ketals and acetals, respectively, upon condensation with glycerol, as discussed above for cyclic acetals. Ketones give primarily 4-hydroxymethyl-1,3-dioxolanes, whereas aldehydes give a mixture of 4-hydroxymethyl-1,3-dioxolanes and 5-hydroxy-1,3-dioxanes. The remaining hydroxyl function is alkoxylated in the presence of a conventional base catalyst.

Ketal-based surfactants have also been prepared in good yields from esters of keto acids by either of two routes, as shown in Figure 11.10. The biodegradation profiles of the dioxolane surfactants given in Figure 11.10 are shown in Figure 11.11. As expected, the degradation rate is very dependent on the alkyl

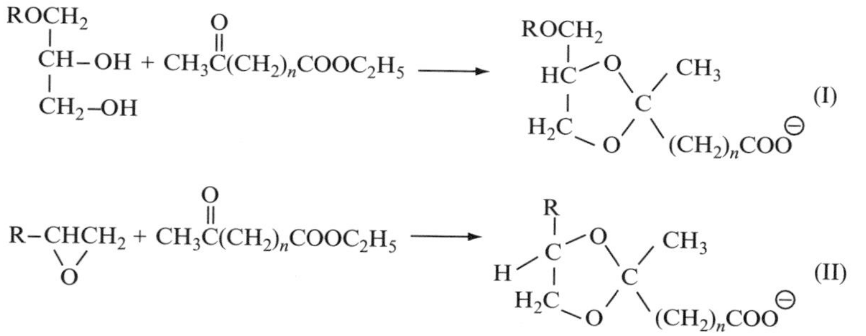

Figure 11.10 Preparation of anionic 1,3-dioxolane surfactants from ethyl esters of keto acids. Reproduced by permission of the American Oil Chemists' Society from A. Sokolowski *et al., J. Am. Oil Chem. Soc.*, **69** (1992) 633

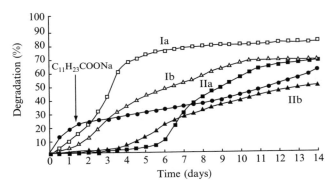

Figure 11.11   Rate of biodegradation versus time for four ketal surfactants and for sodium dodecanoate as a reference. I and II relate to the compounds shown in Figure 11.10: a, R = C$_{12}$H$_{25}$ with $n = 2$; b, R = C$_{16}$H$_{33}$ with, $n = 2$. Reproduced by permission of the American Oil Chemists' Society from D. Ono, *et al., J. Am. Oil Chem. Soc.,* **72** (1995) 853

chain length. The process is markedly faster for the labile surfactants (and particularly for structure I which contains an extra ether oxygen) than for the conventional carboxylate surfactant of the same alkyl chain length used as a reference. Ketal surfactants are, in general, more labile than the corresponding acetal surfactants. As an example, a ketal surfactant kept at pH 3.5 is cleaved to the same extent as an acetal surfactant of similar structure kept at pH 3.0. The relative lability of the ketal linkage is due to the greater stability of the carbocation formed during ketal hydrolysis when compared to the carbocation formed during acetal hydrolysis. (It is noteworthy that biodegradation of an acetal surfactant has been found to be faster than for a ketal surfactant of very similar structure. Evidently, there is no strict correlation between the ease of biodegradation and the rate of chemical hydrolysis.)

Ortho esters

Ortho esters are a new class of surfactants that have recently been described. Surfactant ortho esters are conveniently prepared by the transesterification of a low molecular weight ortho ester (such as triethylorthoformate) with a fatty alcohol and poly(ethylene glycol) (PEG). An example of a structure and a typical method of preparation are given in Figure 11.12. Due to the trifunctionality of the ortho ester, a distribution of species is obtained. Furthermore, if the reactant alcohol is difunctional, cross-linking will occur and a large network may be formed. Such compounds have been shown to be effective foam depressors and an example based on poly(propylene glycol) (PPG) and PEG is also shown in Figure 11.12. By varying the number and types of substituents (fatty alcohol, alkyleneoxy group, end-blocking, etc.), the properties of the

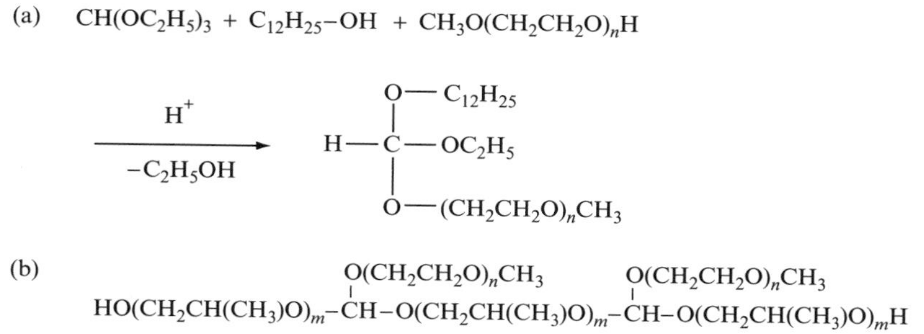

(a) $CH(OC_2H_5)_3 + C_{12}H_{25}-OH + CH_3O(CH_2CH_2O)_nH$

$$\xrightarrow[-C_2H_5OH]{H^+} \quad H-\underset{\underset{O-(CH_2CH_2O)_nCH_3}{|}}{\overset{\overset{O-C_{12}H_{25}}{|}}{C}}-OC_2H_5$$

(b) $O(CH_2CH_2O)_nCH_3 \qquad O(CH_2CH_2O)_nCH_3$

$HO(CH_2CH(CH_3)O)_m-\overset{|}{C}H-O(CH_2CH(CH_3)O)_m-\overset{|}{C}H-O(CH_2CH(CH_3)O)_mH$

Figure 11.12 (a) Synthesis and structure of an ortho-ester-based surfactant and (b) an ortho-ester-based block copolymer. Reproduced by permission of the American Oil Chemists' Society from P.-E. Hellberg, *et al., J. Surf. Detergents,* **3** (2000) 369

ortho-ester-based surfactant or block copolymer can be tailor-made for a specific field of application.

The hydrolysis of ortho esters occurs by a mechanism analogous to that of acetals and ketals and gives rise to one mole of formate and two moles of alcohols. Both formates and alcohols can be regarded as non-toxic substances and recent research has shown that surface active formates (similar to surface active alcohols and esters but in contrast to surface active aldehydes) have no or little dermatological effects, evaluated in terms of sensitizing capacity and irritancy. Ortho-ester-based surfactants undergo acid-catalysed cleavage much more readily than acetal-based surfactants under the same conditions. For instance, a water-soluble ortho ester based on octanol and monomethyl-PEG is hydrolysed to 50% in 2 h at pH 5. The structure of the surfactant has been found to influence the hydrolysis rate and, in general, a more hydrophilic surfactant has a higher decomposition rate.

Ortho ester linkages can also be used to improve biodegradation properties in long-chain ethoxylates or block copolymers. It has been shown that a conventional PEG–PPG copolymer (see Chapter 12) with a molecular weight of 2200 biodegrades to only 3% in 28 days. However, if an equivalent molecule is built up from PEG 350 and PPG 400, connected with ortho-ester links, it will reach 62% biodegradation within 28 days and thus be classified as 'readily biodegradable'.

Alkali Labile Surfactants

Normal ester quats

By the term 'ester quat' we refer to surface active quaternary ammonium compounds that have the general formula $R_4N^+X^-$ and in which the long-chain alkyl moieties, R, are linked to the charged head group by an ester bond.

With normal ester quats we mean surfactants based on esters between one or more fatty acids and a quaternized amino alcohol. Figure 11.13 shows examples of three different ester quats, all containing two long-chain and two short-chain substituents on the nitrogen atom. This figure also shows the 'parent', non-cleavable quat. As can be seen, the ester-containing surfactants contain two carbon atoms between the ester bond and the nitrogen which carries the positive charge. Cleavage of the ester bonds of surfactants II–IV yields a fatty acid soap in addition to a highly water-soluble quaternary ammonium di-or triol. These degradation products exhibit low fish toxicity and they are degraded further by established metabolic pathways. The overall ecological characteristics of esterquats are much superior to those of traditional quats (as represented by compound I of Figure 11.13).

During the last 10–15 years the dialkylester quats have to a large extent replaced the stable dialkyl quats as rinse-cycle softeners which is the single largest application for quaternary ammonium compounds. The switch from stable dialkyl quats to dialkylester quats may represent the most dramatic change of product type in the history of surfactants and it is entirely environment-driven. Unlike stable quats, ester quats show excellent values for biodegradability and aquatic toxicity. Ester quats have also fully or partially replaced traditional quats in other applications of cationics, such as hair care products and various industrial formulations.

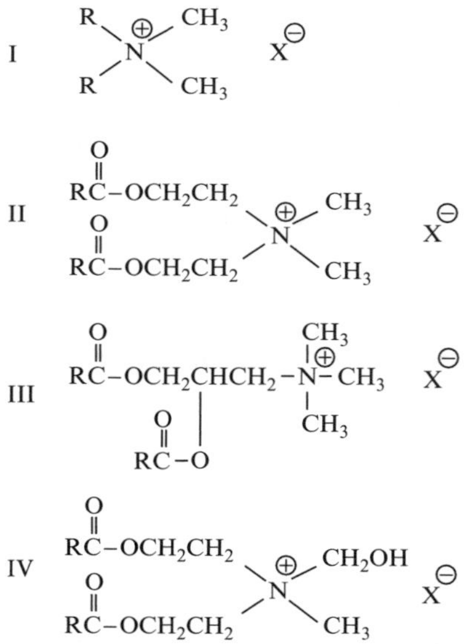

Figure 11.13 Structures of one conventional quaternary ammonium surfactant (I) and three ester quats (II–IV); R is a long-chain alkyl, and X is Cl, Br, or CH_3SO_4

The cationic charge close to the ester bond renders normal ester quats unusually stable to acid and labile to alkali. The strong pH dependence of the hydrolysis can be taken advantage of to induce rapid cleavage of the product. This phenomenon is even more pronounced for betaine esters and the mechanism of hydrolysis is discussed in some detail in the following section. Figure 11.14 illustrates the pH dependence of hydrolysis of an ester quat. As can be seen, the hydrolysis rate is at minimum at pH 3–4 and accelerates strongly above pH 5–6. Evidently, formulations containing ester quats must be maintained at low pH.

Betaine esters

The rate of alkali-catalysed ester hydrolysis is influenced by adjacent electron-withdrawing or electron-donating groups. A quaternary ammonium group is strongly electron-withdrawing. The inductive effect will lead to a decreased electron density at the ester bond; hence, alkaline hydrolysis, which starts by a nucleophilic attack by hydroxyl ions at the ester carbonyl carbon, will be favoured. Compounds II–IV of Figure 11.13 all have two carbon atoms between the ammonium nitrogen and the –O– oxygen of the ester bond. Such esters undergo alkaline hydrolysis at a faster rate than esters lacking the

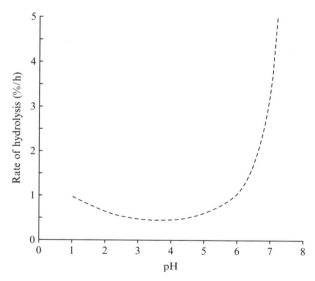

Figure 11.14 Influence of pH on the hydrolytic stability of the dicetylester of bis(2-hydroxyethyl)ammonium chloride at 25°C. From G. Krüger *et al.*, Esterquats, in *Novel Surfactants*, K. Holmberg (Ed.), Surfactant Science Series, 74, Marcel Dekker, New York, 1998

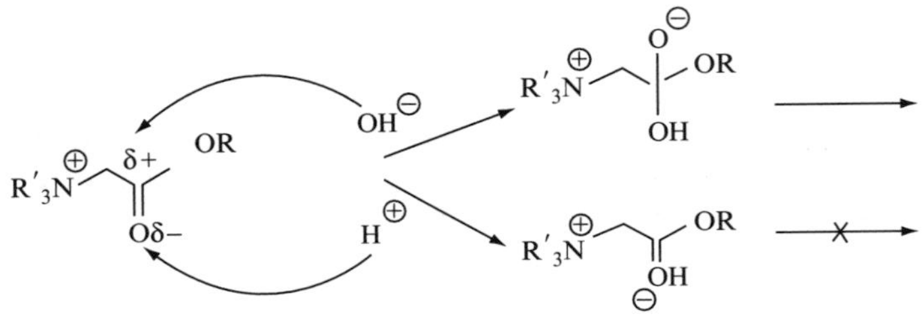

Figure 11.15 Mechanisms for the acid- and base-catalysed hydrolysis of a betaine ester

adjacent charge, although the difference is not very large. If, on the other hand, the charge is at the other side of the ester bond the rate enhancement is much more pronounced. Such esters are extremely labile on the alkaline side but very stable even under strongly acidic conditions. The large effect of the quaternary ammonium group on the alkaline and acid rates of hydrolysis is due to a stabilization/destabilization of the ground state, as illustrated in Figure 11.15. The charge repulsion, involving the carbonyl carbon atom and the positive charge at the nitrogen atom, is relieved by hydroxide ion attack, but augmented by protonation. The net result is that compared with an ester lacking the cationic charge the rate of alkaline hydrolysis is increased 200-fold, whereas the rate of acid hydrolysis is decreased 2000-fold. For surface active betaine esters based on long-chain fatty alcohols, the rate of alkaline hydrolysis is further accelerated due to micellar catalysis, as discussed later in Chapter 23. The presence of large, polarizable counterions, such as bromide, can completely outweigh the micellar catalysis, however.

The extreme pH dependence of surface active betaine esters makes them interesting as cleavable cationic surfactants. The shelf life is long when stored under acidic conditions and the hydrolysis rate will then depend on the pH at which they are used. Single-chain surfactants of this type have been suggested as 'temporary bactericides' for use in hygiene products, for desinfection in the food industry and in other instances where only a short-lived bactericidal action is wanted. Figure 11.16 illustrates the breakdown of a betaine ester with bactericidal action into harmless products, i.e. fatty alcohol and betaine. The patent literature also contains examples of betaine esters containing two long-chain alkyl groups.

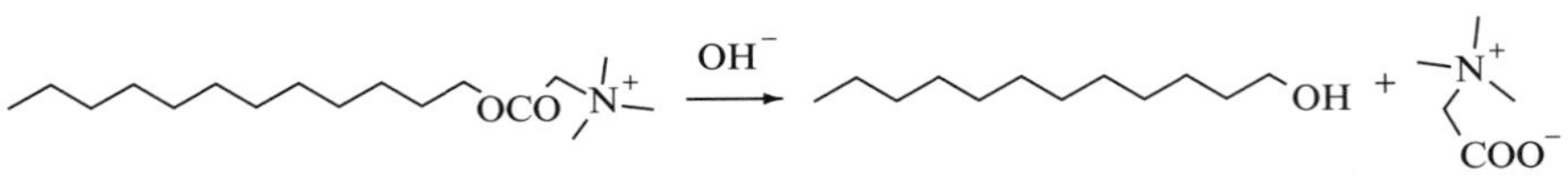

Figure 11.16 Alkaline hydrolysis of a cationic surfactant with biocidal properties into fatty alcohol and betaine

Other esters

As mentioned above, one use of surfactants with limited stability is to have the cleavage product impart a new function. Figure 11.17 illustrates a concept where a surfactant with good detergency properties breaks down under alkaline conditions to a hydrophobic fatty alcohol. Such a surfactant is of interest for combining cleaning and hydrophobization, e.g. when washing tents, rain-clothes, etc. By controlling the pH of the formulation, the textiles will be washed and subsequently hydrophobized in a one-step process.

Concluding Remarks

Amphiphiles with an acid- or alkali-labile bond constitute the most widely explored routes to achieve cleavable surfactants. However, other approaches have also been taken. For instance, several types of surfactants with UV-labile bonds have been synthesized and evaluated. Photochemical cleavage yields non-surface-active species and the concept is attractive because it allows an extremely fast breakdown of the surfactant to occur.

An interesting use of photo-labile surfactants is as emulsifiers in emulsion polymerization. The use of a photo-labile emulsifier opens the possibility to control the latex coagulation process simply by exposing the wet lacquer film to UV irradiation. The ionic head group of the surfactant will be split off by photolysis, thus leading to aggregation of the latex particles. A surfactant with a diazosulfonate head group has been suggested for this application. On UV exposure, the diazo bond is broken, and the surfactant decomposes into a sulfite ion and a hydrophobic residue, both without surface activity. The concept is attractive because UV curing is common in some coatings applications.

Another example of UV-labile amphiphiles relate to double-chain surfactants containing Co(III) as the complexing agent for two single-chain surfactants based on ethylene diamine in the polar head group. UV irradiation, or merely sunlight, causes reduction of Co(III) to Co(II). The latter gives a very labile complex and the double-chain surfactant immediately degrades into two single-chain moieties.

Ozone-cleavable surfactants have also been developed as examples of environmentally benign amphiphiles. These surfactants, which contain unsaturated

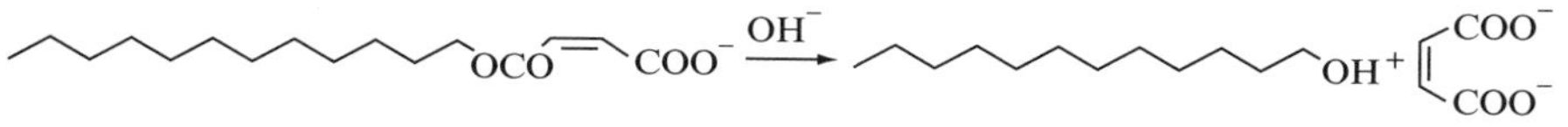

Figure 11.17 Alkaline hydrolysis of an anionic surfactant, a salt of the maleic acid half-ester, into a hydrophobic fatty alcohol and water-soluble maleate

bonds, break down easily during ozonization of water, which is becoming a water purification process of growing importance. It is likely that other mechanisms of surfactant breakdown in the environment will be explored in the future in the design of new surfactants with good environmental characteristics.

Polymerizable Surfactants are of Particular Interest for Coatings Applications

Introduction

Polymerizable surfactants are examples of what is sometimes referred to as 'functional surfactants', i.e. surfactants that possess one characteristic property besides that of pronounced surface activity. The interest in polymerizable surfactants originates from the fact that surfactant action may be needed at some stage of an operation but unnecessary, or even unwanted, at some later stage. The problems with residual surfactants may be environmentally related, such as with slowly biodegradable surfactants in sewage plants. The problems may also be of a technical nature because the presence of surface active agents in the final product may affect the product performance in a negative way.

The paint area is a good example where surfactants are needed at one stage but unwanted at a later stage. Surfactants are used in paints as emulsifiers for the binders, as dispersants for the pigments and to improve wetting of the substrates. In the dried paint film, the presence of a surfactant frequently causes problems, however, since the surfactant acts as an external plasticizer in the film, imparting softness and flexibility. This could be taken advantage of, had the plasticizer been evenly distributed in the coating. However, due to its surface activity, the surfactant will migrate out of the bulk phase and concentrate at the interfaces. It has been shown that surfactant molecules preferably go to the film–air interface, where they align with their hydrophobic tails pointing towards the air (Figure 11.18). As an example, calculations from

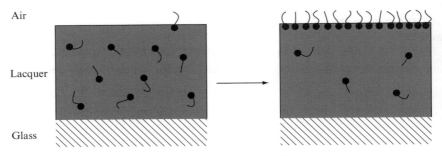

Figure 11.18 Migration of surfactant in a lacquer film leads to enrichment at the surface

ESCA spectra show that the dried film from a lacquer containing 1% surfactant may have an average surface surfactant concentration of 50%. The overall bulk concentration of surfactant is still approximately 1% since the modified surface region is a very thin layer when compared to the thickness of the whole film. The surfactant surface layer constitutes a so-called weak boundary layer when a second coating is applied, leading to the frequently encountered problem in repaintability.

It is also known that during the drying of emulsion paints the surfactant may undergo phase separation, forming lumps, tenths of microns wide, distributed throughout the film. AFM studies have revealed that these surfactant lumps can extend far down into the film. On exposure to water the surfactant is washed out of the film, with the result being that deep cavities appear where the lumps had been. Such film defects are believed to be one of the causes of the poor water resistance of many latex paint films.

One way to overcome the problems associated with the presence of surfactant in the final product is to have the surfactant chemically bound to the latex particle or, alternatively, to make the surfactant polymerize during the setting or curing stage. In principle, the surfactant may either undergo homopolymerization or copolymerize with some other component of the system. In paints and lacquers, the obvious choice of coreactant is the binder.

A completely different way to avoid the problem of residual surfactant in the end product is to use cleavable surfactants, i.e. surfactants that spontaneously break down at some stage. This concept has been discussed in the previous section of this chapter.

Mode of Surfactant Polymerization

Homopolymerization versus copolymerization

In a formulation containing reactive surfactant, homopolymerization of the amphiphile may take place if the concentration is high enough. However, in most technical formulations the surfactant concentration is too low to allow substantial homopolymerization in the bulk phase. A monolayer of surfactant, on the other hand, may homopolymerize when adsorbed at an interface, as will be shown below. The palisade layer may either form by adsorption from an aqueous solution or by migration through a film, as discussed above.

Copolymerization, on the other hand, may take place in a bulk phase. For copolymerization of a monomer, M_1, and a surfactant, M_2, to occur, the reactivity ratios should preferably be $r_1 \leq 1$ and $r_2 \leq 1$, where r_1 and r_2 are defined as $r_1 = k_{11}/k_{12}$ and $r_2 = k_{22}/k_{21}$, where k_{nm} are the rate constants for the four possible ways in which monomer can add:

$$M_1 \bullet + M_1 \longrightarrow M_1 \bullet \quad \text{Rate} = k_{11}[M_1 \bullet][M_1]$$
$$M_1 \bullet + M_2 \longrightarrow M_2 \bullet \quad \text{Rate} = k_{12}[M_1 \bullet][M_2]$$
$$M_2 \bullet + M_1 \longrightarrow M_1 \bullet \quad \text{Rate} = k_{21}[M_2 \bullet][M_1]$$
$$M_2 \bullet + M_2 \longrightarrow M_2 \bullet \quad \text{Rate} = k_{22}[M_2 \bullet][M_2]$$

Autoxidative versus non-autoxidative polymerization

Autoxidation, i.e. oxygen-induced curing, may take place both during copoly-merization in the bulk phase and during homopolymerization of a surface monolayer. Two surfactants capable of undergoing autoxidation are shown in Figure 11.19. The top compound(a), an unsaturated monoethanolamide ethoxylate, was discussed in Chapter 1 (p. 35). Surfactants that can undergo autoxidation are of particular interest in combination with alkyd resins. Aut-oxidation is normally catalysed by cobalt or manganese salts.

Non-autoxidative polymerization includes UV curing or thermally induced curing with the use of free-radical initiators, such as benzoyl peroxide or potassium persulfate. Similar to autoxidation, both bulk and surface curing may occur. Surfactants based on activated vinyl groups, such as acrylate and methacrylate esters, are typical examples of this class. Some examples of surfactants capable of rapid UV curing are shown in Figure 11.20.

Position of polymerizable group

The reactive group may be present in either the polar, hydrophilic or hydro-phobic part of the surfactant molecule.

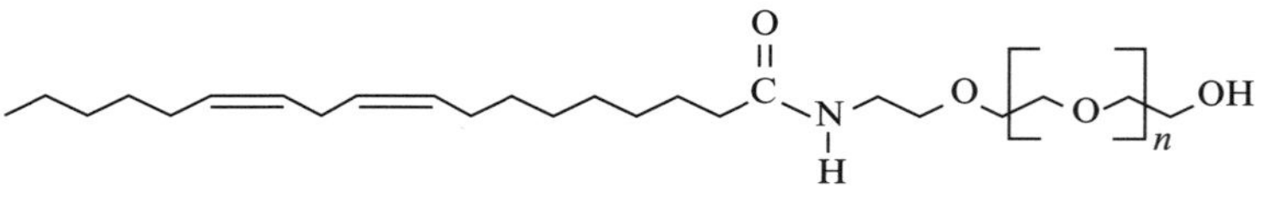

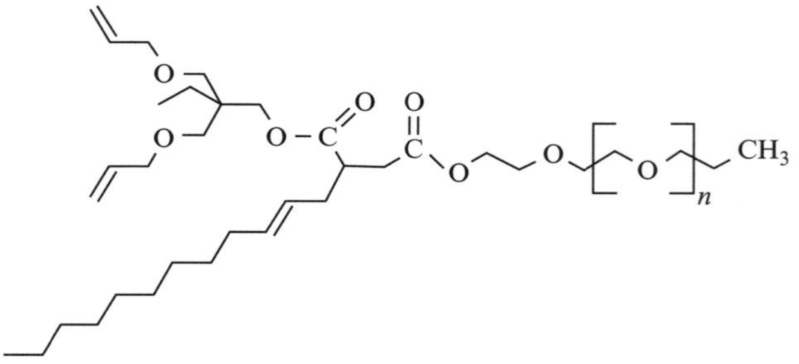

Figure 11.19 Two non-ionic surfactants capable of autoxidation: (a) the ethoxylated monoethanolamide of linoleic acid; (b) the ethoxylated dodecenylsuccinic acid mono-ester of trimethylolpropanediallylether

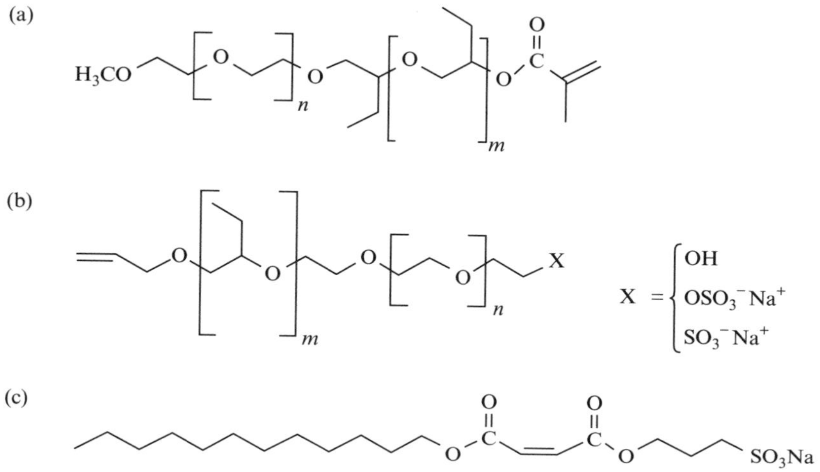

(a)

(b)

$$X = \begin{cases} OH \\ OSO_3^- Na^+ \\ SO_3^- Na^+ \end{cases}$$

(c)

Figure 11.20 Examples of polymerizable surfactants: (a) a methacrylate ester of a methyl-capped block copolymer of ethylene oxide and butylene oxide; (b) an allyl-capped block copolymer of butylene oxide and ethylene oxide with varying end groups; (c) monododecylmonosulfopropylmaleate

In liquid–liquid two-phase systems, the solubility characteristics of the initiator are important. The distribution of the initiator between the oil and water phases should be such that it is predominantly present in the phase where the polymerizable function is located. Thus, surfactants with a polymerizable bond in the polar head group are best served with a water-soluble initiator, whereas a water-soluble initiator is preferred when the polymerizable bond is situated in the hydrocarbon part.

In general, polymerization at the polar end of non-ionic surfactants requires relatively severe conditions and often gives poor yields. When the same functional group is present in the hydrophobic tail, reactivity in a two-phase system is much higher. Cross-linking of the polar groups should also be avoided if the surfactant after polymerization is supposed to provide steric stabilization, for instance, in the stabilization of dispersed systems. The entropy term, which is the main driving force behind steric stabilization, will be reduced if the freedom of motion of the polar head groups is restricted.

Cross-linking of the hydrophobic tails is also natural in those cases where the surfactant polymerizes when adsorbed at a hydrophobic surface. Adsorption at such surfaces occurs with the surfactant hydrophobic chains close together, an orientation which should facilitate the formation of inter-chain bonds.

Applications of Polymerizable Surfactants

Emulsion polymerization

Polymerizable surfactants are of interest in emulsion polymerization, e.g. in the conversion of vinyl chloride to PVC and of acrylates and vinyl acetate to latices for coatings. Use of a reactive surfactant in vinyl chloride polymerization leads to PVC with improved shear stability. In latices, polymerizable surfactants can bring about several advantages such as the following:

- improved stability against shear, freezing and dilution

- reduced foaming

- reduced problems with competitive adsorption (see below)

- improved adhesion properties of the film

- improved water and chemical resistance of the film

Competitive adsorption is a serious problem in many paint formulations, as well as in many other surfactant-containing formulations. A pigmented latex coating contains a variety of interfaces at which surfactants may adsorb, such as binder–water, pigment–water, substrate–water and air–water. In addition, the surfactant molecules may assemble in micelles or form aggregates together with hydrophobic segments of the associative thickener which is normally present in today's latex paints. The situation is illustrated in Figure 11.21. Since different surfactants are normally introduced into the system together with the individual components, e.g. emulsifier (often a mixture of an anionic and a nonionic surfactant) with the binder, pigment dispersant with the pigment, wetting agent added directly to the formulation, etc., the situation becomes very complex and competitive adsorption is a potential problem in all pigmented emulsion paints. The surface active agent used as the binder emulsifier may desorb from the emulsion droplet and adsorb at the pigment surface. The pigment dispersing agent may go the other way. Such an exchange is known to occur and to cause problems in terms of instability and unwanted rheological behavior. Competitive adsorption in general is discussed in Chapter 16.

Compounds of the types shown in Figure 11.20 are useful as reactive surfactants for latex preparation. Out of these, maleic acid derivatives are particularly interesting since they are unable to homopolymerize at ordinary temperature. The maleic-acid-based surfactant shown in the figure can be easily prepared by first reacting maleic anhydride with a fatty alcohol and subsequently treating the monoester formed with propane sultone (which is toxic and should be handled with great care). Extensive homopolymerization is unwanted since the resulting chains of oligomeric or polymeric surfactant will constitute highly

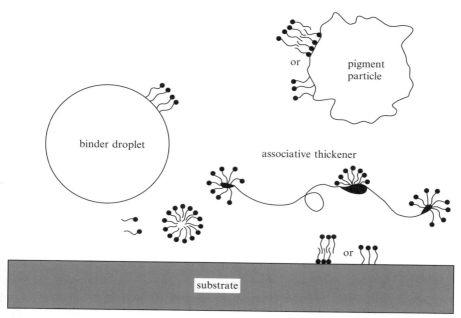

Figure 11.21 A paint formulation contains many different types of surfaces

water-soluble segments on the surface of the latex particles. After drying, these will be distributed throughout the coating and may cause film defects. Instead, the reactive surfactant should preferably copolymerize, i.e. react only with latex monomers, in order to be evenly distributed on the particle surface. It is also important that the reactivity is not too high. If polymerization occurs too rapidly in the water phase, the surfactant-containing polymer may end up in the serum rather than on the particle surface. If the polymerizable surfactant is too rapidly consumed by polymerization in the emulsion droplets, the probability is high that the surfactant will be buried in the interior of the particle.

Polymerizable surfactants for latices are sometimes referred to as 'surfmers', with this name indicating that the surfactant can be seen as a special type of comonomer. Reactive surfactants can also be employed as initiators, in which case they are called 'inisurfs'. Finally, they can be used as chain-transfer agents, i.e. to control the molecular weight of the latex polymer, in which case they are named 'transurfs'. Figure 11.22 shows some representative examples of inisurfs and transurfs, along with the structures of a conventional initiator and a conventional chain-transfer agent for emulsion polymerization. With the use of inisurfs of the type given in this figure, it is possible to prepare latices of high-solids content without the use of extra surfactant.

$$\text{(I)} \quad {}^-O_3S-\phi-(CH_2)_n-OCO-\underset{\underset{CH_3}{|}}{\overset{\overset{CH_3}{|}}{C}}-N=N-\underset{\underset{CH_3}{|}}{\overset{\overset{CH_3}{|}}{C}}-COO-(CH_2)_n-\phi-SO_3{}^-$$

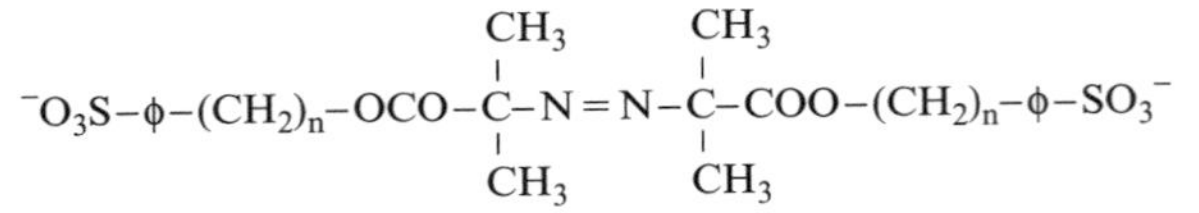

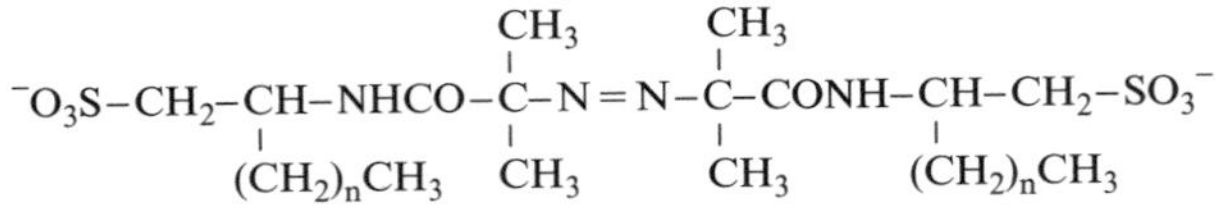

$$\text{(II)} \quad {}^-O_3S-CH_2-\underset{\underset{(CH_2)_nCH_3}{|}}{CH}-NHCO-\underset{\underset{CH_3}{|}}{\overset{\overset{CH_3}{|}}{C}}-N=N-\underset{\underset{CH_3}{|}}{\overset{\overset{CH_3}{|}}{C}}-CONH-\underset{\underset{(CH_2)_nCH_3}{|}}{CH}-CH_2-SO_3{}^-$$

$$\text{(III)} \quad NC-\underset{\underset{CH_3}{|}}{\overset{\overset{CH_3}{|}}{C}}-N=N-\underset{\underset{CH_3}{|}}{\overset{\overset{CH_3}{|}}{C}}-CN$$

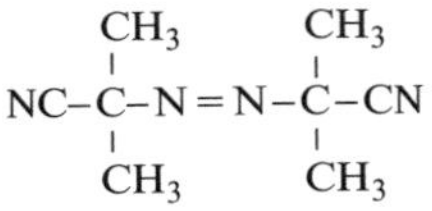

$$\text{(IV)} \quad C_{11}H_{23}-CH=CH-CH_2-SO_3{}^-$$

$$\text{(V)} \quad HS-C_{10}H_{20}-SO_3{}^-$$

$$\text{(VI)} \quad HS-C_nH_{2n}-CH_3$$

Figure 11.22 Structures of two inisurfs (I and II), i.e. polymerizable surfactants that also serve as polymerization initiators, and two transurfs (IV and V), polymerizable surfactants with chain-tranfer capability. The structures of one conventional initiator (III) and one conventional chain-transfer agent (VI) are also shown

Latices are often prepared with a combination of a non-ionic and an anionic surfactant. In addition, cationic surfactants may be employed for latex synthesis, although practical use of positively charged latices is limited. Figure 11.23 shows the structures of three quaternary ammonium surfactants, all containing a reactive styryl group, which have been used as model reactive surfactants for latices.

Figure 11.24 illustrates the effect of the polymerizable surfactants of Figure 11.23 on latex stability. The surfactants were used as the sole emulsifiers for the emulsion polymerization of styrene. For comparison, two non-reactive surfactants were also employed: a cationic surfactant analogous to the middle compound of Figure 11.23, but lacking the vinyl group, and an anionic surfactant.

In the experiment, the latex serum was replaced by continuously pumping distilled water through a cell confining the latex particles by a membrane filter. The procedure results in desorption of weakly bound surfactants. For charged surfactants, desorption leads to a reduction of particle surface charge and to a reduced electrophoretic mobility of the latex. At a certain value of the electrophoretic mobility, positive or negative, the latex will start to agglomerate. As seen in Figure 11.24, the polymerizable surfactants all gave the latex a high positive charge which remained almost constant during the flushing of water

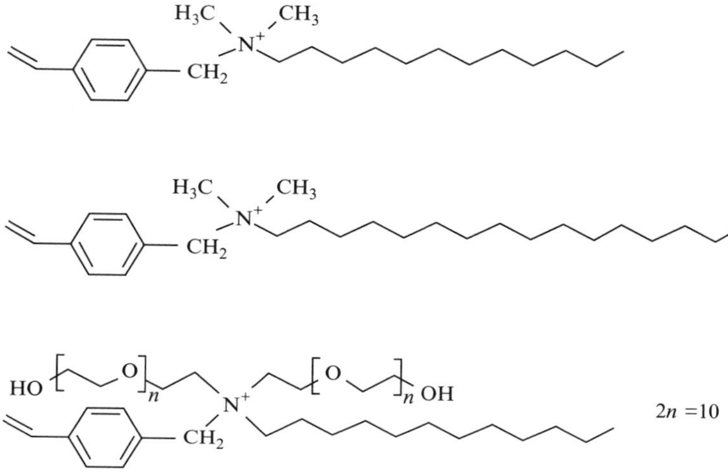

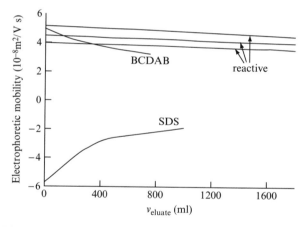

Figure 11.23 Reactive cationic surfactants

Figure 11.24 Electrophoretic mobility as a function of dilution volume of polystyrene latices prepared with the polymerizable cationic surfactants of Figure 11.23 or with the non-polymerizable surfactants, benzylcetyldimethylammonium bromide (BCDAB) or sodium dodecyl sulfate (SDS). Reproduced by permission of the Oil & Colour Chemists Association from K. Holmberg, *Surf. Coatings Int.*, **76** (1993) 481

through the cell. The latices stabilized by the two physically adsorbed surfactants initially had a high absolute electrophoretic mobility. However, as water was passed through the cell the mobilities were continuously decreasing, hence indicating surfactant desorption. The results may be extrapolated to indicate stability on dilution of the different latices.

Alkyd emulsions

Alkyd emulsions are gaining in importance as a consequence of environmental demands. Stable emulsions can be made from most alkyds, provided that the resin viscosity is not too high and sufficient shear forces are applied in the emulsification process. It has been found that by using emulsifiers capable of participating in the autoxidative drying of the binder, the film properties can be considerably improved. The surfactants shown in Figure 11.19, in particular the fatty acid monoethanolamide ethoxylates based on highly unsaturated fatty acid fractions, have proven useful for preparing alkyd emulsions with proper storage stability. It has been found that alkyd lacquer films containing this type of surfactant dry faster and become harder than films containing the same amount of a non-reactive non-ionic surfactant of a similar HLB value. Figure 11.25 depicts schematic representations of the curing process with a conventional emulsifier and with a surfactant capable of copolymerizing with unsaturated acyl chains of the binder.

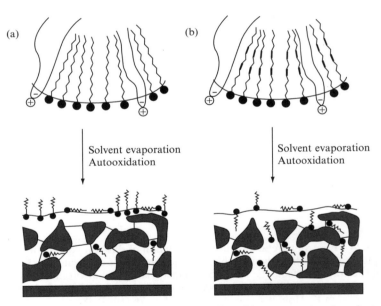

Figure 11.25 Curing of an alkyd emulsion based on (a) a conventional surfactant and (b) a reactive surfactant containing homoconjugated double bonds in the hydrophobic tail. Most of the reactive surfactants become chemically bound to the resin network. Reprinted from K. Holmberg, 'Polymerizable surfactants', *Prog. Org. Coatings*, **20** (1992), 325–337, Copyright (1992), with permission from Elsevier Science

Surface modification

The modification of solid surfaces can be achieved by an adsorbed layer of reactive surfactant on the surface, as illustrated in Figure 11.26. Provided that the surfactant molecules are extensively cross-linked, such a thin surface film will be attached irreversibly. In this way, hydrophobic surfaces can be made hydrophilic or a specific functionality can be introduced.

For example, low-density polyethylene films can be hydrophilized with surfactants having one or two polymerizable groups, such as methacrylate or diacetylene. Adsorption can be made from buffer solutions and the subsequent polymerization of the adsorbed monolayer achieved by UV irradiation. Particularly good results are often obtained with surfactants of the twin-tail type, i.e. having a hydrophobic part consisting of two hydrocarbon chains. This structure gives optimal packing on planar surfaces since such surfactants have a value of the critical packing parameter close to 1 (see Chapter 2). Proper alignment of the surfactants at the surface is believed to be vital for effective cross-linking to occur.

There is experimental evidence that surfactants which contain two polymerizable functions give better result in terms of permanent hydrophilization than surfactants containing only one reactive group. Most likely, surfactants with more than one polymerizable group give a cross-linked networks of higher molecular weight and such a surface layer will be completely water-insoluble and irreversibly attached to the surface.

Paint and lacquer films can also be surface-modified by the migration of a dissolved surfactant to the film–air interface during the drying or curing stage. The principle is shown in Figure 11.27 for an UV-polymerizable surfactant.

As an example, a fluorocarbon surface layer can be obtained by dissolving a small amount, less than 1%, of a polymerizable fluorosurfactant in a lacquer and cross-linking the surfactant monolayer formed at the surface. Figure 11.28(a) shows two fluorocarbon surfactants, one polymerizable (i) and one non-reactive (ii), used in such an experiment. The surfactants were added to a

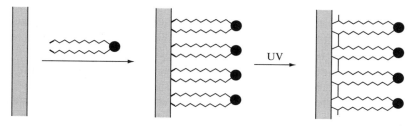

Figure 11.26 Surface modification with UV-curable surfactant

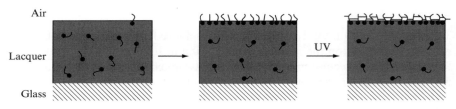

Figure 11.27 Surface modification of a lacquer film through migration of the surfactant to the film–air interface, followed by UV curing

poly(methyl methacrylate) (PMMA) lacquer. PMMA is more polar than the hydrocarbon part of the surfactant so the surfactant orients at the film–air interface with its polar part in the lacquer and the fluorocarbon residue away from it, i.e. pointing into the air. ESCA measurements revealed that a dense fluorocarbon layer formed at the surface. After curing, the films were washed with solvent. Contact angle measurements (Figure 11.28(b)) showed that the lacquer to which the methacrylate surfactant had been added had been properly and permanently hydrophobized. The non-reactive surfactant could be washed away from the surface, i.e. the hydrophobization was not permanent. A pre-formed polymer of the same type (iii in Figure 11.28(a) also gave a substantial and permanent increase in contact angle. Migration of the polymer to the surface was a slow process, however, and several days were required to attain an equilibrium surface composition.

Concluding Remarks

The homopolymerization of reactive surfactants in the form of assemblies, such as micelles or liquid crystals, has been attempted as a way to freeze the structure and prepare various types of nano-sized materials. Polymerization of micelles has *not been entirely successful*, however. With both spherical and rod-like micelles, the polymerized aggregates were of much larger size than the original structures. With liquid crystals and, in particular, with vesicles, the results are more promising. Stable vesicles, of interest for drug administration, have been prepared by free-radical polymerization of preformed vesicles. Such vesicles need not be based entirely on polymerizable surfactants. Incorporation of smaller amounts, 10–30%, of reactive species into a phospholipid-based vesicle formulation leads to vesicles with much improved stability.

Although considerable academic attention is devoted to the homopolymer-ization of surfactant solutions, the main industrial interest in polymerizable surfactants lies in applications, such as those described above, where the surfac-tant is used in the normal way, i.e. as an additive in concentrations of a few percent or less. In such applications, polymerizable surfactants are likely to be of increasing importance in the future.

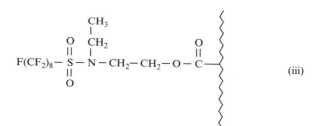

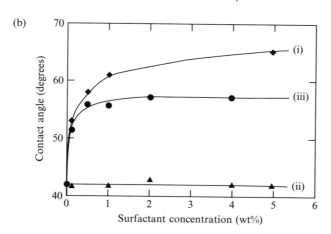

Figure 11.28 Modification of a PMMA lacquer film by surface active fluorocarbons of similar structure. (a) Compounds used: (i) a reactive surfactant containing a methacrylate group; (ii) a non-reactive surfactant; (iii) a surface active polymer. (b) Results obtained from contact angle measurements on the three systems. Reprinted with permission from M. Torstensson, B. Rånby and A. Hult, *Macromolecules*, **23** (1990) 126. Copyright (1990) American Chemical Society

Polymeric Surfactants Constitute a Chapter of their Own

Polymeric surfactants, or surface active polymers, is an area that is currently attracting much interest. Chapter 12 is devoted to this topic.

Special Surfactants Give Extreme Surface Tension Reduction

The vast majority of surfactants have the hydrophobic part of the molecule made up of a hydrocarbon chain, either aliphatic or alkylaryl. With a suitable choice of polar head group, the minimum surface tension that can be achieved with such surfactants is in the order of 26–28 mN/m. Such surface activity is sufficient for most uses of surface active agents. However, there are some applications where a lower surface tension of aqueous solutions is required. The spreading of aqueous foams on top of burning oil and the spreading of water-based formulations on surfaces of polyolefines are two relevant examples.

Two types of speciality surfactants are used to achieve extreme reduction of surface tension, i.e. silicone surfactants and fluorinated surfactants. The former type is based on polydimethylsiloxane as the non-polar group, while the latter contains a fluorocarbon or a combination of a fluorocarbon and a hydrocarbon as the hydrophobic tail. Silicone surfactants are high molecular weight compounds and are treated in Chapter 12. Fluorinated surfactants are briefly introduced below.

Fluorinated surfactants have the following general structure:

$$CF_3-(CF_2)_n-X$$

or

$$CF_3-(CF_2)_n-(CH_2)_m-X$$

where X can be any polar group, charged or uncharged. The fluorocarbon chain is usually rather short, with n typically being 5–9. Fluorinated carboxylates are common. Due to the inductive effect exerted by the electronegative fluorine atoms, these compounds are strong acids and hence relatively insensitive to low pH or hard water. There are also perfluorinated block copolymers, analogous to EO–PO block copolymers, on the market.

Fluorinated surfactants are used for various applications where wetting and spreading of aqueous solutions are difficult. Another use of fluorinated surfactants is to render surfaces, e.g. paper or textiles, both hydrophobic and lipophobic. Fluorinated surfactants, like silicone surfactants but unlike hydrocarbon-based surfactants, are also surface active in organic solvents and are therefore used as surfactants in paints and other non-aqueous formulations.

The main disadvantage with fluorinated surfactants, besides their high price, is their poor biodegradability.

Bibliography

Guyot, A. and K. Tauer, *Adv. Polym. Sci.*, **111** (1994) 44.

Guyot, A., Polymerizable surfactants, in *Novel Surfactants. Synthesis, Applications, and Biodegradability*, K. Holmberg (Ed.), Surfactant Science Series, 74, Marcel Dekker, New York, 1998.

Holmberg, K., *Prog. Org. Coatings*, **20** (1992) 325.

Holmberg, K., Cleavable surfactants, in *Novel Surfactants. Synthesis, Applications, and Biodegradability*, K. Holmberg (Ed.), Surfactant Science Series, 74, Marcel Dekker, New York, 1998.

Holmberg, K., Cleavable surfactants, in *Reactions and Synthesis in Surfactant Systems*, J. Texter (Ed.), Marcel Dekker, New York, 2001.

Menger, F. M. and J. S. Keiper, *Angew. Chem. Int. Ed. Engl.*, **39** (2000) 1906.

Zana, R., Dimeric (gemini) surfactants, in *Novel Surfactants. Synthesis, Applications, and Biodegradability*, K. Holmberg (Ed.), Surfactant Science Series, 74, Marcel Dekker, New York, 1998.

12 SURFACE ACTIVE POLYMERS

Surface active polymers, or polymeric surfactants, have gained in popularity during the last two decades. They are now used commercially in many different applications, among which the stabilization of dispersions and rheology control are probably the most widespread. This chapter will review the most important classes of surface active polymers and also discuss typical properties and uses of the various types.

Surface Active Polymers can be Designed in Different Ways

A polymer with surface active properties can be built along three main routes: with hydrophobic chains grafted to a hydrophilic backbone polymer, with hydrophilic chains grafted to a hydrophobic backbone and with alternating hydrophilic and hydrophobic segments. The three types will be treated individually below.

Surface active polymers are plentiful in nature and the three principal types exist both in the plant and the animal kingdoms. The hydrophilic segment is often a polysaccharide which may be charged or uncharged. For instance, antibodies normally contain carbohydrate residues attached as side chains far away from the antigen binding region, with their main function being to improve the hydrophilicity of the otherwise relatively hydrophobic protein. Other proteins in the body, such as milk and saliva proteins, contain well-defined amino acid sequences very rich in phosphate groups. Such proteins exhibit high surface activities and are excellent stabilizers of fat droplets.

The classification of surface active polymers into the three types mentioned above should not be seen as a clear-cut division. In reality, two or more types may be combined into one product. For example, a surface active macromolecule may have a backbone polymer consisting of alternating hydrophilic and hydrophobic segments and, in addition, contain hydrophilic or hydrophobic side chains, i.e. the molecule may at the same time be both a block and a graft copolymer. A graft copolymer may also contain both hydrophilic and hydrophobic grafts. The important feature from a physicochemical point of view is that the molecule is able to orient itself so as to expose its hydrophilic regions into a polar environment and its hydrophobic segments into a lipophilic phase. By doing so, the interfacial tension will be reduced, i.e. the polymer is by definition a surfactant.

Polymers may have a Hydrophilic Backbone and Hydrophobic Side Chains

This type of graft copolymer is plentiful in nature (Figure 12.1). Different types of lipopolysaccharides are produced in relatively high yields by microorganisms and many attempts have been made to use nature as a source for products for technical use. A particularly well-known example is that of Emulsan. This is the trade name of a polyanionic lipopolysaccharide produced as an extracellular product by the bacterium *Acinetobacter calcoaceticus*. The heteropolysaccharide backbone contains a repeating trisaccharide carrying a negative charge. Fatty acid chains are covalently linked to the polysaccharide through ester linkages. The degree of substitution and the type of fatty acid vary; a representative structure is given in Figure 12.2.

Emulsan, or rather the action of it, was first observed as a spontaneous crude oil–brine emulsion on the sea shore. The bacterium producing the emulsifying agent was isolated and later the surface active macromolecule was characterized. More in-depth studies revealed that the bacterium actually produces two components vital to the emulsion-forming process, i.e. a low molecular weight peptide, which is very surface active and which is an excellent emulsifier, and the lipopolysaccharide of molecular weight around 10^6, which

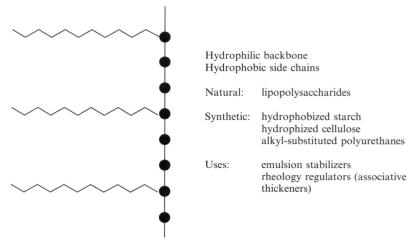

Hydrophilic backbone
Hydrophobic side chains

Natural: lipopolysaccharides

Synthetic: hydrophobized starch
             hydrophized cellulose
             alkyl-substituted polyurethanes

Uses: emulsion stabilizers
             rheology regulators (associative
             thickeners)

Figure 12.1 Some examples of polymers containing a hydrophilic backbone and hydrophobic side chains

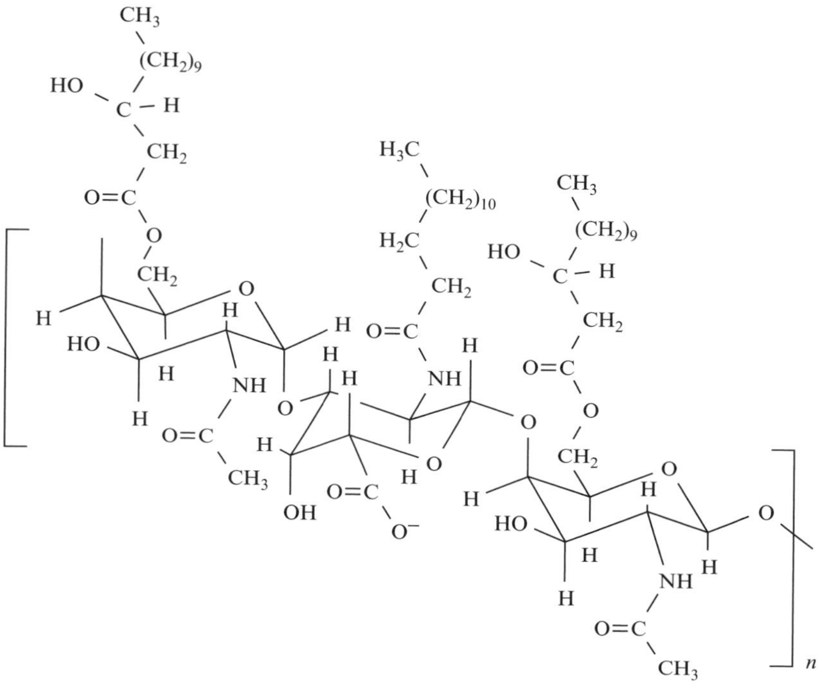

Figure 12.2 Structure of Emulsan

is remarkably effective in stabilizing the newly formed emulsion. The weight ratio between the two is typically 1:9. The term 'Emulsan' is sometimes used for the lipopolysaccharide–peptide combination and sometimes only for the lipopolysaccharide; in the following, Emulsan refers only to the latter. This is a good illustration of how nature often uses surface active polymers—not to create emulsions but to stabilize those that are formed by low molecular weight surfactants. In fact, polymeric surfactants are usually not suitable as emulsifying agents since they are unable to diffuse rapidly to newly created interfaces.

Some characteristic properties of Emulsan are as follows:

(1) It gives only moderate reductions in the surface and interfacial tensions.

(2) It has a very strong tendency to go to oil–water interfaces.

(3) It is, by itself, not an efficient emulsifying agent.

(4) It is an extremely good stabilizer for emulsions of specific oils in water (but not water in oil).

(5) It is very 'substrate specific' and functions best in the presence of divalent cations.

It is interesting to note that the molecule is only moderately surface active. For instance, it reduces the water–hexadecane interfacial tension from 47 to around 30 mN/m, which is not a very low value. The combination of lipopolysaccharide and peptide reduces the same interfacial tension to 14 mN/m. Emulsan, however, is very insoluble in both oil and water; hence, its driving force for the interface is very strong, which is an important property of an emulsion stabilizer.

The 'substrate specificity' of Emulsan is striking. Stable emulsions are only formed with specific oils, i.e. with specific hydrocarbon combinations. The best results in terms of emulsion stability are obtained with an oil consisting of a mixture of aliphatic hydrocarbons and alkylaryl compounds, i.e. the typical composition of heavy crude oil. This is the type of substrate which the bacterium that produces Emulsan has become adapted to. Excretion of Emulsan is the bacterium's way of helping to increase the crude out–water interface where it lives.

Great expectations were put on Emulsan as a crude oil emulsifier. The production of crude oil-in-water emulsions of high internal ratio were seen as a means to make extremely viscous heavy oil mobile enough to be pumped in pipelines. Of particular interest at the time when Emulsan was developed was the transport of Alaskan oil to mainland USA. Crude oil-in-water emulsions of around 75% oil content, which are feasible with Emulsan as the emulsion stabilizer, would enable direct burning without water evaporation, in analogy with the burning of concentrated coal slurries.

The efforts to use Emulsan as a crude oil emulsifier for transport in pipelines were thwarted by the observation that Emulsan could be enzymatically hydrolysed. Breakdown of the emulsion stabilizer would lead to emulsion coalescence during transport which would be very serious. The main use of Emulsan today is for cleaning oil tanks—an application of much smaller potential.

Emulsan is the most well-known example of biologically produced surface active polymers for industrial use. However, considerable efforts are today directed towards development of applications and improved production methods for other lipopolysaccharides excreted by microorganisms, both bacteria and fungi. In recent years, applications in the cosmetic field have come in focus. The work-up process involved in the production of these biosurfactants is still a rather tedious matter, however, and the product prices are comparatively high. It is reasonable to believe that the price gap between biosurfactants and synthetic surfactants will decrease in the future.

Natural polysaccharides may also be chemically modified into the equivalent of lipopolysaccharides by attachment of long alkyl or alkylaryl chains. One such route of derivatization is shown in Figure 12.3. The starting hydrophilic polymer is starch, which is a mixture of linear amylose and highly branched amylopectin. The latter component may be selectively degraded at the crosslinks by an enzyme acting only on the 1,6–glucosidic linkages. The linear

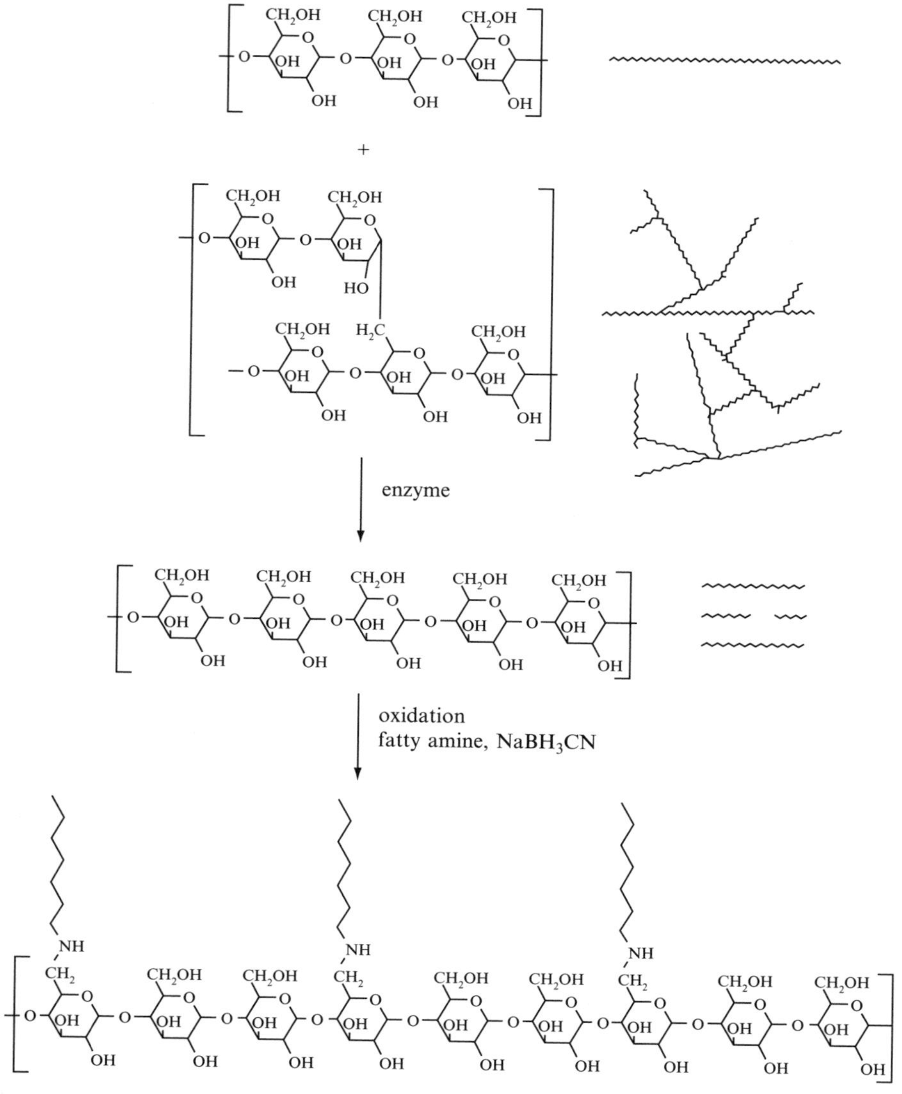

Figure 12.3 Derivatization of starch into a surface active polymer. In the scheme, oxidation occurs only at the 6–carbon of the anhydroglucose units. In reality, oxidation may also lead to ring cleavage between carbon atoms 2 and 3, so generating aldehyde groups at these positions. These aldehydes may also undergo reductive amination with the fatty amine

polysaccharide formed is oxidized to create aldehyde groups (and possibly ketones) and then reacted with a fatty amine. The degree of substitution, which is governed by the ratio of fatty amine to anhydroglucose rings, should be low, i.e. below 10%, or otherwise solubility problems will arise.

A similar type of derivatization has also being camed out for cellulose (Figure 12.4). A common practice is to swell cellulose in strong alkali (mercerization) and react the semi-soluble material with ethylene oxide and alkyl chloride. If the alkyl group is short, e.g. ethyl, a product with moderate surface activity is obtained. If some of the ethyl groups are replaced by long-chain alkyls, a polymer with a higher surface activity is obtained. Such graft copolymers are commercially available and are referred to as 'associative thickeners'. They are used as rheology control agents for aqueous formulations, such as water-borne paints. In addition, only a few per cent of the anhydroglucose rings should carry a long-chain substituent, or otherwise the product will become insoluble in water. Depending on the reaction conditions, the hydrophobic substituents could be more or less randomly distributed along the polysaccharide backbone. The substitution pattern is of importance for the physicochemical properties. It seems that domains of highly substituted anhydroglucose rings, followed by domains of a low degree of substitution, give a more surface active polymer. However, it is not easy to govern the substitution pattern, particularly in large-scale syntheses.

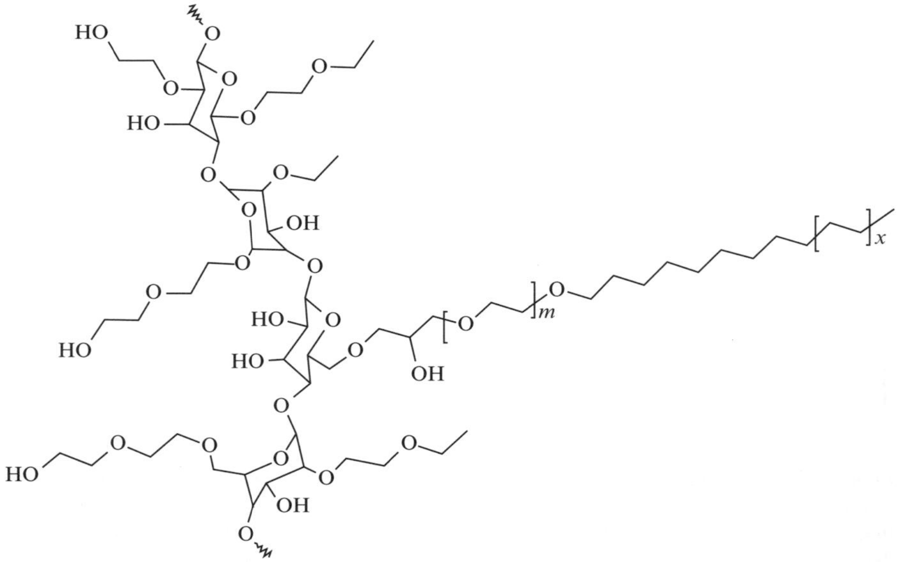

Figure 12.4 Structure of cellulose which has been modified with ethylene oxide and alkyl chloride. Reproduced by permission of Leif Karlson, Akzo Nobel Surface Chemistry, Stockholm, Sweden

Polymers may have a Hydrophobic Backbone and Hydrophilic Side Chains

Glycoproteins can be said to represent a natural product class of this type although the polypeptide backbone is, of course, not entirely hydrophobic. In fact, many glycosylated proteins can be seen as a combination of graft and block copolymers since the polypeptide chain often contains distinct hydrophobic and hydrophilic segments.

Several types of synthetic graft copolymers of this type are shown in Figure 12.5. There is considerable current interest in copolymers with poly(ethylene glycol) (PEG) tails. These are effective steric stabilizers for various kinds of dispersions. Figure 12.6 shows three different ways to prepare PEG-substituted polyacrylates. In principle, all three routes are feasible for large-scale production. Ethoxylated acrylate monomers are commercially available.

As mentioned above, this type of graft copolymer has found use as a steric stabilizer of dispersions, for instance, in the paint field. Another interesting application of these surface active polymers is for modification of solid surfaces in order to prevent adsorption of proteins and other biomolecules. Polymers of this type can be made to adsorb in a monolayer at hydrophobic surfaces and adsorption invariably occurs via interaction between the hydrophobic backbone and the solid surface, i.e. the PEG chains are oriented towards the aqueous phase. This type of PEG coating has been found to be an efficient way of obtaining low protein adsorption and low cell adhesion characteristics. For instance, PEG coatings are reported to give a marked suppression of

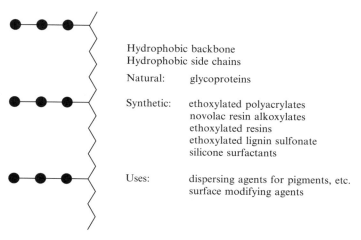

Figure 12.5 Some examples of polymers containing a hydrophobic backbone and hydrophilic side chains

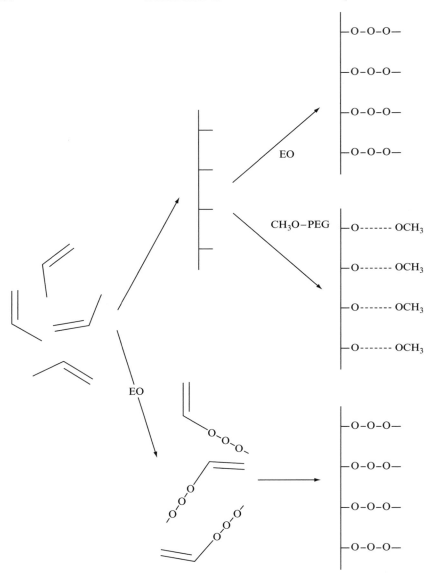

Figure 12.6 PEG-modified polyacrylates can be prepared by different types of processes. The upper route illustrates ethoxylation of a polyacrylate containing hydroxyethyl groups along the chain. The middle route shows reaction of a polyacrylate containing methyl ester groups with PEG monomethyl ether. Continuous removal of methanol during the reaction will lead to effective transesterification. The lower route illustrates polymerization of ethoxylated acrylate monomers. The ethoxylated monomer is copolymerized with conventional monomers, such as acrylic acid and methacrylic acid

plasma protein adsorption and platelet adhesion, thus leading to reduced risk of thrombus formation, as demonstrated both *in vitro* and *in vivo*.

The inert character of PEG surfaces is believed to be due to the solution properties of the polymer and to the fact that it is completely uncharged. The polymer has a high dipole moment and is strongly solvated in water. Interestingly, its homologues, polyoxymethylene and polyoxypropylene, as well as its isomer, polyacetaldehyde, are insoluble in water and, thus, not useful as hydrophilic polymer grafts.

The ability of PEG to prevent proteins and other biomolecules from approaching the surface can be considered as a steric stabilization effect (Figure 12.7). Such a stabilization usually has two contributions, i.e. an elastic term and an osmotic term. The elastic, or volume restriction, component results from the loss of conformational entropy when two surfaces approach each other, caused by a reduction in the available volume for each polymer segment. Thus, when a protein approaches the PEG-modified surface, a repulsive force develops due to the loss of conformational freedom of the polyoxyethylene chains.

The osmotic interactions, or mixing interactions, arise from the increase in polymer concentration on compressing two surfaces. When a protein or another large molecule in water solution approaches the surface, the number of available conformations of the PEG segments is reduced due to compression or interpenetration of polymer chains. In addition, an osmotic repulsive force develops. Whether interpenetration or compression, or both, occur depends on the density of the PEG chains. If the PEG grafting is dense, it is probable

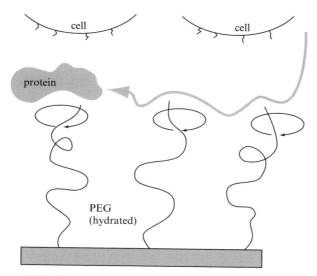

Figure 12.7 Protein rejection by PEG attached to a solid surface

that compression is preferred to interpenetration, while if the grafting is less dense then interpenetration is likely to dominate.

Uncharged polysaccharides are also very efficient in preventing protein adsorption to solid surfaces. For instance, a dextran coating results in a very inert surface, and the mechanism by which biomolecules are rejected is probably similar to that discussed above for PEG. The two types of polymers have very different temperature dependences, however. Whereas PEG and PEG derivatives are temperature-sensitive, giving less efficient steric stabilization at higher temperatures, polysaccharides are relatively unaffected by temperature.

Ethoxylated or ethoxylated and propoxylated phenol–formaldehyde resins, particularly of the novolac type, are a product class that has found use as a steric stabilizer in coatings. The EO–PO block copolymer products are also widely used as demulsifiers, e.g. in oil production. The alkylaryl segment constitutes a strongly hydrophobic backbone which binds firmly to most hydrophobic surfaces and allows the hydrophilic chains to be long (often, 50–100 oxyethylene units) without desorbing from the surface. The backbone, i.e. the alkylphenol–formaldehyde condensate, is usually of relatively low molecular weight, typically 1000–3000. Figure 12.8 shows the product class.

Comb copolymers based on poly(12–hydroxystearic acid) as grafts are frequently used as dispersants in non-aqueous formulations, e.g. paints. The poly(12–hydroxystearic acid) chains, which are of low molecular weight, provide steric stabilization analogous to how PEG behaves in aqueous solution. The backbone polymer contains specific groups which function as anchoring sites for the polymer to the surface of the particles to be dispersed. Proper anchoring of the polymer, which usually is based on acid–base interactions, is often the critical issue, determining dispersion efficiency. Silicone surfactants constitute another type of comb polymer with polar side chains. Two important types of silicone surfactants are shown in Figure 12.9.

The backbone of silicone surfactants is practically always polydimethylsiloxane, which in itself is highly hydrophobic and completely insoluble in water. Water-soluble substituents, charged or non-charged, will impart surface activity in aqueous systems. Poly(ethylene glycol) or poly(ethylene glycol)–poly

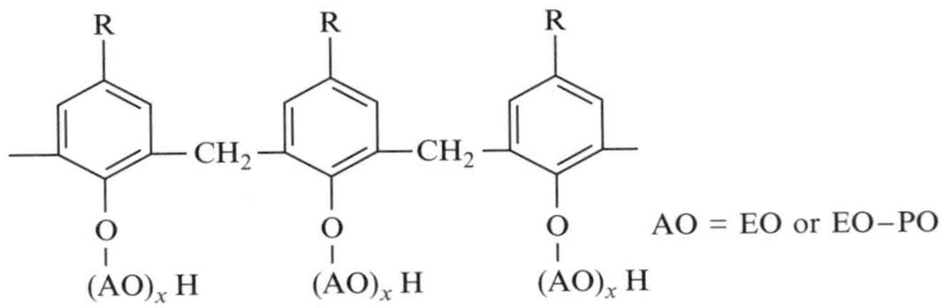

Figure 12.8 Phenol–formaldehyde resin ethoxylates; R is a lower alkyl, typically propyl or butyl

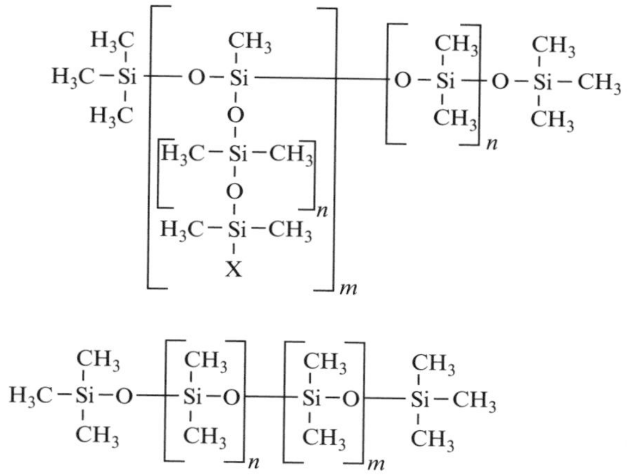

Figure 12.9 Some examples of silicone surfactants. X is usually an ionic or non-ionic polar group, most often an EO–PO copolymer, although it may also be a weakly polar group, such as an ester, amide, epoxy, etc.

(propylene glycol) block copolymers are by far the most common substituents. The linkages between Si and the polyether chain may be either Si–O–C or Si–C. The Si–O–C link is made by esterification of chloropolysiloxanes with hydroxyl-functional organic compounds such as an EO–PO copolymer. The bond is not very stable to hydrolysis and such products are unsuitable for use under either acid or alkaline conditions. The Si–C linkage, where a carbon of the EO–PO copolymer is directly linked to the silicon atom, is stable. Such a linkage is usually made by a Pt-catalysed hydrosilylating addition of an Si–H function in the polysiloxane to a terminal olefinic bond in the substituent polymer.

The substituent X of Figure 12.9 may also have a weakly polar character, in which case the product exhibits surface active properties in organic solvents. Silicone surfactants are, together with fluoro surfactants, the only products suitable for use in non-aqueous systems.

Silicone surfactants are typical speciality surfactants used in niche situations in which they perform better than conventional surfactants. Table 12.1 lists some characteristic properties of silicone surfactants.

Table 12.1 Properties of silicone surfactants

1. Very effective in lowering the surface tension (down to around 20 mN/m).
2. Excellent wetting on low-energy surfaces.
3. Powerful antifoamers.
4. Poorly biodegradable.
5. Relatively expensive. However, due to high efficiency, cost-performance may not be unfavourable when compared with conventional surfactants.

Silicone surfactants are used in a variety of applications, such as the following:

(1) Cell control additives in polyurethane foams.

(2) Antifoams in many types of aqueous systems (EO–PO substituted products) and in non-aqueous systems (usually non-substituted polysiloxanes).

(3) Additive in paints to prevent:
     (a) floating of pigments;

     (b) film defects (cratering, orange peel, etc.).

(4) Wetting agents on polyolefines and other hard-to-wet materials, and also as wetting agents in non-aqueous systems, e.g. in lubrication.

(5) Emulsifiers for silicone oil emulsions.

Polymers may Consist of Alternating Hydrophilic and Hydrophobic Blocks

Many proteins contain regions with distinct hydrophilic and hydrophobic characters (see Figure 12.10). Casein, the milk protein, and many salivary proteins are examples of surface active proteins of this type; these contain polar segments with a high concentration of phosphate groups, along with regions dominated by hydrophobic amino acids.

There is a variety of man-made surface active block copolymers. By far the most common and well-known are the poly(alkylene glycol) copolymers. The hydrophilic segment is almost invariably poly(ethylene glycol) (PEG), obtained by ethylene oxide (EO) polymerization. The hydrophobic segment is usually poly(propylene glycol) (PPG) but poly(butylene glycol)-based products also exist. Since propylene oxide (PO) and butylene oxide (BO) are the starting

Block copolymer

Natural: certain proteins

Synthetic: EO–PO block copolymers
              copolymers between EO and
              12-hydroxystearic acid

Figure 12.10 Some examples of polymers containing alternating hydrophilic and hydrophobic blocks

materials for the two latter polymer segments, all three poly(alkylene glycol)s will have the same –O–C–C–backbone structure, i.e. the repeating units are:

$$-OCH_2CH_2 - \quad \text{for poly(ethylene glycol)}$$

$$-OCH_2CH(CH_3) - \quad \text{for poly(propylene glycol)}$$

and

$$-OCH_2CH(CH_2CH_3) - \quad \text{for poly(butylene glycol)}$$

As mentioned before, the small difference in structure between the three types of repeating units lends to a surprisingly large difference in their physicochemical properties. Whereas PEG is water-soluble regardless of molecular weight, the two other poly(alkylene glycol)s are water-insoluble and, thus, act as hydrophobic segments in block copolymers.

In the literature, the term poly(alkylene oxide) is sometimes used instead of poly(alkylene glycol), e.g. poly(ethylene oxide) instead of poly(ethylene glycol). PEGs are also sometimes referred to as polyoxyethylenes (POEs) or polyoxiranes. In general usage, poly(ethylene glycol) refers to polymers of molecular weight below about 20 000, poly(ethylene oxide) to higher molecular weight polymers, and POE and polyoxirane are not specific in this regard. The segments used in the block copolymers are always of relatively low molecular weight; poly(ethylene glycol) (PEG) is therefore the appropriate term.

There are many possible variations of EO–PO block copolymers and the patent literature is rich in suggestions. However, the number of commercially available types is more limited and the most important ones are shown earlier in Figure 4.14. Products of this type are often referred to as Pluronics (EO/PO/EO) or inverse Pluronics (PO/EO/PO), with Pluronic being the Wyandotte trade name for this surfactant class.

Figure 4.16 above shows a phase diagram of a triblock EO–PO–EO copolymer–water system. As can be seen from this, the phase diagram exhibits a multitude of liquid crystalline phases, more than one normally finds in surfactant–water systems. In addition, the ternary phase diagram is very rich in phases, as is evident from Figure 12.11. It is unusual to find practically all possible liquid crystalline phases, ranging from micellar cubic to reverse micellar cubic, in one single system. Evidently, the block copolymers can attain a very wide range of curvatures in solution.

There are also EO–PO copolymers that are claimed to contain segments with random distribution of the monomers. These are made by the copolymerization of EO and PO. However, EO is much more reactive than PO; therefore, such mixtures give products of the considerable block type. Various methods are being used to affect the monomer distribution in these products, e.g. specific

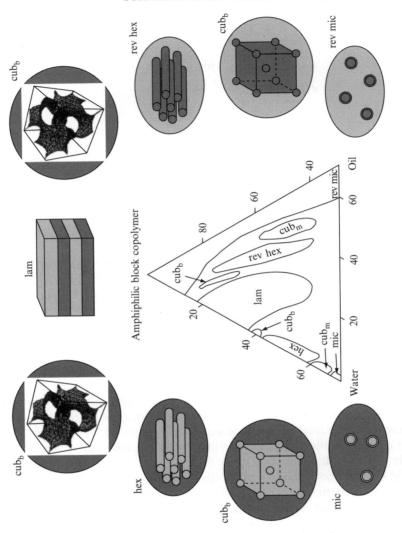

Figure 12.11 Phase diagram for a ternary system consisting of *p*-xylene, water and an EO–PO–EO block copolymer

procedures for monomer feeding. These procedures differ from one producer to another. Since the product properties are very dependent on the monomer distribution, random EO–PO copolymers with the same monomer ratio obtained from different sources may have large differences in properties.

Table 12.2 Properties of EO–PO block copolymers

---

1. They show reversed solubility vs. temperature dependence, i.e. they are more soluble in cold than warm water, and exhibit a cloud point.
2. Products with low EO contents have low foaming properties. (The best antifoam is obtained for EO/PO ratios of 1:4 to 1:9.) Reverse products, i.e. PO/EO/PO, give lowest foam.)
3. High molecular weight products with high PO contents have good wetting properties.
4. Products with high EO contents have good dispersing properties.
5. Biodegradability is slow, particularly for products with high PO contents.

Table 12.2 lists some important properties of EO–PO block copolymers. In general, the products are versatile and inexpensive and their physicochemical properties can be tailor-made more easily than is the case for most other surfactant types since both the hydrophobic and hydrophilic segments can be varied at will. In Chapter 5, the solution properties of these block copolymers are discussed in some detail. Poor degradability is a negative characteristic which limits their use at the present time and which can be expected to do so even more in the future.

Some typical uses of EO–PO block copolymers include the following:

(1) Foam control agents in:

  (a) machine dishwashing powders;

  (b) the textile industry (dyeing and finishing);

  (c) oil production;

  (d) emulsion paints.

(2) Wetting agents in:

  (a) machine dishwashing, i.e. rinse aids;

  (b) lubricants.

(3) Dispersing agents for pigments.

(4) Emulsifiers or coemulsifiers for herbicides and insecticides.

(5) Demulsifiers, e.g. in oil production (products containing 20–50% EO are used for w/o emulsions, and 5–20% EO products for o/w emulsions).

(6) Personal care products.

(7) Pharmaceutical formulations.

Polymeric Surfactants have Attractive Properties

The growing interest in polymeric surfactants (or surface active polymers) can be said to emanate from two characteristic features:

(1) They have a very strong driving force to go to interfaces, with this tendency to collect at interfaces being not as dependent on the physical variables as that observed for normal, low molecular weight surfactants. This means that:

    (a) the products are effective at low total concentrations;

    (b) the products show little sensitivity to salts, temperature changes, etc.

(2) They can have very long polyoxyethylene (or polysaccharide) chains and still be retained at interfaces. (Low molecular weight surfactants with long hydrophilic chains tend to desorb from the interface and dissolve in the aqueous phase.) Thus, such products are very efficient steric stabilizers for dispersed systems and effective non-fouling agents on solid surfaces.

Bibliography

Alexandridis, P. and B. Lindman (Eds), *Amphiphilic Block Copolymers: Self-Assembly and Applications*, Elsevier, Amsterdam, 1999.

Kosaric, N. (Ed.), *Biosurfactants*, Surfactant Science Series, 48, Marcel Dekker, New York, 1993.

Piirma, I., *Polymeric Surfactants*, Surfactant Science Series, 42, Marcel Dekker, New York, 1992.

Porter, M. R., *Handbook of Surfactants*, Blackie & Sons, London, 1991.

Tadros, Th. F., Polymeric surfactants, in *Handbook of Applied Surface and Colloid Chemistry*, K. Holmberg (Ed.), Wiley, Chichester, 2001.

13 SURFACTANT–POLYMER SYSTEMS

Surfactants and water-soluble polymers have very broad ranges of applications. This has been described in different chapters of this book, together with the underlying physicochemical mechanisms. Reviewing the compositions of various products, we learn that in the large majority of cases one or more polymers are present, together with one or more surfactants. In a typical situation, they are employed to achieve different effects—colloidal stability, emulsification, flocculation, structuring and suspending properties, rheology control—but in some cases a synergistic effect is addressed. The combined occurrence of polymers and surfactants is found in such diverse products as cosmetics, paints, detergents, foods, polymer synthesis and formulations of drugs and pesticides.

In this chapter we will broadly investigate the interactions between different types of polymers, in particular water-soluble homopolymers and graft copolymers, with the different classes of surfactants. It will be found that important starting points in a discussion are other mixed solute systems, in particular surfactant–surfactant and polymer–polymer mixed solutions.

Polymers can Induce Surfactant Aggregation

One of the most significant aspects of a surfactant is the ability to lower the interfacial tension between an aqueous solution and some other phase. In particular, for an ionic surfactant this is modified by the presence of a polymer in the solution. As illustrated in Figure 13.1, the effect of a polymer on the surface tension of an aqueous solution is different for different surfactant concentrations. At low concentrations, there may or may not be, depending on the surface activity of the polymer, a lowering of the surface tension. However, at some concentration there is a break in the surface tension curve and a more or less constant value is attained. There is then a concentration region, roughly proportional in extension to the polymer concentration, with a constant γ value. Finally, there is a decrease towards the value obtained in the absence of polymer.

We may interpret the concentration dependence of γ in the presence of a polymer as follows: At a certain concentration, often termed the critical

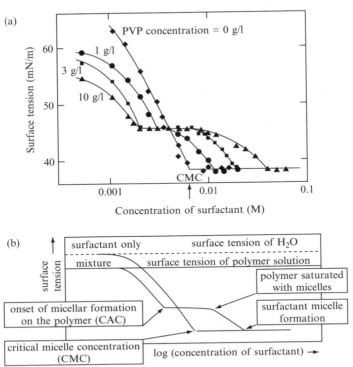

Figure 13.1 (a) Surface tension of SDS solutions as a function of surfactant concentration in the presence of different concentrations of poly(vinyl pyrrolidone) (PVP). From M.M. Breuer and I. D. Robb, *Chem Ind.*, (1972) 530535. (b) A schematic plot of the concentration dependence of the surface tension for mixed polymer–surfactant solutions. For comparison, the corresponding curve for the system which only contains surfactant is also shown

association concentration (CAC), there is an onset of association of surfactant to the polymer. Because of this there is no further increase in surfactant activity and thus no further lowering of γ. As the polymer is saturated with surfactant, the surfactant unimer concentration and the activity start to increase again and there is a lowering of γ until the unimer concentration reaches the CMC, after which γ is constant and normal surfactant micelles start to form.

This picture is confirmed if the association of the surfactant is monitored directly (e.g. by surfactant selective electrodes, by equilibrium dialysis, by self-diffusion or by some spectroscopic technique). As illustrated by the binding isotherm shown in Figure 13.2, there is, at low surfactant concentrations, no significant interaction. At the CAC, a strongly co-operative binding is indicated. At higher concentrations, we see a plateau level, and then a further increase of the free surfactant concentration until the surfactant activity or

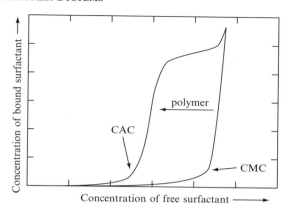

Figure 13.2 The binding isotherm of a surfactant to a polymer without distinct hydrophobic moieties, giving the concentration of bound surfactant as a function of the free surfactant concentration, can be interpreted as a lowering of the surfactant CMC by the polymer, or a strongly co-operative binding

unimer concentration joins the curve obtained in the absence of polymer. We note from Figure 13.2 the strong analogy with micelle formation and the interpretation of the binding isotherm in terms of a depression of the CMC.

Such a description is supported by solubilization studies, as illustrated in Figure 13.3. We note that the solubilization curves in the presence of polymer

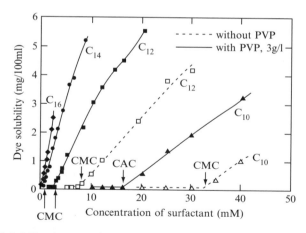

Figure 13.3 Solubilization experiments support the notion of a polymer-induced micellization. The amount of a dye, Orange OT, solubilized in mixtures of sodium alkyl sulfates of different chain lengths ($C_{10} - C_{16}$) and poly(vinyl pyrrolidone) (PVP) is given as a function of the surfactant concentration. From H. Lange, *Kolloid-Z. Z. Polym.*, **243** (1971) 101

are shifted to lower surfactant concentrations but are otherwise essentially the same as those without polymer. From the break-points, we can deduce CMC/CAC values which decrease by the same factor as those without polymer as the alkyl chain is lengthened.

Experimental binding studies of mixed polymer-surfactant solutions can be summarized as follows (cf. Figure 13.4):

(1) CAC/CMC is only weakly dependent on polymer concentration over wide ranges.

(2) CAC/CMC is, to a good approximation, independent of polymer molecular weight down to low values. For very low molecular weights, the interaction is weakened.

(3) The plateau binding increases linearly with the polymer concentration.

(4) Anionic surfactants show a marked interaction with most homopolymers while cationic surfactants show a weaker but still significant interaction. Non-ionic and zwitterionic surfactants only rarely show a distinct interaction with homopolymers.

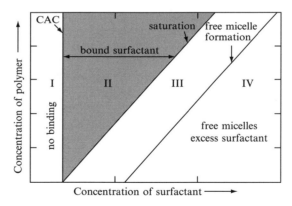

Figure 13.4 Association between a homopolymer and a surfactant in different concentration domains. (I) At low surfactant concentrations, there is no significant association at any polymer concentration. (II) Above the CAC, association increases up to a surfactant concentration which increases linearly with the polymer concentration. (III) Association is saturated and the surfactant unimer concentration increases. (IV) There is a co-existence of surfactant aggregates at the polymer chains and free micelles. This picture is schematic but gives a good description for aqueous mixtures of an ionic surfactant and a non-ionic homopolymer. From B. Cabane and R. Duplessix, *J. Phys. (Paris)*, **43** (1982) 1529

Attractive Polymer–Surfactant Interactions Depend on both Polymer and Surfactant

There are thus two alternative pictures of mixed polymer–surfactant solutions, one describing the interaction in terms of a (strongly co-operative) association or binding of the surfactant to the polymer and one in terms of a micellization of surfactant on or in the vicinity of the polymer chain. Both descriptions are useful and are largely overlapping. However, we will see that for polymers with hydrophobic groups the binding approach is preferred, while for hydrophilic homopolymers the micelle formation picture has distinct advantages.

As regards aggregate structure in these systems, the 'pearl-necklace model' (Figure 13.5), with the surfactant forming discrete micellar-like clusters along the polymer chain, has received wide acceptance for the case of mixed solutions of ionic surfactants and homopolymers. The micelle sizes are similar with polymer present and without, and the aggregation numbers are typically similar or slightly lower than those of micelles forming in the absence of a polymer.

In the presence of a polymer, the surfactant chemical potential is lowered with respect to the situation without polymer (Figure 13.6). There are several interactions that can be responsible for surfactant binding or a polymer–induced micellization. We note that in many respects (variation with surfactant alkyl chain length, solubilization, micelle structure and dynamics) there is a close similarity to the micellization of the surfactant alone. The normal hydrophobic

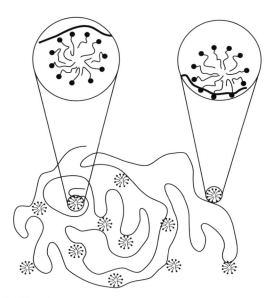

Figure 13.5 'Pearl-necklace model' of a surfactant–polymer association

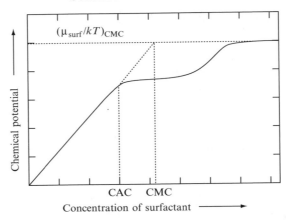

Figure 13.6 In the presence of a polymer, the chemical potential may be lowered, leading to micelle formation at a lower concentration. This figure gives the chemical potential (μ) of the surfactant divided by kT as a function of surfactant concentration on a logarithmic scale. In the presence of a polymer, the chemical potential is lowered and micelle formation on the polymer starts at a lower concentration (CAC) than in the absence of polymer (CMC). Reproduced by permission of John Wiley & Sons, Inc., from D. F. Evans and H. Wennerström, *The Colloidal Domain. Where Physics, Chemistry, Biology, and Technology Meet*, VCH, New York, 1994, p. 312

interaction between the alkyl chains must therefore still be a dominating contribution to the free energy of association. However, it is modified by mainly one of the following two factors.

For polymers that have hydrophobic portions or groups, there will be a hydrophobic attraction between the polymer molecules and the surfactant molecules. Such interactions will be particularly strong for block copolymers, with hydrophobic and hydrophilic blocks, and with graft copolymers, where hydrophobic groups have been grafted onto a hydrophilic polymer backbone. However, homopolymers can also have hydrophobic groups, strong as in poly(styrene sulfonate) or weak as in poly(ethylene glycol). The polarity of such groups (as we have seen in Chapter 4 for poly(ethylene glycol)) can decrease with increasing temperature.

Electrostatic interactions are obvious if both surfactant and polymer are charged; then in the case of opposite signs of the charges we can expect a quite strong association. However, we must also take into account the repulsive interactions between charged polymer molecules or between charged surfactant molecules. In particular, we learnt in our discussion of ionic surfactant self-assembly that the entropy loss associated with the increased concentration of counterions at the aggregate surface compared to the bulk is highly unfavourable for self-assembly and explains, *inter alia*, why ionic surfactants have orders-of-magnitude-higher CMCs than non-ionics.

A polymer may modify this entropy contribution in a number of different ways. If it is ionic and has a similar charge, then we have a simple and relatively moderate electrolyte effect. If its charge is opposite, and it acts as a multivalent electrolyte, then the interaction becomes very strong since an association between the polymer and micelle leads to a release of the counterions of both the micelles and the polymer molecules; a very similar effect will be obtained in mixtures of two oppositely charged polymers. Indeed, there is for such a case a lowering of the CMC by orders of magnitude.

In addition, non-ionic cosolutes may decrease the CMC for ionics, as we learnt from the example of addition of different alcohols (Chapter 2). If the cosolute is slightly amphiphilic it will be located in the micelle surface, will lower the charge density, and hence will decrease the entropic penalty in forming micelles. This is believed to be the mechanism behind the moderate depression of the CMC for ionic surfactants produced by poly(ethylene glycol) and several non-ionic polysaccharides.

We note from these arguments that ionic surfactants would be expected to interact broadly with different types of water-soluble polymers. This is true, in particular, for anionics; the considerably weaker interaction of cationics is due to a higher degree of counterion binding. Non-ionic surfactants, on the other hand, should have little tendency to interact with hydrophilic homopolymers since no further stabilization of the micelles can be expected; we noted above that non-ionic surfactants only exceptionally associate to homopolymers. The situation will, of course, be different for polymers with hydrophobic parts to which non-ionics will associate by hydrophobic interaction.

Surfactant Association to Surface Active Polymers can be Strong

The modification of water-soluble homopolymers by grafting a low amount of hydrophobic groups (of the order of 1% of the monomers reacted is a typical figure), like alkyl chains, leads to amphiphilic polymers which have a tendency to self-associate by hydrophobic interaction. This weak aggregation (Figure 13.7) leads to an increase in viscosity, and in other rheological characteristics— hence the use of these 'associative thickeners' as rheology modifiers in paints and other products.

An added surfactant will interact strongly with the hydrophobic groups of the polymer, leading to a strengthened association between polymer chains, and thus to an increased viscosity. We exemplify this behaviour with a hydrophobically modified non-ionic cellulose ether (HM-EHEC) (Figure 13.8). As can be seen in Figure 13.9, sodium dodecyl sulfate increases the viscosity dramatically for HM-EHEC but very slightly for unmodified EHEC. At a higher surfactant content, the viscosity effect is lost. As we will see, we can best understand these systems in terms of a mixed micelle formation between the surfactant and the amphiphilic polymer. In order to have cross-linking

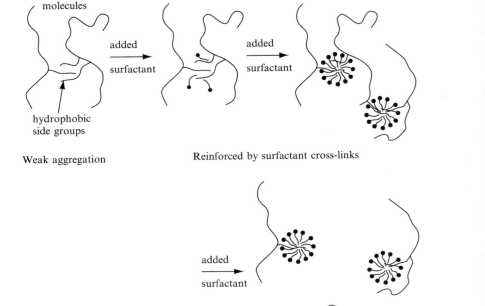

Figure 13.7 The self-association of a hydrophobically modified (HM) water-soluble polymer can be strengthened or weakened by a surfactant, depending on the stoichiometry

and thus a viscosity effect, there must be a sufficiently high number of polymer hydrophobes per micelle. At higher surfactant concentrations, there will be only one polymer hydrophobe in a micelle and all cross-linking effects are lost.

These viscosity effects of the addition of surfactant to a solution of a hydrophobically modified water-soluble polymer are general, but the effect will be modified by other interactions such as electrostatic ones. As exemplified in Figure 13.10, addition of an oppositely charged surfactant to a solution of a hydrophobically modified polyelectrolyte gives much larger viscosity effects than for a non-ionic or a similarly charged surfactant; however, even in the

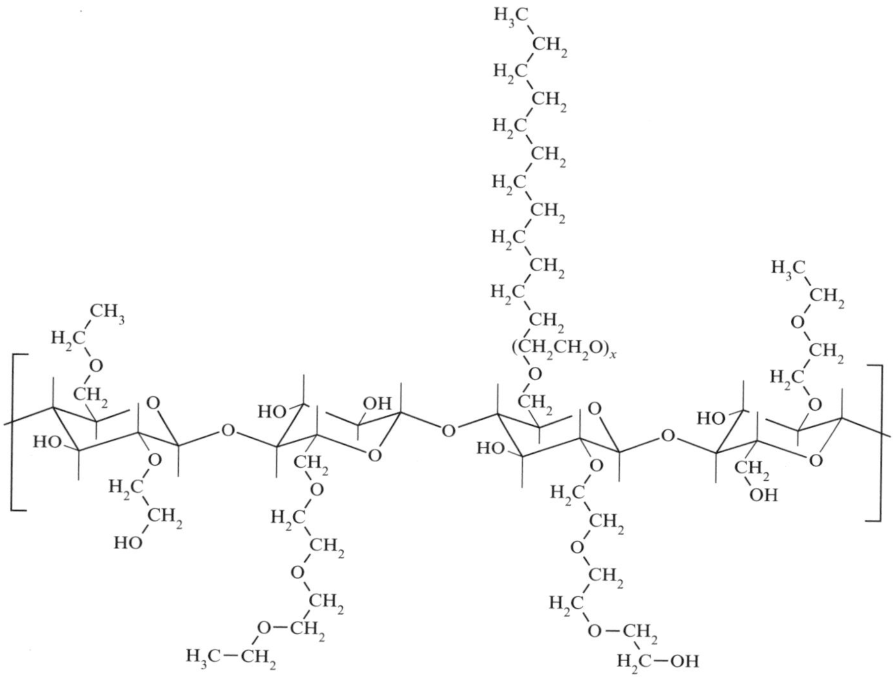

Figure 13.8 Cellulose can be modified by a relatively random substitution of hydro-xyethyl and ethyl groups to give ethyl(hydroxyethyl) cellulose (EHEC). In HM-EHEC, a low fraction of hydrophobic groups is inserted

latter case there are substantial effects, especially if the hydrophobic grafts are larger.

In mixtures of surfactants and clouding polymers, which become less polar at increasing temperatures, thermoreversible gels may be formed. On increasing the temperature, gelation is induced and on cooling the gel melts. Such an effect may also be obtained in mixtures of a non-ionic surfactant and a hydrophobi-cally modified water-soluble polymer, when there is an important temperature-induced micelle growth or a transition from micelles to vesicles or some other self-assembly structure.

The Interaction between a Surfactant and a Surface Active Polymer is Analogous to Mixed Micelle Formation

A hydrophobically modified water-soluble polymer (HM-polymer) can be viewed as a modified surfactant. It forms micelles, or hydrophobic microdo-mains, on its own at very low concentrations (intramolecularly at infinite dilution) and these micelles can solubilize hydrophobic molecules. Furthermore, an HM-polymer and a surfactant in general have a strong tendency to

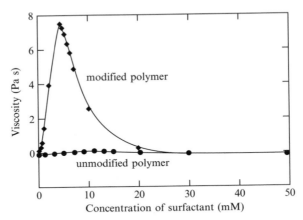

Figure 13.9 Addition of sodium dodecyl sulfate to a solution of a hydrophobically modified polymer (EHEC) gives a strong increase in viscosity at first and then a decrease to a low value. For the unmodified polymer, changes in viscosity are small. Reproduced by the courtesy of K. Thuresson

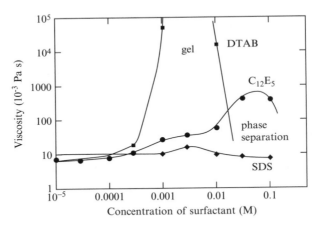

Figure 13.10 Addition of a surfactant to a solution of a hydrophobically modified polyelectrolyte (in this case, hydrophobically modified polyacrylate) gives very different viscosity effects for different types of surfactants. The effect is much larger for an oppositely charged surfactant (dodecyltrimethylammonium bromide) (DTAB) than for a non-ionic ($C_{12}E_5$) or a similarly charged one (sodium dodecyl sulfate) (SDS). From B. Magny, I. Iliopoulos, R. Audebert, L. Piculell and B. Lindman, *Prog. Colloid Polym. Sci.*, **89** (1992) 118

form mixed micelles in a similar way as two surfactants. Two stoichiometries are important for HM-polymer-surfactant systems, namely the alkyl chain stoichiometry and the charge stoichiometry.

Since the mixed aggregates dominate, we have a low concentration of free surfactant and essentially no free micelles until the concentration of micelles exceeds the concentration of polymer hydrophobic groups. From this, we can understand the binding isotherm of a surfactant to an HM-polymer, as schematized in Figure 13.11. The binding isotherm of an ionic surfactant to a non-ionic HM-polymer will be very similar to its binding to non-ionic surfactant micelles. We can distinguish between three concentration regions. In the first region, there is a high affinity non-co-operative binding of individual ionic surfactant molecules to micelles of the non-ionic surfactant or the HM-polymer. As the number of ionic surfactant molecules per micelle exceeds one, the binding becomes anti-co-operative since the binding of surfactant to a similarly charged micelle is unfavourable. Finally, as the free surfactant concentration equals the CAC, i.e. the CMC in the presence of the homopolymer, there is a self-association of the ionic surfactant to micelles at the polymer. This is seen as a co-operative binding region, which is the only binding seen for the corresponding homopolymer.

The peak in the plot of viscosity versus surfactant concentration in general occurs in the vicinity of the CAC. Here, the composition of the mixed HM-polymer-surfactant micelles changes strongly with concentration to become dominated by the surfactant; thus the cross-linking effect is lost.

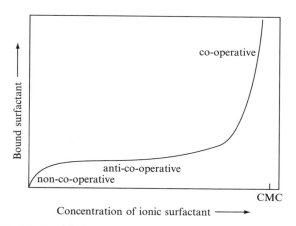

Figure 13.11 The binding isotherm (giving the concentration of bound surfactant as a function of the free surfactant concentration) for an ionic surfactant to a non-ionic HM-polymer is similar to the binding to a non-ionic micelle, i.e. with (a) one non-co-operative, (b) one anti-co-operative, and (c) one co-operative region. With the parent homopolymer, only the co-operative region is seen

Phase Behaviour of Polymer-Surfactant Mixtures Resembles that of Mixed Polymer Solutions

General Aspects and Non-ionic Systems

For the case of mixed solutions of a surfactant and an HM-polymer we can see a large tendency to association between the two cosolutes. Above we also learnt that many homopolymers facilitate micelle formation of an ionic surfactant. However, an associative interaction is by no means evident since for most polymer-surfactant pairs there is no net attractive interaction.

For two polymers in a common solvent, the entropic driving force of mixing is weak and we typically encounter a segregation into two solutions, with one rich in one polymer and one rich in the other; the tendency to phase-separate increases strongly with the molecular weights of the polymers. Since a micelle is also characterized by a high molecular weight, we would expect segregative phase separation to be a common phenomenon (Figure 13.12).

For polymer solutions we can distinguish between two types of phase separation, i.e. one segregative and one associative (Figure 13.13). If there is no attractive interaction, a segregative phase separation is obtained, while if there

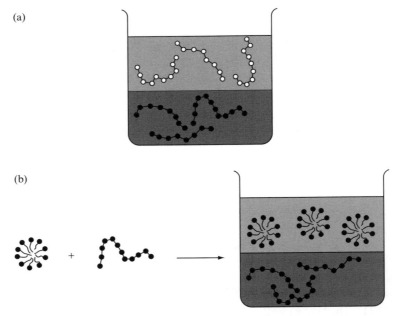

Figure 13.12 Because of the high molecular weight of a micelle we expect a mixed polymer–surfactant solution, in the absence of significant attractive interactions, to display (a) the incompatibility typical of polymer mixtures, i.e. (b) segregation into one polymer-rich and one micelle-rich solution

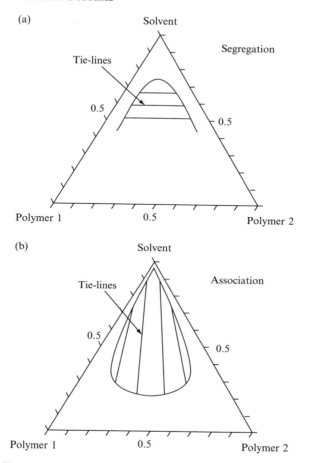

Figure 13.13 Ternary phase diagrams for two polymers in a common solvent showing segregation and association. In the two-phase regions, tie-lines are given to show the compositions of co-existing solution phases. Reprinted from *Adv. Colloid Interface Sci.*, **41**, L. Piculell and B. Lindman, 'Association and segregation in aqueous polymer/ polymer, polymer/surfactant and surfactant/surfactant mixtures. Similarities and differences', 149–178, Copyright (1992), with permission from Elsevier Science

is a moderately strong attraction complete miscibility may result. In the case of strong attraction between the two polymers, we will have an associative phase separation, with one phase concentrated in both polymers and one dilute solution. The degree of phase separation will in both cases increase with the molecular weight of the polymers.

For a mixed polymer–surfactant system the behaviour is completely analogous. One difference is that the 'degree of polymerization' of a micelle, unlike a polymer, is not fixed but may vary with the conditions (temperature,

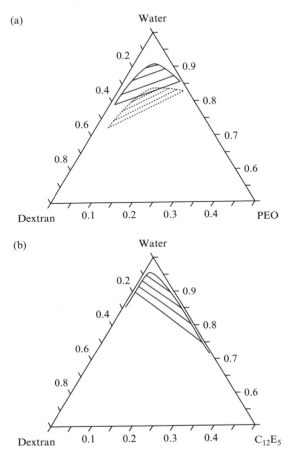

Figure 13.14 Phase diagrams of (a) aqueous mixtures of dextran (molecular weight 23 000) and poly(ethylene glycol) (molecular weight 2300 (dotted curve), and 18 000 (continuous curve)) compared with (b) a mixture of dextran (40 000) with $C_{12}E_5$. Above the curves there is miscibility, while below the curves there is phase separation into two solutions. (a) Reprinted from *Adv. Colloid Interface Sci.*, **41**, L. Piculell and B. Lindman, 'Association and segregation in aqueous polymer/polymer, polymer/surfactant and surfactant/surfactant mixtures. Similarities and differences', 149–178, Copyright (1992), with permission from Elsevier Science. (b) Reprinted with permission from L. Piculell, K. Bergfeldt and S. Gerdes, *J. Phys. Chem.*, **100** (1996) 3675. Copyright (1996) American Chemical Society

electrolyte concentration, etc.), as we learnt in our discussion of micellar growth (Chapter 3).

In Figure 13.14, we compare a polymer-surfactant mixture, a non-ionic polyoxyethylene surfactant and dextran, with a related polymer-polymer mixture, poly(ethylene glycol) and dextran. We can see that the two mixtures phase-separate in qualitatively the same way.

Changing the surfactant produces major changes in the phase diagram. Thus, $C_{12}E_5$ gives a much stronger segregation than $C_{12}E_8$ and with the former segregation increases strongly with increasing temperature. We note that this is exactly what we would expect from our discussion of micelle sizes for these surfactants (Chapter 4): $C_{12}E_8$ forms small roughly spherical micelles at all temperatures, while $C_{12}E_5$ forms large micelles which grow strongly with temperature.

The segregative phase separation is not the only one observed for mixtures of a non-ionic polymer and a non-ionic surfactant. For the case of a less polar polymer, the phase separation can be associative, especially at higher temperatures, due to hydrophobic association.

Introduction of Charges

The introduction of charged groups in the solutes has a profound influence on phase separation phenomena. We have already learnt that polyelectrolytes are much more soluble than the corresponding uncharged polymer, which we attribute to the entropy of the counterion distribution. Confining polymer molecules to part of the system costs little entropy due to the small number of entities. On the other hand, there is a large entropy loss on confining the (much more numerous) counterions. In mixed polymer systems we see many consequences of the electrostatic interactions due to net charges. One is the low tendency to phase separation in a mixed solution of one non-ionic and one ionic polymer; in the presence of added electrolyte, this inhibition of phase separation is largely eliminated and typical polymer incompatibility is observed.

Completely analogous effects are observed for mixed polymer-surfactant solutions. Even a slight charging up of either the polymer (by introducing ionic groups) or the micelles (by adding ionic surfactant) strongly enhances polymer-surfactant miscibility. This is illustrated for the system of Figure 13.14(b) in Figure 13.15, where we also see that addition of electrolyte tends to eliminate these effects of the charges. When the polymer molecules are similarly charged, there is also a return towards the incompatibility of the parent non-ionic mixture.

Ionic surfactants tend broadly to associate with non-ionic polymers. The lower the polarity of the polymer, then the stronger the association. For clouding polymers, which are on the limit of being water soluble, the association is strong and leads (as we saw in Chapter 4) to a strong increase in the cloud point. This decreased tendency to phase separation is illustrated in Figure 13.16, where we also see that quite low concentrations of electrolyte change the behaviour completely. As we have seen above, there is polymer-induced micellization in these cases and thus a formation of polymer-surfactant complexes. Because of the charging up of the polymer, its solubility increases, again a result

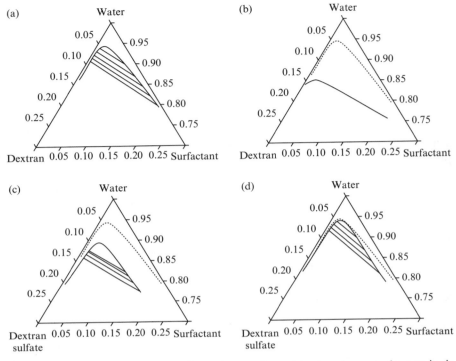

Figure 13.15 Introducing charges in a mixture of a non-ionic polymer and a non-ionic surfactant, illustrated for the mixture of dextran and a polyoxyethylene surfactant (reference system in (a)). Both on introducing (b) a low fraction of ionic surfactant in the micelles, or (c) a low amount of sulfate groups in dextran, the mutual miscibility is strongly enhanced. This can be eliminated (d) either if electrolyte is added or if both the polymer molecules and the micelles are charged. Above the curves there is miscibility, while below there is phase separation into two solutions. The dotted curves give the miscibility limit for the reference system. Reprinted with permission from K. Bergfeldt and L. Piculell, *J. Phys. Chem.*, **100** (1996) 5935. Copyright (1996) American Chemical Society

of the entropic penalty of confining the counterions to one phase. However, this is eliminated on addition of low concentrations of electrolyte and there is a dramatic lowering of the cloud point.

Mixed Ionic Systems

A mixture of two oppositely charged polyelectrolytes shows a strongly associative behaviour, as demonstrated by a strong tendency to phase separation. A mixture of a polyelectrolyte and an oppositely charged surfactant will also associate strongly. As exemplified in Figure 13.17, there is a lowering of the

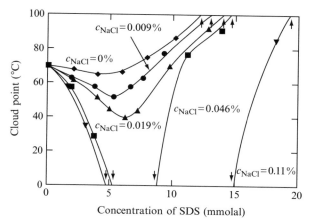

Figure 13.16 Addition of an ionic surfactant to a solution (0.9 wt%) of a clouding polymer (illustrated by ethylhydroxyethylcellulose (EHEC)) raises the cloud point in the absence of added electrolyte but decreases it in the presence of (low amounts of) electrolyte. The change in the cloud point on addition of SDS is given in the absence of added electrolyte and in the presence of different concentrations of added NaCl; from top to bottom the curves refer to 0, 0.01, 0.02, 0.04 and 0.10 wt% of salt. Reprinted with permission from A. Carlsson, G. Karlström and B. Lindman, *Langmuir*, **2** (1986) 536. Copyright (1986) American Chemical Society

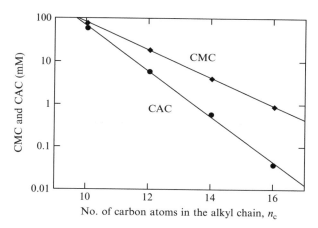

Figure 13.17 The CAC is typically orders of magnitude lower than the CMC for an ionic surfactant in the presence of an oppositely charged polymer. The logarithm of the CMC and the CAC of alkyltrimethylammonium bromides in the presence of an anionic polysaccharide, sodium hyaluronate, is plotted versus the number of carbons in the alkyl chain. Reprinted with permission from K. Thalberg and B. Lindman, *J. Phys. Chem.*, **93** (1989) 1478. Copyright (1989) American Chemical Society

CMC by orders of magnitude for long-chain surfactants and, as shown in Figure 13.18, there is a strong associative phase separation. An aqueous mixture of a polyelectrolyte and an oppositely charged surfactant phase separates into one dilute phase and one, typically highly viscous, phase concentrated in both polymer and surfactant.

The extent of phase separation in the system of Figure 13.18 increases strongly with the surfactant alkyl chain length and polymer molecular weight. On addition of electrolyte, the phase separation is eliminated but at a higher electrolyte content there is a phase separation again. However, as we can see in Figure 13.19, this is of a different nature. This behaviour, which is very similar to what we observe for mixtures of two oppositely charged polymers, can best be understood from a combination of polymer incompatibility and electrostatic effects. We note that the concentration of counterions is very high and, therefore, unlike for the non-ionic polymers, phase separation with a polyelectrolyte in one phase leads to confinement of the counterions and a very significant entropy loss. Therefore, polyelectrolytes are highly soluble when compared to the corresponding uncharged polymers. At high electrolyte concentrations, this entropy contribution is eliminated and phase separation will be similar to that of uncharged polymer systems. In the case of Figure 13.19, we have an intrinsically segregating system, as can be seen from the phase diagram at high electrolyte concentrations.

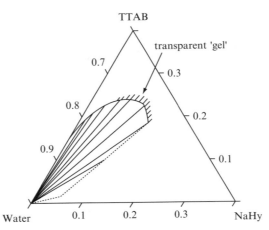

Figure 13.18 A mixture of an ionic surfactant and an oppositely charged polyelectrolyte typically gives an associative phase separation, as here exemplified by tetradecyltrimethyl ammonium bromide and sodium hyaluronate. Reproduced by permission of CRC Press from B. Lindman and K. Thalberg, in *Interactions of Surfactants with Polymers and Proteins*, E. D. Goddard and K. P. Ananthapadmanabhan (Eds), CRC Press, Boca Raton, FL, 1993, p. 252

The associative phase separation at low salt contents is also understood from the entropy of the counterion distribution. The highly charged micelles and polyelectrolyte molecules are enriched with counterions at their surface due to the Coulombic attraction. On association, counterions of both cosolutes are transferred into the bulk with a concomitant entropy gain; therefore, there is a strong tendency to associative phase separation in the absence of added salt.

When the surfactant of this system is replaced with one that has a similar charge as the polymer, the entropic loss is essentially absent and a segregative phase separation is the rule, as illustrated in Figure 13.20. At a high salt content, phase separation is enhanced, which can be referred to the growth of sodium dodecyl sulfate micelles.

Phase Behaviour of Polymer–Surfactant Mixtures in Relation to Polymer–Polymer and Surfactant–Surfactant Mixtures

We may summarize the phase behaviour of mixtures of polymer (P) and surfactant (S) depending on charge (superscript $+$, $-$ or 0) as follows:

P^+S^-, P^-S^+	Association without added electrolyte, miscibility at intermediate electrolyte and segregation at high electrolyte concentration
P^-S^-, P^+S^+	Segregation
P^0S^0	Segregation; association may occur for less polar polymers, especially at higher temperatures
P^0S^+, P^0S^-	Phase separation inhibited; association or segregation may be
P^+S^0, P^-S^0	induced by added salt

PS systems are closely analogous to PP systems. The degrees of polymerization of both polymer and surfactant aggregate determine the extent of phase separation. Since the micelle size is not fixed but may vary with the conditions, phase separation will be variable.

The phase behaviour of PS systems is also affected by specific interactions between the two cosolutes, like hydrophobic interactions for the case of the HM-polymer. This may enhance phase separation for non-ionic systems but decrease it for ionics. For a mixture of oppositely charged polymer and surfactant, the formation of a concentrated phase with charge stoichiometric amounts of polymer and surfactant and a dilute phase containing any excess of either polymer or surfactant becomes unfavourable if the polymer is hydrophobically modified. The reason is the tendency of the polymer to associate hydrophobically with the micelles in the concentrated phase. Then, this phase will lose the charge stoichiometry and have a tendency to swell. Therefore, associative phase separation will occur only over a restricted concentration region for a mixture of a hydrophobically modified polyelectrolyte and an oppositely charged surfactant.

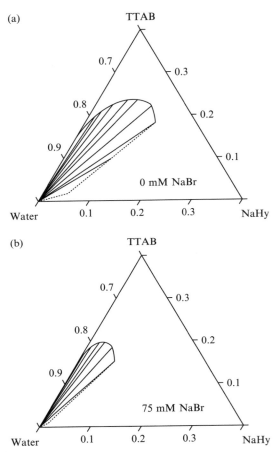

Figure 13.19 Phase separation in mixtures of a polyelectrolyte and an oppositely charged surfactant changes from associative to no phase separation and finally to segregative as electrolyte is added. The example shows mixtures of a cationic surfactant, tetradecyltrimethyl ammonium bromide, and an anionic polysaccharide, sodium hyaluronate. Reproduced by permission of CRC Press from B. Lindman and K. Thalberg, in *Interactions of Surfactants with Polymers and Proteins*, E. D. Goddard and K. P. Ananthapadmanabhan (Eds), CRC Press, Boca Raton, FL, 1993, p. 254

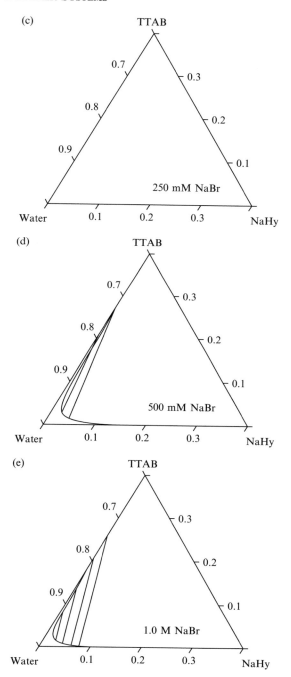

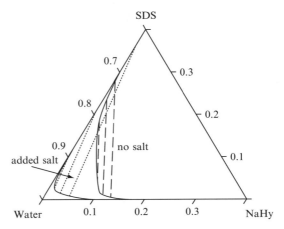

Figure 13.20 Phase separation in mixtures of a polyelectrolyte and a similarly charged surfactant is typically segregative and may be enhanced on salt addition if there is an electrolyte-induced micellar growth. The example refers to mixtures of sodium dodecyl sulfate and the anionic polysaccharide sodium hyaluronate. Reproduced by permission of CRC Press from B. Lindman and K. Thalberg, *Interactions of Surfactants with Polymers and Proteins*, E. D. Goddard and K. P. Ananthapadmanabhan (Eds), CRC Press, Boca Raton, FL, 1993, p. 257

Mixing two surfactants, on the other hand, does not give rise to segregative phase separation (Chapter 5). The reason is the strong tendency to form mixed aggregates which gives an important additional contribution to the entropy of mixing. It is common, however, for a mixture of two oppositely charged surfactants to display an associative phase separation.

Polymers may Change the Phase Behaviour of Infinite Surfactant Self-Assemblies

The larger the surfactant self-assembly aggregate, then the larger will be the tendency for phase separation of any kind. For polymers mixed with infinite surfactant aggregates, as in a bicontinuous microemulsion or a lamellar phase, weak repulsive and attractive interactions will have a profound influence on phase behaviour. As described in Chapter 6, a bicontinuous microemulsion has oil and water domains, which are separated by monolayer surfactant films which are connected over macroscopic distances. The structure is similar to that of the sponge phase, with the surfactant bilayer replaced by a monolayer and every second water channel replaced by an oil channel. In the phase diagram of a surfactant–oil–water system, there is (under balanced conditions) a large three-phase region where the bicontinuous microemulsion is in equilibrium with the oil and water phases.

(a)

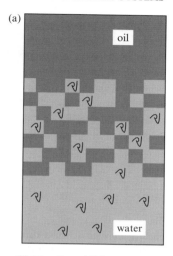

(b)

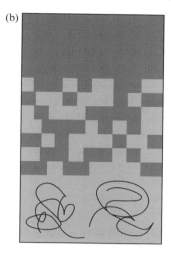

Figure 13.21 On addition of a water-soluble non-associating polymer to a system of a bicontinuous microemulsion in equilibrium with excess water and oil, the distribution of the polymer will be very different for different molecular weights. (a) A low molecular weight polymer is soluble in the microemulsion phase, whereas (b) coils that are larger than the pore size of the microstructure are insoluble in the microemulsion phase and confined to the excess water phase (or excess oil phase in the case of an oil-soluble polymer). From A. Kabalnov, B. Lindman, U. Olsson, L. Piculell, K. Thuresson and H. Wennerström, *Colloid Polym. Sci.*, **274** (1996) 297

For the case of adding a polymer to a lamellar phase or a bicontinuous phase, where there is no net attraction between the polymer and surfactant films, the polymer may enter the narrow water slits or channels only if it has a low molecular weight and a low radius of gyration (Figure 13.21). For a higher molecular weight it will stay outside (Figure 13.22 and 13.23) and thereby destabilize the surfactant phase. However, only a moderate hydrophobic modification of the polymer will induce an association to the surfactant films and increase the stability of the surfactant phase (Figure 13.24).

There Are Many Technical Applications of Polymer–Surfactant Mixtures

The use of a polymer and a surfactant in combination may be based on different effects such as controlling phase behaviour, controlling the interfacial properties or controlling the formation of networks due to association. The most important and well-understood use of polymers and surfactants together is to achieve a suitable rheology, i.e. thickening and gelation effects. It is also possible to design systems, based on hydrophobically modified water-soluble polymers or

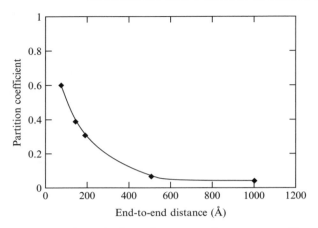

Figure 13.22   Dextran and non-ionic surfactants tend to segregate. For a three-phase system of $C_{12}E_5$ microemulsions and dextran, the partitioning of the polymer into the bicontinuous microemulsion phase is high for low molecular weights but very low for high molecular weights. The partition coefficient for dextran is plotted versus the unperturbed end-to-end distance. This microemulsion has a pore size of *ca.* 300 Å. Reproduced by permission of Springer-Verlag from A. Kabalnov, B. Lindman, U. Olsson, L. Piculell, K. Thuresson and H. Wennerström, *Colloid Polym. Sci.*, **274** (1996) 297

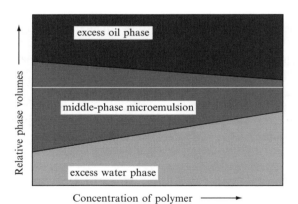

Figure 13.23   The partitioning of high molecular weight polymers into the lower aqueous phase of a three-phase system leads to a destabilization and decreased volume of the microemulsion phase. The relative phase volumes as a function of the polymer concentration are illustrated. (A corresponding effect is obtained on adding a high molecular weight oil-soluble polymer.) Reproduced by permission of Springer-Verlag from A. Kabalnov, B. Lindman, U. Olsson, L. Piculell, K. Thuresson and H. Wennerström, *Colloid Polym. Sci.*, **274** (1996) 297

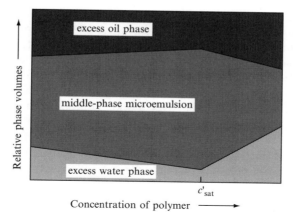

Figure 13.24 While a hydrophilic polymer enriches in the lower aqueous phase of a three-phase system and destabilizes the microemulsion phase, the corresponding hydrophobically modified polymer, which associates to the surfactant films, is incorporated in the microemulsions and stabilizes them. The relative phase volumes as a function of the polymer concentration are illustrated. The initial swelling of the middle phase is reversed at higher polymer concentrations when the middle phase has been saturated with polymer (c_{sat}). Reproduced by permission of Springer-Verlag from A. Kabalnov, B. Lindman, U. Olsson, L. Piculell, K. Thuresson and H. Wennerström, *Colloid Polym. Sci.*, **274** (1996) 297

homopolymers, which respond to different external factors. One example of a responsive system is gelation induced on an increase in temperature.

Phase behaviour effects include the solubilization of water-insoluble polymers. One example is the depression of clouding (increase of the cloud point) of a polymer solution achieved by the addition of an ionic surfactant. The polymer-induced micellization leads to a lowering of the surfactant unimer concentration and activity. This can be significant for the elimination of irritation caused by the surfactant.

The interfacial behaviour of surfactant–polymer mixtures, utilized for example, in the stabilization of suspensions, depends on a complex interplay between different pair interactions. Addition of a polymer can either remove the surfactant from a surface or enhance its adsorption, and vice versa, depending on the relative stability of polymer–surfactant complexes in solution and at the interface.

DNA is Compacted by Cationic Surfactants, which gives Applications in Gene Therapy

Double-stranded DNA is a highly charged and stiff polyanion. Because of its high charge density, DNA interacts strongly with cationic surfactants. The surfactant binding isotherms show a strongly co-operative association, while the phase diagrams display a very strong associative phase separation. Since DNA has very high molecular weight, it is possible to directly monitor the interactions on a single molecular level by using microscopy. As cationic surfactant is added to a DNA solution, the DNA molecules change their conformation from an extended 'coil' state to compact 'globules' (Figure 13.25). The DNA molecules are compacted individually, and over a wide concentration range there is a co-existence of coils and globules.

Native DNA is highly extended due to the electrostatic repulsions between different parts of the polymer molecule (driven by the energy of the counterion distribution). In the presence of multivalent counterions, there are attractive forces, due to ion–ion correlation effects (see Chapter 8), between different parts of a DNA molecule, thus leading to contraction. DNA induces self-assembly of a cationic surfactant and the surfactant aggregates act as multivalent counterions. DNA compaction by a surfactant can also be viewed as an associative phase separation at the molecular level.

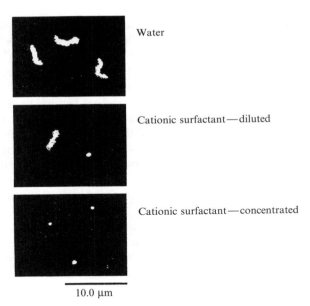

Water

Cationic surfactant—diluted

Cationic surfactant—concentrated

10.0 μm

Figure 13.25 By adding a fluorescent probe molecule, DNA molecules can be directly visualized by fluorescent microscopy. Addition of a cationic surfactant leads to compaction of individual DNA molecules. Reproduced by permission of John Wiley & Sons, Ltd., from B. Lindman, Surfactant-Polymer Systems, in Handbook of Applied Surface and Colloid Chemistry, Volume 1, K. Holmberg (Ed), Figure 20.27, page 460

Many deceases have a genetic origin and are cured by modifying the DNA sequence. However, in order to allow DNA to be transferred into cells, it needs to be compacted. Cationic cosolutes, such as surfactants, are efficient transfection agents of great interest for developments in gene therapy.

Bibliography

Kabalnov, A., B. Lindman, U. Olsson, L. Piculell, K. Thuresson and H. Wennerström, *Colloid Polym. Sci.*, **274** (1996) 297.

Kwak, J. C. T. (Ed.), *Polymer–Surfactant Systems*, Marcel Dekker, New York, 1998.

Lindman, B. and K. Thalberg, Polymer–Surfactant interactions—Recent developments, in *Interactions of Surfactants with Polymers and Proteins*, E. D. Goddard and K. P. Ananthapadmanabhan (Eds), CRC Press, Boca Raton, FL, 1993, Ch. 5.

Lindman, B., Surfactant–Polymer Systems, in *Handbook of Applied Surface and Colloid Chemistry Vol. 1*, K. Holmberg (Ed), John Wiley & Sons, Ltd., Chichester, 2002, Ch. 20.

Piculell, L. and B. Lindman, *Adv. Colloid Interface Sci.*, **41** (1992) 149.

Piculell, L., F. Guillemet, K. Thuresson, V. Shubin and O. Ericsson, *Adv. Colloid Interface Sci.*, **63** (1996) 1.

Piculell, L. K. Thuresson and B. Lindman, *Polym. Adv. Technol.*, **12** (2001) 44.

Rubingh, D. and P. M. Holland (Eds), Interaction between polymers and cationic surfactants, in *Cationic Surfactants: Physical Chemistry*, Marcel Dekker, New York, 1991, pp. 189–248.

14 SURFACTANT–PROTEIN MIXTURES

Proteins are Amphiphilic

While our general treatise of surfactant–polymer interactions also applies to proteins, these have some features, which are different from those of most other macromolecules. Proteins are copolymers built up of amino acids that contain polar and non-polar groups; the polar groups may be ionic or non-ionic, which may also vary with the pH of the solution. Thus, we can characterize a typical protein as an amphoteric polyelectrolyte with some hydrophobic groups or as an amphiphilic polymer with a variable charge density. The important differences when compared with the polymers which we have discussed so far are that proteins are monodisperse and that there is much less conformational freedom and variation for proteins. Many proteins are compact macromolecules with a well-defined structure on different levels. Rather different from these globular proteins (such as enzymes, haemoglobin and albumins) are fibrous proteins (collagen, keratin, etc.) with extended linear structures. Most fibrous and globular proteins become disordered on heating.

Proteins are copolymers built up of a large number of monomer units (amino acids) which have a prescribed sequence. In addition to this primary structure, there is also a three-dimensional arrangement, described in terms of secondary and tertiary structures; many proteins are composed of separate sub-units arranged in a distinct way with respect to each other, i.e. the quaternary structure. The three-dimensional arrangement of the monomers is such that the hydrophobic and hydrophilic amino acids are segregated to a considerable extent: globular proteins have a predominantly hydrophobic interior and a hydrophilic surface, with smaller hydrophobic patches.

Proteins, which are perturbed extensively away from the native structure, can not easily find their way back to this. Therefore, the unfolding of a protein can lead to (reversible or irreversible) denaturation. Various physical and chemical treatments can lead to denaturation of proteins in solution, with one example being the addition of certain surfactants at higher concentrations.

It must be pointed out that proteins do not constitute a homogeneous group of macromolecules; for example, highly soluble proteins, such as serum proteins, are quite different from insoluble proteins, such as membrane proteins. It is, therefore, not possible to provide a generally applicable picture of proteins, or indeed surfactant–protein systems. Furthermore, studies of such systems have still a limited coverage and there are no generally accepted models to describe them.

Surfactant–Protein Interactions have a Broad Relevance

Since surfactants and proteins are both amphiphilic, it is natural that they interact. These interactions show similarities to other surfactant–macromolecule systems, in particular to systems of amphiphilic polymers, such as hydrophobically modified water-soluble polymers. Globular proteins differ from the other polymers we have discussed in that they are monodisperse and in that they are compact and essentially lack conformational freedom. One important difference relates to the reversibility: the surfactant binding to a protein at higher stages of binding, for ionic surfactants in particular, may lead to conformational changes which are irreversible, i. e. the equilibrium conformation is not recovered on removing the surfactant. Protein denaturation by a surfactant is an important aspect in many applications. In studies of interactions, it is a complicating factor since mixed solutions of surfactants and proteins may involve protein molecules in different conformational states. Complexes involving a denatured protein have close analogies to complexes between surfactants and partially hydrophobic, flexible polymers.

The interactions between small amphiphilic molecules—lipids and surfactants—and proteins have a broad significance in industrial, biological, food, pharmaceutical and cosmetic systems. Examples of these include the following: lipid–membrane protein interactions in cell membranes; surfactant-induced skin irritation, related to the interaction of surfactant with stratum corneum proteins; the exposure of human hair and wool proteinacious substrates to surfactants on washing and other processes; protein–surfactant and protein–lipid combinations in edible products; the interactions of surfactants with gelatin in industrial, pharmaceutical and photographic applications; sodium dodecyl sulfate–polyacrylamide gel electrophoresis as an important technique for the determination of the molecular weight of proteins; enzyme-containing washing powders; renaturation of proteins; extraction and purification of proteins assisted by non-ionic surfactants; enzyme catalysis in microemulsions.

Surface Tension and Solubilization give Evidence for Surfactant Binding to Proteins

The normal surface tension versus surfactant concentration curve (see Chapter 2) is modified by a protein in a way which resembles that observed for weakly polar homopolymers. Thus, Figure 14.1, which gives the surface tension of

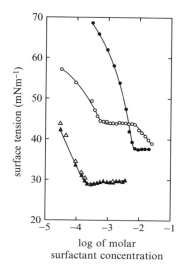

Figure 14.1   The effect of gelatin on the surface tension of solutions of sodium dodecyl sulfate (circles) and Triton X-100 (deca(ethylene glycol)monooctylphenol ether) (triangles): filled symbols, without protein; open symbols, in the presence of protein. From W. L. Knox, Jr and T. O. Parshall, *J. Colloid Interface Sci.*, **33** (1970) 16

solutions of sodium dodecyl sulfate in the presence of gelatin, shows clear analogies with that shown earlier in Figure 13.1: the surface tension curve has a break far below the CMC, followed by a plateau region and then a further decrease down to values approaching those of the solution of surfactant alone. In the case of a non-ionic surfactant, no effect of the protein is seen, as shown in Figure 14.1. However, for other cases, a clear interaction between a protein and a non-ionic surfactant is indicated from surface tension curves (Figure 14.2). As discussed in Chapter 13, most polymers do not associate with non-ionic surfactants, with the general exception being strongly amphiphilic polymers, such as hydrophobically modified water-soluble polymers.

The solubility of a water-insoluble non-polar substance, such as a dye, starts to be important above the CMC for a surfactant solution (see Chapter 2). In the presence of a polymer, solubilization starts at a lower concentration, thus signifying a polymer-induced surfactant micellization (see Figure 13.3). An analogous effect is observed for many protein solutions, as illustrated in Figure 14.3. This indicates that a water-soluble protein can also induce surfactant self-assembly and the formation of hydrophobic microdomains. However, the behaviour is not as clear-cut as for many homopolymer systems, and it is difficult to identify a critical association concentration (CAC).

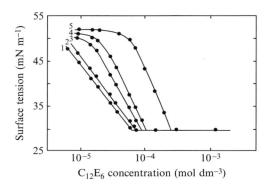

Figure 14.2 Surface tension curves for $C_{12}E_6$ in the presence of bovine serum albumin, of the following concentrations: (1) 0; (2) 10^{-6}; (3) 5×10^{-6}; (4) 1×10^{-5}; (5) 5×10^{-5}. Reproduced by permission of the Chemical Society of Japan from N. Nishikido *et al.*, *Bull. Chem. Soc. Jpn*, **55** (1982) 3085

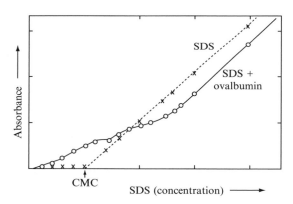

Figure 14.3 The solubilization (measured by the increase in UV absorbance) of dimethyl aminoazobenzene (probe) by ovalbumin–sodium dodecyl sulfate complexes. Taken from 'Protein-surfactant solutions' by K.P. Ananthapadmanabhan, Fig. 4, Ch. 8, pp 320–365 in *Interaction of Surfactants with Polymers and Proteins*. Edited by E.D. Goddard and K.P. Ananthapadmanabhan, copyright CRC Press, Boca Raton, 1993.

The Binding Isotherms are Complex

The association of surfactants to proteins is directly illustrated in binding isotherms obtained by equilibrium dialysis and other techniques for a few proteins (Figures 14.4 and 14.5). The important features of the binding isotherms are the combination of a low degree of non-co-operative binding at low surfactant concentrations and a massive co-operative binding at higher concentrations. Often binding is described in terms of a low amount of high affinity binding to specific sites, followed in the second stage by co-operative binding.

The interesting analogy with surfactant binding to hydrophobically modified polymers (described in Chapter 13) is illustrated in Figure 14.5(b) for the case of sodium dodecyl sulfate interacting with lysozyme; at low concentrations, there is a low amount of non-co-operative binding, which is followed by first an anti-co-operative region and then an extensive co-operative binding. In considering binding isotherms for oppositely charged protein–surfactant pairs, precipitation phenomena must be noted; so-called 'high-affinity binding' at low surfactant concentrations has been shown to involve precipitation of a stoichiometric complex (see below). In the same way as for non-ionic polymers,

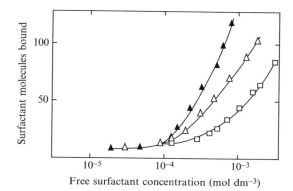

Figure 14.4 Surfactant binding (expressed as the number of surfactant molecules per protein molecule) to bovine serum albumin for sodium dodecyl sulfate (triangles) and sodium octylbenzene sulfonate (squares). The open symbols refer to a NaCl concentration of 10^{-2} M, and the filled symbols to a concentration of 10^{-1} M. From H. M. Rendall, *J. Chem. Soc., Faraday Trans. 1*, **72** (1976) 481. Reproduced by permission of the Royal Society of Chemistry

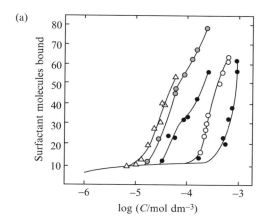

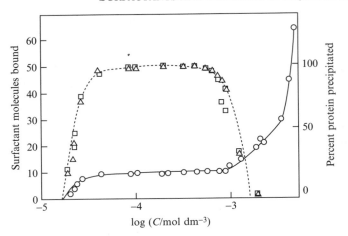

Figure 14.5 (a) Surfactant binding (expressed as the number of surfactant molecules per protein molecule) to bovine serum albumin for sodium dodecyl sulfate at the following pH levels: from left, 3.8, 4.1, 4.8, 5.6 and 6.8. From J. A. Reynolds *et al.*, *Biochemistry*, **9**, (1970) 1232. (b) Surfactant binding and protein precipitation in dilute mixtures of sodium dodecyl sulfate and lysozyme. The number of bound SDS molecules per lysozyme molecule (continuous curve), and the fraction of the protein in the precipitated phase (dashed curve), are shown as a function of the free surfactant concentration. Reproduced by permission of the Chemical Society of Japan from K. Fukushima, Y. Murata, G. Sugihara and M. Tanaka, *Bull. Chem. Soc. Jpn*, **54** (1981) 3122

anionic surfactants associate much more strongly than cationic ones, while non-ionics often show no association. Anionics may also associate to proteins above the isoelectric point, where the protein has a net negative charge.

Structural and mechanistic descriptions of protein–surfactant complexes are few but it appears that the first step is due to hydrophobic interactions between the surfactant hydrophobic tails and non-polar residues/patches of the protein, while the second stage shows analogies with polymer-induced micellization. In the second stage of binding, there is an unfolding of the protein to expose the hydrophobic interior. Therefore, many globular proteins are gradually converted into random coils and a 'pearl necklace' structure, similar to that adopted for simple polymers (see Figure 13.5), is frequently used as a model; thus, there are strings of globular micelles decorating the polypeptide backbone. The micelles formed appear to have considerably lower aggregation numbers (and, therefore, a lower solubilization capacity) than those micelles formed by the surfactant alone or in the presence of hydrophilic polymers.

Protein–Surfactant Solutions may have High Viscosities

Addition of a surfactant to a solution of gelatin may, for higher protein concentrations, show a marked viscosity increase with a maximum as a function of the surfactant concentration (Figure 14.6). Similar observations have been made for mixed solutions of globular proteins and ionic surfactant. The behaviour has analogies with that displayed by solutions of mixtures of surfactants and hydrophobically modified water-soluble polymers (but is not at all so pronounced). This is in line with the amphiphilic nature of the protein. The viscosity effect is related to the formation of micelle-like aggregates cross-linking the protein molecules.

Another interesting manifestation of the amphiphilic character of proteins is when a protein is added to a solution of a hydrophobically modified polymer. As can be seen in Figure 14.7 for the case of hydrophobically modified polyacrylate and lysozyme, protein addition leads to a very significant viscosity increase; at higher protein concentrations, the viscosity decreases to low levels. There is a very interesting similarity to the effect of surfactant micelles (see Figures 13.9 and 13.10), which is in line with both surfactant micelles and proteins having a hydrophobic interior and a hydrophilic surface.

Protein–Surfactant Solutions may give rise to Phase Separation

Precipitation of a protein with net positive charge on binding of an anionic surfactant is a common phenomenon; at higher surfactant concentrations, there is a redissolution. The behaviour shows analogies with that of mixed systems of an ionic polymer and oppositely charged surfactant in general (cf. Figure 13.18), but oppositely charged protein–surfactant pairs are more

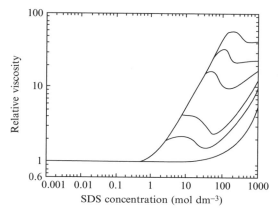

Figure 14.6 Relative viscosity for gelatin solutions as a function of the concentration of added sodium dodecyl sulfate; from bottom to top, the gelatin concentrations are 0, 2.0, 3.0, 5.0, 7.0 and 10.0 wt%. Reprinted with permission from J. Greener, *et al.*, *Macromolecules*, **20** (1987) 2490. Copyright (1987) American Chemical Society

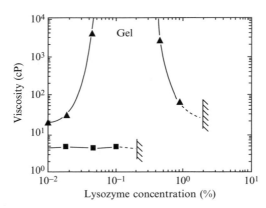

Figure 14.7 Variation of viscosity as a function of lysozyme concentration for 1% solutions of polyacrylate (squares) and hydrophobically modified polyacrylate (3% modification by octadecyl groups) (triangles). From R. Petit, R. Audebert and I. Iliopoulos, *Colloid Polym. Sci.*, **273** (1995) 777

complex as we will illustrate with mixed solutions of lysozyme and sodium dodecyl sulfate (SDS).

As SDS is added to a solution of lysozyme, there is, at low concentrations of SDS, formation of a solid-like neutral complex with excess protein in solution (Figure 14.8). The amount of precipitate reaches a maximum as the number of surfactant molecules per protein is 8, i. e. the number of charges on lysozyme. Excess of surfactant leads to redissolution; for surfactants with shorter alkyl chain lengths, more surfactant is needed for redissolution. The latter occurs in the co-operative part of the binding isotherm, when the free surfactant concentration reaches the CMC (cf. Figure 14.5(b)).

The low amount of surfactant needed for precipitation, as well as the low amount of water in the precipitate, and the stoichiometry of the complex, is related to the globular nature of the protein. A well-defined tertiary structure favours the formation of well-defined complexes and the water content is low for a neutral precipitate formed by rigid particles; this is in contrast to the gel-like coacervates formed in mixtures of surfactants and hydrophobically modified water-soluble polymers.

The cooperative binding and the precipitation of a neutral complex at low surfactant concentrations can be described in terms of a solubility product, K_s, of the neutral compound, according to the following:

$$PS_n(s) \longleftrightarrow P^{n+} + nS^-; \quad K_s = [P^{n+}][S^-]^n \qquad (14.1)$$

when considering the case of a negatively charged surfactant (S^-) and a positively charged protein (P^{n+}). The critical surfactant concentration for

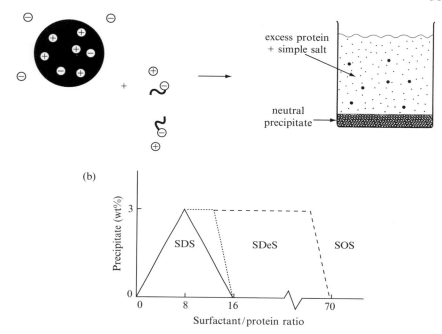

Figure 14.8 (a) Mixed solutions of a cationic protein and an anionic surfactant give, over wide composition ranges, a neutral precipitate and a solution with excess protein. (b) The amount of precipitate in solutions of lysozyme and a sodium alkyl sulfate (dodecyl (SDS), decyl (SDeS) and octyl (SOS)) increases linearly with the surfactant/ protein stoichiometry until it reaches 8, the charge number of the enzyme. Redissolution at higher surfactant concentrations is more efficient the more strongly associating the surfactant. (Redrawn from L. Piculell, A. K. Morén and A. Khan, unpublished)

binding is that where the solubility product is just exceeded. Since the complex involves many surfactant molecules, the critical surfactant concentration depends only weakly on the protein concentration and the binding becomes markedly co-operative.

The phase behaviour is complex, with not only solid and solution phases but also a (blue) isotropic gel phase, occurring over a narrow range around a well-defined surfactant/protein ratio. The effect of progressive surfactant addition is shown in Figure 14.9(a), together with a schematic phase diagram (Figure 14.9(b)). While the associative phase separation shows similarities with that of other polyelectrolyte–ionic surfactant systems, we underline the important additional features. One is the minute amount of surfactant needed for the onset of phase separation, while another is the occurrence of a novel gel phase in which a three-dimensional network is due to partially denatured protein molecules.

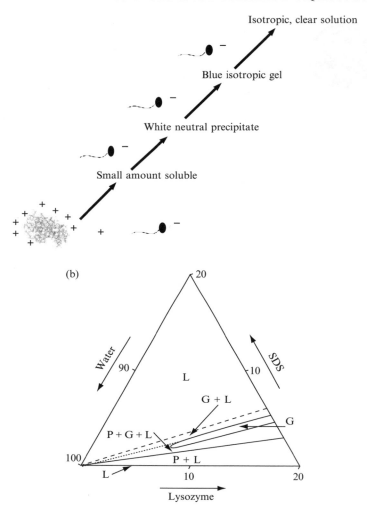

Figure 14.9 (a) There is a sequence of sample characteristics as SDS is progressively added to a solution of lysozyme. Reproduced by courtesy of Anna Stenstam. (b) The phase diagram (concentrations in wt%) shows schematically the regions for the occurrence of solution (L), gel (G) and precipitate (P), as well as two- and three-phase regions. Reprinted with permission from A. K. Morén and A. Khan, *Langmuir*, **11** (1995) 3636. Copyright (1995) American Chemical Society

Surfactants may Induce Denaturation of Proteins

As we have seen, ionic surfactants in general bind to proteins and initiate unfolding of the tertiary structure. Anionic surfactants such as SDS denature proteins more than cationics. Non-ionic surfactants, on the other hand, do not alter the tertiary structure of the protein. The unfolding of the protein typically occurs in the region of the binding isotherm, where a significant increase in surfactant binding by non-specific co-operative interactions starts. SDS is a strong protein denaturant, but interestingly, insertion of oxyethylene groups reduces the effect considerably. Denaturation by SDS is also reduced by the addition of non-ionic or zwitterionic surfactants. At low binding ratios, surfactants can stabilize proteins against thermally induced unfolding.

Bibliography

Anantapadmanaban, K. P., Protein–surfactant interactions, in *Interactions of Surfactants with Polymers and Proteins*, E. D. Goddard and K. P. Anantapadmanaban (Eds), CRC Press, Boca Raton, FL, 1993, Ch. 8.

Bos, M., T. Nylander, T. Arnebrant and D. C. Clark, in *Food Emulsifiers and their Applications*, G. L. Hasenhuettl and R. W. Hartel (Eds), Chapman & Hall, New York, 1997, Ch. 5.

Dickinson, E., Proteins in solution and at interfaces, in Interactions of Surfactants with Polymers and Proteins, E. D. Goddard and K. P. Anantapadmanaban (Eds), CRC Press, Boca Raton, FL, 1993, Ch. 7.

Kwak, J. C. T. (Ed.), *Polymer–Surfactant Systems*, Marcel Dekker, New York, 1998.

Nylander, T., Protein–lipid interactions, in *Proteins at Liquid Interfaces*, D. Möbius and R. Miller (Eds), Elsevier, Amsterdam, 1998, pp. 385–431.

Tanford, C., *The Hydrophobic Effect: Formation of Micelles and Biological Membranes*, Wiley, New York, 1980.

Watts, A. (Ed.), *Protein–Lipid Interactions*, New Comprehensive Biochemistry, Vol. 25, Elsevier, Amsterdam, 1993.

15 AN INTRODUCTION TO THE RHEOLOGY OF POLYMER AND SURFACTANT SOLUTIONS

The flow properties are in focus in most applications of polymer and surfactant solutions as well as for mixed systems. In a number of previous chapters we have already touched on the rheological aspects of different types of systems. While it is not possible to go any deeper into the advanced topic of rheology, we wish to give some central aspects and thereby achieve a comparison between the rheological properties of different systems.

Rheology Deals with how Materials Respond to Deformation

Solid bodies, liquids and gases can respond in different ways to a deformation; they can be elastic, they can flow or they can fracture. It is a large and important subject not least for colloidal systems; here, we will be interested in both the flow properties of dispersions as well as simple and complex solutions, and the elasticity of solutions and gels. The steady-state behaviour will be in focus but it should be recalled that time-dependent effects are most significant for applications.

Rheology is not an easy subject. Experiments that give relevant information need to be carefully designed, the mathematics are quite difficult to master, and even minute changes in composition (as we saw *inter alia* in Chapter 13) and in aggregation can dramatically change the rheological properties.

Here we will mainly introduce basic relationships and definitions, as well as consider the rheology of simple colloidal dispersions and solutions.

The Viscosity Measures how a Simple Fluid Responds to Shear

Here, we will only consider one simple type of deformation, *shear*, as introduced in Figure 15.1. Under so-called 'non-slip' conditions, the layer of the liquid closest to the wall is stationary with respect to the latter. Different

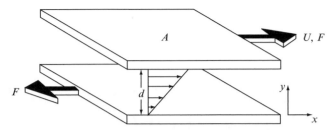

Figure 15.1   Here are shown two parallel planes, each of area A, located at $y = 0$ and $y = d$. The space between the planes is filled with a sheared fluid. The upper plane moves with a viscosity U relative to the lower one and the lengths of the arrows between the planes are proportional to the local velocity in the fluid. The applied shearing force is F. Reprinted from *An Introduction to Rheology*, H. A. Barnes, J. F. Hutton and K. Walters, Figure 1.1, Copyright (1989) with permission from Elsevier Science

systems respond very differently to shear, e.g. a fluid (liquid or gas) by viscous flow, or a solid by an elastic deformation. For a fluid, the deformation is irreversible, while for a solid it is reversible. For many of the systems we are interested in here the situation is intermediate (viscoelastic): there is both a viscous and an elastic response.

We will only consider simple laminar (i. e. non-turbulent) flow and first we look at the simplest (so-called Newtonian) behaviour. With reference to Figure 15.1, we can define the following:

- Shear stress, $\tau = F/A$ (often also denoted σ) (units of Pa)

- Shear strain, dx/dy (alternatively, γ)

- Flow rate, $v = dx/dt$

- Shear rate, $D = dv/dy = (d/dt)(dx/dy)$ (units of s^{-1})

For two parallel plates at a constant separation, there is a linear variation in the flow rate in the y-direction, i. e. D is the same in all layers. For a cylindrical tube (with Poiseuille flow), v is a parabolic function of r, largest at the centre and zero at the wall, while D is zero at the centre and largest at the wall (Figure 15.2).

For the simplest case, which we call a Newtonian fluid, the shear stress and the shear rate are proportional to each other (Figure 15.3); the proportionality constant is the viscosity, η:

$$F/A = \eta\, dv/dy, \text{ or } \tau = \eta D \tag{15.1}$$

The viscosity, which measures the resistance against deformation for a fluid, can vary widely between different materials, as illustrated in Table 15.1. The

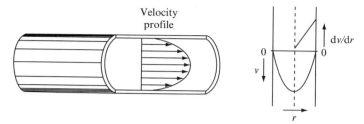

Figure 15.2 For the flow of a fluid in a cylindrical tube, the velocity is a parabolic function of the radius r, with a maximum in the centre. The shear rate is a linear function of r, largest at the wall and zero at the centre. Reprinted from *An Introduction to Rheology*, H. A. Barnes, J. F. Hutton and K. Walters, Figure 2.14, Copyright (1989) with permission from Elsevier Science

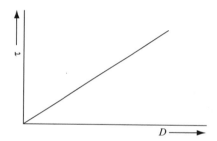

Figure 15.3 For a Newtonian fluid, the shear stress (τ) is proportional to the shear rate (D)

Table 15.1 The viscosity of some familiar materials. Reprinted from *An Introduction to Rheology*, H. A. Barnes, J. F. Hutton and K. Walters, Table 2.1, Copyright (1989) with permission from Elsevier Science

Liquid	Approximate viscosity (Pa s)
Glass	10^{40}
Molten glass (500°C)	10^{12}
Bitumen	10^{8}
Molten polymers	10^{3}
Golden syrup	10^{2}
Liquid honey	10^{1}
Glycerol	10^{0}
Olive oil	10^{-1}
Bicycle oil	10^{-2}
Water	10^{-3}
Air	10^{-5}

larger the viscosity, then the larger is the shear stress required for a certain shear flow.

Simple liquids are generally Newtonian, while colloidal systems, dispersions or solutions, may or may not be Newtonian. The nature may also change with concentration and shear rate; the latter is very different for different processes (Table 15.2).

For solutions, it is often convenient to discuss different derived functions, such as relating the viscosity to that of the neat solvent or normalizing with respect to concentration. It will be relevant here to discuss both the relative viscosity (viscosity ratio), the specific viscosity, the reduced viscosity (viscosity number), and the intrinsic viscosity (limiting viscosity number) (Table 15.3.)

The viscosity can be measured by a number of different shear experiments, suitable for different systems and purposes. In the capillary, or Ostwald viscosimeter the measured flow time gives the viscosity. The flow time (t) depends on the viscosity as well the density (ρ) of the liquid and generally the instrument is calibrated by measurements with liquid of known viscosity. The following relationship applies for two liquids 1 and 2:

Table 15.2 Shear rates typical of some familiar materials and processes. Reprinted from *An Introduction to Rheology*, H. A. Barnes, J. F. Hutton and K. Walters, Table 2.2, Copyright (1989) with permission from Elsevier Science

Situation	Typical range of shear rates (s^{-1})	Application
Sedimentation of fine powders in a suspending liquid	$10^{-6} - 10^{-4}$	Medicines, paints
Levelling due to surface tension	$10^{-2} - 10^{-1}$	Paints, printing inks
Draining under gravity	$10^{-1} - 10^{1}$	Painting and coating, toilet bleaches
Extruders	$10^{0} - 10^{2}$	Polymers
Chewing and swallowing	$10^{1} - 10^{2}$	Foods
Dip coating	$10^{1} - 10^{2}$	Paints, confectionary
Mixing and stirring	$10^{1} - 10^{3}$	Manufacturing liquids
Pipe flow	$10^{0} - 10^{3}$	Pumping, blood flow
Spraying and brushing	$10^{3} - 10^{4}$	Spray-drying, painting, fuel atomization
Rubbing	$10^{4} - 10^{5}$	Application of creams and lotions to the skin
Milling pigments in fluid bases	$10^{3} - 10^{5}$	Paints, printing inks
High-speed coating	$10^{5} - 10^{6}$	Paper
Lubrication	$10^{3} - 10^{7}$	Gasoline engines

Table 15.3 Important derived functions used to characterize viscosity behaviour, including the International Union for Pure and Applied Chemistry (IUPAC) names. From, P. C. Hiemenz and R. Rajagopalan, *Principles of Colloid and Surface Chemistry*, 3rd Edn, Marcel Dekker, New York, 1997

Functional form	Symbol	Common name	IUPAC name	$\lim\limits_{c\to 0}$
—	η	Viscosity	—	η_0
η/η_0	η_r	Relative viscosity	Viscosity ratio	1
$\eta/\eta_0 - 1$	η_{sp}	Specific viscosity	—	0
$(\eta/\eta_0 - 1)/c$	η_{red}	Reduced viscosity	Viscosity number	$[\eta]$
$c^{-1}\ln(\eta/\eta_0)$	η_{inh}	Inherent viscosity	Logarithmic viscosity number	$[\eta]$
$\lim\limits_{c\to 0}\eta_{red}$ or $\lim\limits_{c\to 0}\eta_{inh}$	$[\eta]$	Intrinsic viscosity	Limiting viscosity number	—

$$\eta_1/\eta_2 = \rho_1 t_1/\rho_2 t_2 \tag{15.2}$$

For dilute solutions or dispersions, there is no difference in density between the solution and the neat solvent so that we have:

$$\eta/\eta_0 = t/t_0 \tag{15.3}$$

In the concentric-cylinder and cone-and-plate viscometers illustrated in Figure 15.4, the viscous fluid is enclosed in the gap between two surfaces. As one of them moves, for example, by rotation of the outer cylinder in the concentric-cylinder viscometer, the viscous resistance will be transmitted through the fluid and produces a torque on the stationary surface, which is related to the viscosity. By varying either the width of the gap or the angular velocity, the shear rate can be varied.

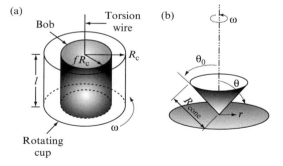

Figure 15.4 Schematic representations of (a) the concentric-cylinder, and (b) the cone-and-plate viscometers. From P. C. Hiemenz and R. Rajagopalan, *Principles of Colloid and Surface Chemistry*, 3rd Edn, Marcel Dekker, New York, 1997

The Presence of Particles Changes the Flow Pattern and the Viscosity

If colloidal size particles are introduced in a liquid, the flow pattern is changed, as illustrated in Figure 15.5. If a non-rotating particle is inserted, the liquid must be slowed down since the layers on opposite sides of the particle must have the same velocity, which must be the same as that of the particle. The overall velocity gradient is thus reduced and, since the shear is the same, this corresponds to an increased viscosity. For a rotating particle there will also be a reduced velocity gradient.

The increase in viscosity due to the particles increases with their concentration and the relationship can for a Newtonian colloidal system be expressed as a power series in the particle concentration. For low concentrations and low shear rates, we therefore have the following:

$$\eta/\eta_0 = 1 + k_1 c + k_2 c^2 + k_3 c^3 \ldots \tag{15.4}$$

where η_0 is the viscosity of the medium (absence of particles) and c the concentration of particles. This is the famous Einstein equation for the laminar flow of a dilute dispersion of spherical particles. The constant k_1 depends on particle shape and k_2 on particle–particle pair interactions, with numerical values depending on the concentration units chosen (k_1 equals the intrinsic viscosity or limiting viscosity number defined above).

It is fundamental to choose the volume fraction, ϕ, of particles as the concentration unit. For rigid, spherical particles, we have:

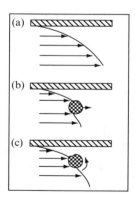

Figure 15.5 Illustration of the flow pattern near a stationary wall for (a) a neat liquid, and in the presence of particles, (b) non-rotaing, or (c) rotating. From P. C. Hiemenz and R. Rajagopalan, *Principles of Colloid and Surface Chemistry*, 3rd Edn, Marcel Dekker, New York, 1997

$$\eta/\eta_0 = 1 + 2.5\phi \tag{15.5}$$

which is known as Einstein's equation of viscosity of dispersions, i.e. the intrinsic viscosity $[\eta]_\phi = 2.5$.

The relative viscosity and the reduced viscosity as a function of the volume fraction of spherical particles is illustrated in Figure 15.6. In the former case, the slope gives the intrinsic viscosity, while in the latter it is given by the intercept on extrapolation to $\phi = 0$. Another experimental verification for a wide range of particles is illustrated in Figure 15.7.

The Einstein theory is based on a model of dilute, unsolvated spheres. Deviations can be due to a high concentration, non-spherical shape of particles and particle swelling due to solvation. For rigid non-spherical particles, $[\eta]_\phi > 2.5$, with theoretical equations available for simple ellipsoidal shapes. Absorption of solvent and solvation leads to an increase of $[\eta]_\phi$, but the Einstein equation is still valid if the effective volume of the particle, including absorbed solvent, is considered. This effect becomes very important for flexible polymers. If solute adsorption occurs, there is an analogous effect and the effective volume of the particle, including adsorbed molecules, needs to be considered.

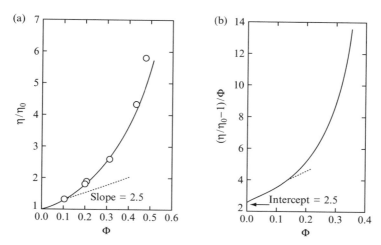

Figure 15.6 (a) Relative, and (b) reduced viscosity versus volume fraction of spherical particles; the slope and intercept, respectively, give the intrinsic viscosity. The continuous line in (a) represents the theoretical prediction of the effect of particle crowding. From P. C. Hiemenz and R. Rajagopalan, *Principles of Colloid and Surface Chemistry*, 3rd Edn, Marcel Dekker, New York, 1997

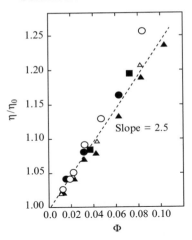

Figure 15.7 Experimental verification of Einstein's law of viscosity for spherical particles of different sizes and types (yeast particles, fungus spores and glass particles). From P. C. Hiemenz and R. Rajagopalan, *Principles of Colloid and Surface Chemistry*, 3rd Edn, Marcel Dekker, New York, 1997

The Relationship between Intrinsic Viscosity and Molecular Mass can be Useful

An important generalization about the intrinsic viscosity of polymer solutions is given by the Staudinger–Mark–Houwink relationship, as follows:

$$[\eta] = kM^{\alpha} \tag{15.6}$$

Different particle shapes and different solvency conditions give different exponents, as exemplified by rigid spheres giving 0, rigid long rods 1.8, flexible chains in a theta (θ) solvent 0.5 and in an athermal solvent 0.8. (For the athermal case, the χ parameter is zero, cf. Chapter 9.) For the molecular weight determination of polymers, k and α are experimentally determined constants, as exemplified in Table 15.4.

The Rheology is often Complex

We have considered above the simple Newtonian behaviour, where there is proportionality between the shear stress and the shear rate. For colloidal systems, this simple relationship generally breaks down and we have a *non-Newtonian* liquid. However, we can then define an *apparent viscosity* as follows:

$$\tau = \eta_{\text{app}} D \tag{15.7}$$

Table 15.4 Mark – Houwink coefficients for some polymer – solvent systems. From P. C. Hiemenz and R. Rajagopalan, *Principles of Colloid and Surface Chemistry*, 3rd Edn, Marcel Dekker, New York, 1997

Polymer	Solvent	Temperature (°C)	$k \times 10^3$ (cm^{3}g^{-1})	a
Polypropylene	Cyclohexanone	$92 = \Theta$	172	0.50
Poly(vinyl alcohol)	Water	80	94	0.56
Polyoxyethylene	Carbon tetrachloride	25	62	0.64
Poly(methyl methacrylate)	Acetone	30	7.7	0.70
Polystyrene	Toluene	34	9.7	0.73
Natural rubber	Benzene	30	18.5	0.74
Polyacrylonitrile	Dimethylformamide	20	17.7	0.78
Poly(vinyl chloride)	Tetrahydrofuran	20	3.63	0.92
Poly(ethylene terephthalate)	*m*-Cresol	25	0.77	0.95

For a non-Newtonian liquid, the value of η_{app} depends on the shear stress (or shear rate), a dependence that will interest us now.

However, let us first mention another complication. The apparent viscosity generally varies with time at a given shear rate (or stress); thus, η_{app} depends on the history of the solution. Above, we have assumed that a certain shear rate produces a certain shear stress which does not change as long as a constant shear rate is applied. After some time, at constant D or τ a *stationary state* often develops, where η_{app} is constant and independent of the history of the system.

The relationship between shear rate and shear stress is different for different systems, as exemplified in Figure 15.8. For a shear-thinning (or pseudo-plastic) system, η_{app} decreases with increasing shear rate; plastic liquids are characterized by a finite yield value, i.e. a minimal yield stress needed before they start to flow. A shear-thickening (or dilatant system) is characterized by an apparent viscosity that increases with increasing shear rate.

Non-Newtonian behaviour can arise from many different mechanisms, some of which are illustrated in Figure 15.9. For dilute systems, shear-thinning can be due to flow orientation of the particles or the change in conformation of the polymer molecules. Thus, when the shear rate is larger than the rate of thermal reorientation of the particles, they will be aligned in the flow direction. For concentrated systems, shear-thinning appears when the shear rate is larger than the rate of build-up of the equilibrium supra-particulate structures.

The behaviour of shear-thinning systems is illustrated in Figure 15.10, and as can be seen there are two Newtonian plateaus where η_{app} is independent of shear rate and shear stress. The plateau value for low shear rates is called the zero shear viscosity. The shear-thinning behaviour of polymer solutions depends strongly on concentration, as illustrated in Figure 15.11.

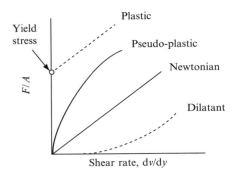

Figure 15.8   Different relationships between shear stress and shear rate for the station-ary state. From P. C. Hiemenz and R. Rajagopalan, *Principles of Colloid and Surface Chemistry*, 3rd Edn, Marcel Dekker, New York, 1997

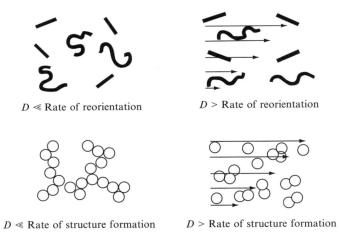

Figure 15.9   Examples of the origin of shear-thinning for dilute and concentrated systems, showing schematic structures for low and high shear rates (D). Reproduced by permission of Lennart Piculell, University of Lund, Sweden

The viscosity can either increase or decrease with time and these changes can be reversible or irreversible. A common case is *thixotropy*, where the viscosity decreases with time. Thixotropy typically occurs when the solution is shear-thinning, with the opposite applying for shear-thickening situations. For such cases, the flow history of the sample must be taken into account. The opposite case to thixotropy is called *rheopexy*.

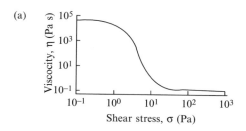

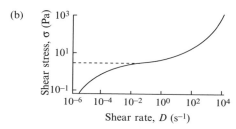

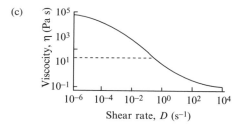

Figure 15.10 Illustrations of the typical behaviour of a non-Newtonian liquid: (a) viscosity versus shear stress; (b) shear stress versus shear rate (the dotted line representing the ideal yield stress, or Bingham plastic, behaviour); (c) viscosity versus shear rate. Reprinted from *An Introduction to Rheology*, H. A. Barnes, J. F. Hutton and K. Walters, Figure 2.3, Copyright (1989), with permission from Elsevier Science

Viscoelasticity

In defining the viscosity from the simple shear experiment, we saw that for the simple case of a Newtonian liquid, the shear stress is proportional to the shear rate. For a solid body subjected to shear, there will be deformation rather than flow; the shear stress is proportional to the shear or deformation rather than the shear rate:

$$\tau = G \mathrm{d}x/\mathrm{d}y \qquad (15.8)$$

where G (units of Pa) is called the shear modulus (a solid is elastic). A characteristic of many surfactant and polymer systems is that they have both

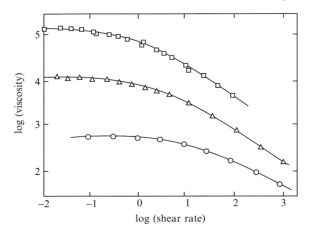

Figure 15.11 Shear-thinning for concentrated polysaccharide solutions, illustrated for λ- carrageenan at concentrations (from bottom to top) of 1.5, 2.75 and 5.0 wt%. The viscosity (mPa s) is plotted versus the shear rate (s^{-1}). From E. R. Morris, A. N. Cutler, S. B. Ross-Murphy and D. A. Rees, *Carbohydr. Polym.*, **1** (1981) 5

types of responses simultaneously; they are both viscous and elastic, i.e. *visco-elastic*.

Viscoelasticity can be investigated in oscillating measurements. Considering Figure 15.4, the outer cylinder (or the plate) is oscillated sinusoidally; the torque experienced by the inner cylinder (giving the shear stress) is measured as a function of time.

For an elastic body, the shear stress is largest at the maximum deformation, i. e. at the extreme points of the oscillation, while it is zero when the deformation is zero; the shear stress varies in phase with the deformation, i. e. the phase angle $\delta = 0$.

For a viscous liquid, the shear stress is largest at the maximum shear rate, and therefore this occurs for zero deformation. When the deformation is at a maximum, the shear rate is zero and also the shear stress. The shear stress is thus out-of-phase with respect to the deformation, and the phase angle is 90°.

For a viscoelastic liquid the phase angle has an intermediate value, $0° < \delta < 90°$. It is characterized by the in-phase and out-of-phase components of the modulus:

$$\text{Shear storage modulus, } G' = G_0 \cos \delta \tag{15.8a}$$

$$\text{Shear loss modulus, } G'' = G_0 \sin \delta \tag{15.8b}$$

Viscoelastic systems are also characterised by two other parameters, namely:

$$\text{Dynamic shear modulus,} \quad G^* = [(G')^2 + (G'')^2]^{1/2} \qquad (15.9a)$$

$$\text{Dynamic viscosity,} \quad \eta^* = G^*/\omega = [(G')^2 + (G'')^2]^{1/2}/\omega \qquad (15.9b)$$

The limiting types of behaviour are an elastic gel and a Newtonian liquid. For an elastic gel, the elastic component dominates, and thus $G' \gg G''$ and G' is independent of frequency. The dynamic viscosity, η^*, approaches G'/ω, which is inversely proportional to the frequency. For a Newtonian system, the viscous component dominates and $G'' \gg G'$ and $G'' = \eta\omega$. The (dynamic) viscosity is independent of frequency.

The Rheological Behaviour of Surfactant and Polymer Solutions Shows an Enormous Variation: Some Further Examples

Surfactant Systems

The viscosity of surfactant solutions is only weakly affected by surfactant self-assembly into spherical aggregates. On the other hand, as we investigated in Chapter 3, growth into cylindrical or thread-like aggregates is accompanied by major increases in viscosity.

In Figure 3.1, the viscosity is shown for the case of spherical aggregates. At low volume fractions, the viscosity follows the Einstein equation for hard spheres. For higher concentrations, there is agreement with models that take into account crowding and other particle–particle interactions.

An illustration for the case of microemulsions composed of spherical droplets is given in Figure 15.12, again showing the weak increase in viscosity with droplet volume fraction.

The increase in viscosity due to uni-dimensional growth of micelles was illustrated earlier in Figure 3.2. The dramatic increase in viscosity with concentration is due to a combination of micellar growth and micelle–micelle interactions; the viscosity enhancement relative to neat water may exceed 10^7. Such solutions are typically viscoelastic. Cationic surfactants with certain organic counterions may give particularly strong effects, as illustrated in Figures 15.13 and 15.14 for hexadecylpyridinium in the presence of salicylate. Figure 15.13 shows the zero shear viscosity data, while Figure 15.14 shows the storage and loss moduli data.

There are also cases of reversed micellar growth leading to similar effects. For lecithin–water–oil systems, addition of water to a solution of lecithin in oil can lead to viscosity increases by a factor of 10^6 because of this.

Bicontinuous, 'sponge' or L_3 phase, surfactant solutions are characterized by a Newtonian behaviour and moderate viscosities, even at relatively high volume fractions.

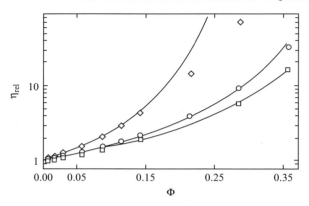

Figure 15.12 Relative viscosity as a function of the droplet volume fraction, Φ (volume fraction of surfactant plus oil), for oil-in-water (o/w) microemulsions composed of mixed surfactant, oil and water. The continuous lines correspond to the predictions of a theoretical model for spherical particles. From M. Gradzielski and H. Hoffmann, in *Handbook of Microemulsion Science and Technology*, P. Kumar and K. L. Mittal (Eds), Marcel Dekker, New York, 1999

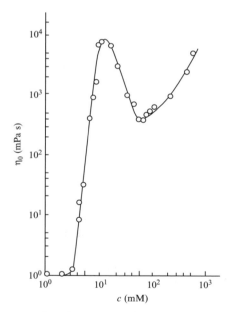

Figure 15.13 The concentration dependence of the viscosity extrapolated to zero shear rate for equimolar concentrations of hexadecylpyridinium chloride and sodium salicylate. From H. Hoffmann and H. Rehage, in *Surfactant Solutions. New Methods of Investigation*, R. Zana (Ed.), Marcel Dekker, New York, 1987

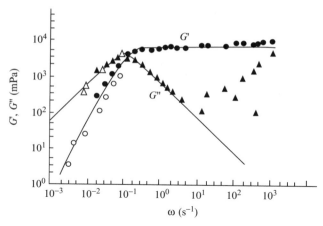

Figure 15.14 Storage and loss moduli (G' and G'', respectively) for a solution of 80 mM hexadecylpyridinium salicylate. From H. Hoffmann and H. Rehage, in *Surfactant Solutions. New Methods of Investigation*, R. Zana (Ed.), Marcel Dekker, New York, 1987

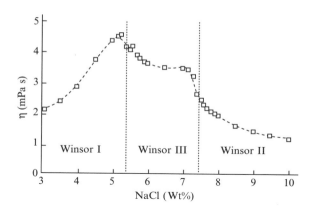

Figure 15.15 Viscosity (at low shear rates) as a function of salinity for a system of brine, toluene, sodium dodecyl sulfate and butanol. As the salinity increases there is a change over from a Winsor I (o/w) to a Winsor III (bicontinuous) and finally to a Winsor II (w/o) behaviour. From M. Gradzielski and H. Hoffmann, in *Handbook of Microemulsion Science and Technology*, P. Kumar and K. L. Mittal (Eds), Marcel Dekker, New York, 1999

Similarly, the structurally related bicontinuous microemulsions also show low viscosities and simple Newtonian flow behaviour. In Figure 15.15, the relationship between microstructure and viscosity is illustrated for a microemulsion system where a transition from an oil-in-water (o/w) to a bicontinuous and finally to a water-in-oil (w/o) structure can be induced by the addition of salt. Irrespective of the microstructure, the viscosity is low and there are only moderate changes on changes in microstructure. In microemulsion systems, where one can pass between different types of structures, it is typically observed that there is a maximum in viscosity around the transition from the droplet to the bicontinuous microstructure.

For liquid crystalline phases there is a relationship between rheology and composition but the most important factor is phase structure. Of the common phases, the lamellar one has the lowest viscosity, while the cubic ones (both discrete micellar and bicontinuous types) have the highest viscosities. In fact, cubic phases are characterized by very high viscosities and may even possess a yield value. They exhibit elastic properties, and have a solid-like appearance. When subjected to shear, the liquid crystalline phases present complex rheological responses, which are due to microstructural rearrangements; for lamellar phases, there is a complex interplay between extended bilayer stacks, bilayer fragments and multilamellar vesicles, 'onions'.

Polymer Solutions

As we have already touched upon above in this present chapter, as well as in Chapters 4 and 9, polymers have a broad range of applications for rheology control. We have also seen that the viscosity depends on a number of factors, such as molecular weight, concentration, solvency conditions and extension of polymer molecules. Special attention will be paid here to the properties of associating polymers.

The discussion above regarding the exponent in the Staudinger–Mark–Houwink equation illustrates the role of solvency, with another illustration being given earlier in Figure 9.14. The extension of an ionic polymer is strongly influenced by salt, as well as polymer concentration. Due to electrostatic screening, polyions adopt a more compact conformation in the presence of salt as well as at higher polyion concentration; this explains the decrease in reduced viscosity.

Associating amphiphilic polymers can give rise to a wide range of rheology behaviour depending on the polymer architecture and we have already alluded to two cases in Chapters 4 and 13. Graft copolymers, with hydrophobic grafts on a water-soluble backbone, often referred to as hydrophobically modified water-soluble polymers have wide applications as 'associative thickeners'; a typical system is illustrated above in Figure 13.8. Due to interpolymer association between the grafts (cf. Figure 13.7), there is a very significant increase in viscosity due to hydrophobic modification; for the case referred to, this is approximately a factor of 10 for a 1 wt% solution.

The block copolymer case of Figure 4.15 is of a very different nature. Here self-assembly into the same type of structures as for surfactant systems controls the rheological behaviour. Thus, the dramatic increase in viscosity with increasing temperature (sometimes referred to as 'thermal gelation') is due to a phase transition from a micellar solution to a cubic liquid crystalline phase.

The shear-thinning behaviour of polysaccharide solutions was exemplified above in Figure 15.11. An illustration of the widely different types of behaviour that can be encountered for chemically similar aqueous polysaccharide systems is provided in Figure 15.16. This figure presents the 'mechanical spectra' of G', G'' and $\eta*$ plotted logarithmically against the frequency. Figure 15.16(a) shows the data obtained for a polymer (dextran) solution, which approximates a purely viscous system, while Figure 15.16(b) illustrates a gel (agarose), responding, to a good approximation, as a an elastic solid. For the solution case, $G'' \gg G'$, both increasing with frequency, and the viscosity is independent of frequency over a wide range. In the gel case, $G' \gg G''$, both independent of ω, with the dynamic viscosity decreasing with frequency.

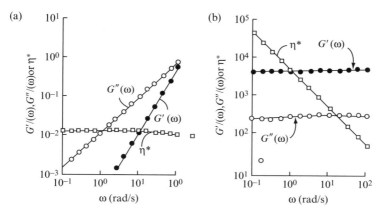

Figure 15.16 Storage and loss moduli (G' and G'', respectively) and complex viscosity (η^*) as a function of frequency for (a) a dextran solution, and (b) an agarose gel. The solution shows a Newtonian behaviour, while the gel displays strongly elastic properties. From E. R. Morris and S. B., Ross-Murphy, in *Techniques in the Life Sciences*, Vol. B3, Elsevier/North Holland, Amsterdam, 1981

Mixed Polymer–Surfactant Systems

The rheological behaviour of mixed polymer–surfactant solutions is as for polymer solutions in general strongly dependent on polymer concentration; however, for the mixed systems much more dramatic changes are seen.

The case of dilute solutions is illustrated in Figure 15.17 for the case of a non-ionic polymer associating with an ionic surfactant. The reduced viscosity increases with the polymer molecular weight, but in particular, with the binding

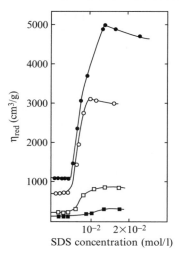

Figure 15.17 Reduced viscosity data obtained for solutions of poly(ethylene glycol) as a function of the concentration of added sodium dodecyl sulfate for four different molecular weights; from the bottom, 7×10^4, 2×10^5, 1×10^6 and 2×10^6. Reprinted from *Eur. Polym. J.*, **21**, J. Francois, J. Dayantis and J. Sabbadin, 'Hydrodynamic behaviour of the poly(ethylene oxide) – sodium dodecylsulphate complex', 165 – 174, Copyright (1985), with permission from Elsevier Science

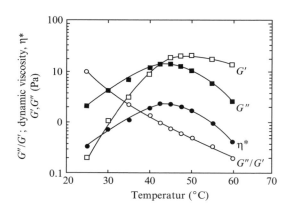

Figure 15.18 Viscoelastic properties as a function of temperature for a solution (1 wt%) of a non-ionic polymer (ethylhydroxyethyl cellulose (EHEC)) and a cationic surfactant (hexadecyltrimethylammonium bromide). Reprinted from *Colloid. Surf.*, **47**, A. Carlsson, G. Karlström and B. Lindman, 'Thermal gelation of nonionic cellulose ethers and ionic surfactants in water', 147 – 165, Copyright (1990), with permission from Elsevier Science

of the surfactant (starting at the critical association concentration (CAC)). As the ionic surfactant binds, the polymer is effectively transformed into a polyelectrolyte with a more extended conformation.

The dramatic viscosity changes for mixed semi-dilute solutions of a hydrophobically modified water-soluble polymer and a surfactant was discussed in some detail in Chapter 13 in terms of mixed aggregate formation. The viscosity is very sensitive to the stoichiometry of the micellar-type self-assemblies. Similar cross-linking effects and very efficient gelation has been been observed with vesicles instead of micelles.

Mixed polymer–surfactant solutions are typically viscoelastic; for certain systems there may be a change-over from the dominance of viscous effects to the dominance of elastic properties as the temperature increases. This may occur for non-ionic polymers, which become less polar at higher temperature, and show a clouding behaviour, as illustrated in Figure 15.18; this system shows a temperature-induced gelation.

Bibliography

Barnes, H. A., J. F. Hutton and K. Walters, *An Introduction to Rheology*, Elsevier, Amsterdam, 1989.

Bergström, L., Rheology of concentrated suspensions, in *Surface and Colloid Chemistry in Advanced Ceramics Processing*, R. J. Pugh and L. Bergström (Eds), Marcel Dekker, New York, 1994.

Gradzielski, M. and H. Hoffmann, in *Handbook of Microemulsion Science and Technology*, P. Kumar and K. L. Mittal (Eds), Marcel Dekker, New York, 1999.

Hiemenz, P. C. and R. Rajagopalan, *Principles of Colloid and Surface Chemistry*, 3rd Edn, Marcel Dekker, New York, 1997.

Hoffmann, H. and H. Rehage, in *Surfactant Solutions. New Methods of Investigation*, R. Zana (Ed.), Marcel Dekker, New York, 1987.

Larson, R. G., *The Structure and Rheology of Complex Fluids*, Oxford University Press, New York, 1999.

Ross-Murphy, S. B., Rheological methods, in *Biophysical Methods in Food Research*, H. W.-S. Chan (Ed.), Blackwell Scientific Publications, Oxford, UK, 1984, pp. 138–199.

16 SURFACE TENSION AND ADSORPTION AT THE AIR–WATER INTERFACE

Surface Tension is due to Asymmetric Cohesive Forces at a Surface

The surface tension of liquids causes the formation of drops and is related to the attractive forces between the molecules, as illustrated in Figure 16.1(a). These attractive forces are the source of condensation of vapours into liquids and they originate from dispersion, dipole–dipole and dipole-induced–dipole forces and hydrogen bonding. In the bulk liquid, a molecule senses the same

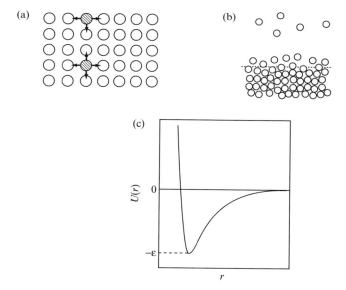

Figure 16.1 (a) The origin of the surface tension is found in the imbalance of the attractive forces on a molecule at the surface. (b) Close to the surface the molecules are at larger separations and hence they have a higher energy, as depicted in (c) the energy versus distance diagram

attractive forces in all directions, while for a molecule at the surface this attraction is lacking in one direction. This asymmetry is the origin of the surface energy and is manifested in the surface tension. Thus, the surface tension is a reflection of the cohesive forces in a liquid.

A second molecular explanation for the surface tension is depicted in Figure 16.1(b). Close to the surface the molecules are at larger separations and hence they have a larger energy, as depicted in the energy versus distance diagram in Figure 16.1(c). Also in this molecular picture, the surface tension correlates with the cohesive energy of the liquid.

Examples of the surface tension for some neat liquids are shown in Table 16.1. The surface tension is expressed in units of dyn/cm or mN/m. Fortunately, the conversion factor is unity in going from one unit to the other and hence they can be used interchangeably. Since the surface tension is equivalent to a surface energy, or to be more correct, a surface free energy, the surface tension can also be expressed in energy units, erg/cm^2 or mJ/m^2, here also with a conversion factor of unity. From Table 16.1, we note that the surface tension correlates with the cohesion energy in the liquid, where diethyl ether has the smallest and mercury the largest cohesion energy.

The cohesion energy is in turn a function of the strength of the dispersion forces in the liquid, as exemplified by comparing the surface tension of chloroform and bromoform; the latter is a more polarizable molecule and hence shows a higher surface tension. The cohesion energy is also correlated with the free volume in the liquid, which is realized when comparing the surface tension of the series of normal alkanes—the higher the free volume, then the lower the cohesion energy and hence the lower the surface tension.

The interfacial tension is the surface tension, or the surface free energy, which appears at the interface between two immiscible, or partially miscible liquids. The reason for immiscibility is normally to be found in a large difference in cohesion forces between the molecules in the two liquids. Thus, at the interface there will be a net force, as depicted in Figure 16.2. The larger this difference in

Table 16.1 Surface tension (in mN/m) of some neat liquids at 25°C

Water	72
Ethanol	22
Chloroform	27
Bromoform	45
n-Hexane	18
n-Octane	22
n-Dodecane	25
n-Hexadecane	27
Diethyl ether	17
Mercury	480

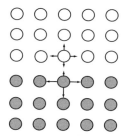

Figure 16.2 As for the surface tension, the interfacial tension originates from an imbalance of the attractive forces on molecules at the interface

cohesion forces, then the larger the interfacial tension. If this difference is small enough, however, the two liquids are miscible. The problem of miscibility and immiscibility is satisfactorily described by the regular solution theory, as outlined earlier in Chapter 10.

The units of interfacial tension are the same as for surface tension. Besides being dependent on the difference in cohesion forces, the interfacial tension also depends on the specific interactions between the molecules in the two liquids. Water/octane, for example, has an interfacial tension of 51 mN/m, while water/octanol has an interfacial tension of only 8 mN/m. Clearly, the hydroxyl group of the octanol faces the aqueous phase and thus reduces the interfacial tension considerably. It is this phenomenon that is utilized in surfactant molecules, having one very polar part and one non-polar part, thus reducing the interfacial tension between organic liquids and water.

Solutes Affect Surface Tension

In aqueous systems, an added component can affect the surface tension in three different ways, as depicted in Figure 16.3. Organic water-soluble materials, such as ethanol, normally decrease the surface tension monotonically with increasing concentration. This is due to a preferential adsorption of the organic molecules at the liquid–air surface, as will be discussed further below. Surfactants, on the other hand, show a very large reduction in surface tension at very low concentrations up to the critical micelle concentration (CMC), whereafter the surface tension is practically constant. This large reduction in surface tension is due to a strong adsorption of the surfactants at the liquid–air surface. At concentrations higher than the CMC, all additional surfactant will form new micelles, thus keeping the surfactant unimer activity (concentration) more or less constant. Hence, the surface tension will not change with surfactant concentration above the CMC. Electrolytes normally increase the surface tension. The reason is that the electrolytes are depleted from the surface i.e. there is a negative adsorption of the ions at the liquid–air surface. The relationship between surface tension

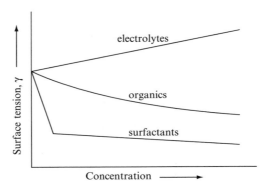

Figure 16.3 In aqueous systems, an added component affects the surface tension in one of three different ways

and adsorption is discussed later in this chapter in terms of the regular solution theory and is given by equations (16.22a,b).

Dynamic Surface Tension is Important

Dynamic surface tension is the change in surface tension before equilibrium conditions are obtained. At the instant when a new liquid/air surface is formed, the surface composition is the same as the bulk composition and the surface tension is close to the arithmetic mean of the surface tensions of the neat components. At equilibrium, the component with the lower surface tension will be preferentially adsorbed at the surface and hence the surface composition will be in excess of the component with the lower surface tension and therefore the surface tension of the solution will be lowered. This process is time-dependent and depends on the diffusion of the components. An example is shown in Figure 16.4(a) for the water–ethanol system. If a new surface suddenly is created in this system at a solution composition of 40% alcohol, for example, the surface tension will be close to the arithmetic mean between the two pure component surface tensions, i.e. proportional to the composition. Due to the preferential adsorption of the component with the lower surface tension (alcohol), the surface tension will be lowered to the equilibrium value, as shown by the arrow in the figure.

The time dependence of this change in surface tension normally follows the path depicted in Figure 16.4(b) for the case of a surfactant in aqueous solution. Here, the surface tension at the limit of infinitely short times is close to that of pure water. Dynamic surface tension is important in many technical applications, with one example being the printing process. Here, the equilibrium surface tension will never be attained and hence characterization of such systems using equilibrium surface tension is not meaningful.

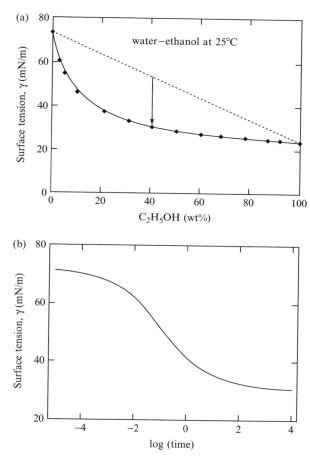

Figure 16.4 (a) The dynamic surface tension is the change from the arithmetic mean of the surface tensions of the pure components to the equilibrium surface tension, shown here for the system water–ethanol. Reproduced by permission of John Wiley & Sons, Inc. from A.W. Adamson, *Physical Chemistry of Surfaces*, 5th Edn, Wiley, New York, 1990. (b) After the formation of a new surface, the system relaxes to the equilibrium surface tension due to adsorption of the more surface active species

Besides the time taken to diffuse to the liquid–air interface, there can be two other reasons for the surface tension changing with time. The first is due to a change in configuration of larger molecules, such as polymers, at the surface. Polymers, in general, take a very long time (of the order of hours) to reach equilibrium at any surface. The second reason is found when the system contains more than one solute. In such cases, long times to reach equilibrium are found because of a competitive adsorption of the solutes at the liquid–air interface. This will be discussed further below.

The Surface Tension is Related to Adsorption

A detailed discussion on the relationship between the adsorption and the surface tension, in the terminology of the regular solution theory, will be given later in this chapter. In discussing the adsorption of a solute (surfactant) at the liquid–air surface, one first needs to define the surface. This was first came out by Gibbs, who identifyed the surface to be the position where the solvent concentration is half between that in the solvent and in the vapour. The adsorption of a surfactant is normally written as $\Gamma^{(1)}$, where the (1) indicates that the surface, and hence the adsorption, is defined by the solvent.

For non-ionic surfactants, the adsorption is related to the surface tension through the Gibbs equation, as follows:

$$\Gamma^{(1)} = -\frac{1}{RT}\frac{d\gamma}{d\ln a} \tag{16.1a}$$

where a is the activity of the solute in the bulk solution. Since the surfactant concentration in bulk solutions below the CMC is normally very low, the surfactant activity can be replaced by the surfactant concentration, C, and thus we have:

$$\Gamma^{(1)} = -\frac{1}{RT}\frac{d\gamma}{d\ln C} \tag{16.2a}$$

Thus, the surfactant adsorption is obtained from the slope of a plot of the surface tension versus the logarithm of the concentration. Figure 16.5 shows such plots for a series of medium-chain alcohols in water. The curves have the same slope, indicating that the adsorption, or surface excess, is the same irrespective of the chain length of the alcohol. Note the approximate linearity of the curves at high concentrations, i.e. the lower part of the curves, indicating a constant adsorption according to equation (16.2).

Assuming monolayer adsorption, one realizes that the adsorbed amount is inversely proportional to the cross-sectional area per adsorbed molecule, A. The relationship can be shown to be as follows:

$$A(\overset{\circ}{A}{}^2/\text{molecule}) = \frac{1000}{6.023\Gamma} \tag{16.3}$$

where Γ is expressed in $\mu\text{mol/m}^2$.

The identical slopes of the curves in Figure 16.5 therefore mean that the alcohols have the same surface area per molecule, irrespective of chain length. A constant surface area per molecule for these medium-chain alcohols is in accordance with a molecular ordering at the interface, where the polar part is directed towards the water and the non-polar hydrocarbon chain is directed towards the air.

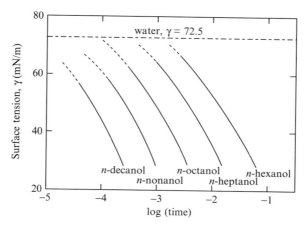

Figure 16.5 Surface tension versus concentration for aqueous solutions of a series of long-chain alcohols. Reproduced by permission of Pearson Educational from R. Defay, I. Prigogine, A. Bellemans and D.H. Everett, *Surface Tension and Adsorption*, Longmans, London, 1966

For ionic surfactants, one has to take into account the fact that there is a counterion associated with the surfactant and that the surface as a whole must be electrically neutral. Equations (16.1) and (16.2) then have the forms:

$$\Gamma^{(1)} = -\frac{1}{2RT}\frac{d\gamma}{d\ln a} \tag{16.1b}$$

and:

$$\Gamma^{(1)} = -\frac{1}{2RT}\frac{d\gamma}{d\ln c} \tag{16.2b}$$

Note, however, that in the presence of excess salt, equations (16.1a) and (16.2a) should be used. This is because the excess salt swamps the effect of the counterion at the surface and hence the surfactant adsorbs as a single species, without the company of a counterion.

Surfactant Adsorption at the Liquid–Air Surface is Related to the Critical Packing Parameter

Like the medium-chain alcohols, surfactants adsorb at the liquid–air surface with their polar parts directed towards the aqueous solution and the hydrocarbon parts directed towards the air. Figure 16.6 shows the surface tension of an alkyl ether sulfate at various salt concentrations. This figure shows that with

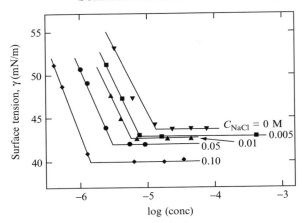

Figure 16.6 Surface tension of $C_{16}E_3SO_4^-$ Na^+ at various salt concentrations. From D. Attwood, *Kolloid-Z.*, **232** (1969) 788

increasing salt content, (a) the CMC decreases, (b) the surface tension above the CMC decreases, and (c) the slope of the curves just below the CMC increases, indicating that the adsorption increases, according to equation (16.2a,b). All three observations are indicative of an increasing critical packing parameters (CPP) of the system, which is due to the shielding of the surfactant head group repulsion by the salt (see Chapter 13). From the slopes of the curves below the CMC, we find that the surfactants pack closer at higher salt concentrations, which is not surprising considering the increasing shielding effect of the salt.

Figure 16.7 shows some surface tension data obtained for the non-ionic series $C_{16}E_n$. We find that the surfactants with shorter polyoxyethylene chains pack better at the liquid–air interface, i.e. the slopes at concentrations just below the CMC are higher. This is in line with the CPP concept where the latter increases as the polyoxyethylene chain decreases in length. We also note that the surface tension at and above the CMC is lowered as the polyoxyethylene chains are shortened. Note, however, that the effect on the CMC is rather small.

Figure 16.8 shows the temperature dependence of the surface tension for $C_{12}E_6$. It is well known that as the temperature is increased, the polyoxyethylene chain compresses, thus leading to an increased CPP value (Chapter 4). The figure shows that this results in a lowering of the surface tension as well as the CMC. Careful analysis also shows that the adsorption at concentrations below the CMC increases with temperature, i.e. the surface area per molecule decreases with temperature.

Finally we need to make a note of the effect of impurities on surface tension data. Figure 16.9 shows the surface tension of a sodium dodecyl sulfate (SDS) solution as a function of concentration. The minimum in surface tension is due

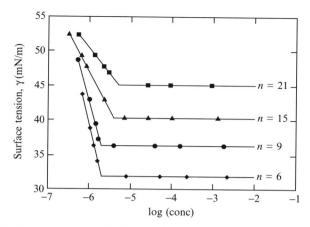

Figure 16.7 Surface tension data for the non-ionic series $C_{16}E_n$ with the indicated values of n. From P. Elworthy and C. MacFarlane, *J. Pharm. Pharmacol.*, **14** (1962) 100T

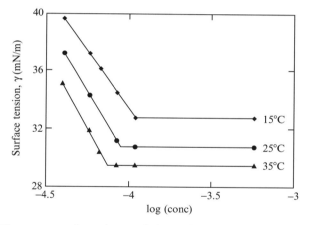

Figure 16.8 Temperature dependence of the surface tension for $C_{12}E_6$. From J. M. Corkhill, J. F. Goodman and R. H. Ottewill, *Trans. Faraday Soc.*, **57** (1961) 1927. Reproduced by permission of The Royal Society of Chemistry

to the presence of dodecyl alcohol, which is formed from hydrolysis of the SDS molecules. Dodecyl alcohol is more surface active than SDS and is thus preferentially adsorbed at the liquid–air surface, so lowering the surface tension. When micelles form in the solution, however, the dodecyl alcohol molecules become solubilized in the micelles and hence desorb from the liquid–air interface, thus resulting in an increase in surface tension. Therefore, whenever a minimum in surface tension versus surfactant concentration is found,

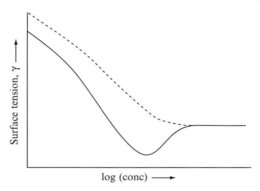

Figure 16.9 Surface active impurities in a surfactant sample cause a minimum in the surface tension versus surfactant concentration curve

the surfactant sample contains an impurity that is more surface active than the surfactant itself.

Two important observations are made for such systems. The first is that the surface tension will usually decrease slowly for a very long time before equilibrium is obtained. This is because the impurity may be present in very low concentrations and the diffusion from the bulk to the surface is the rate-determining factor. The other observation is that the equilibrium surface tension is dependent on the surface/volume ratio of the sample. The larger the bulk volume, then the more impurity is available for the adsorption at the liquid–air interface and hence the lower the surface tension.

Polymer Adsorption can be Misinterpreted

The measurement of surface tensions of polymer solutions has sometimes led to confusion. Figure 16.10 shows the surface tension of polydimethylsiloxane in tetralin. Inspection of this figure reveals that as the polymer molecular weight is increased the surface tension drop at low polymer concentrations becomes more strongly pronounced. The reason for this is to be found in the poor compatibility between the polymer and the solvent. As discussed in Chapter 9, the solubility of a polymer decreases as the molecular weight increases. Thus, the higher molecular weight polymers are 'pushed out' towards the surface where the number of solvent–polymer segment contacts are less than in the solution (see equations (16.4) and (16.6) below). Due to the low solubility of the higher molecular weight species, the polymer will be found at the surface and hence the surface tension is drastically lowered. This phenomenon has in the literature sometimes been mistaken for a micellization of the polymer, since the surface tension curves resemble those of micellar solutions.

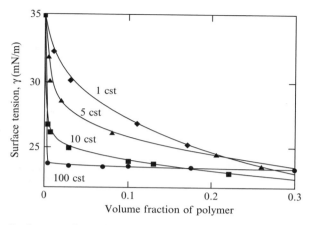

Figure 16.10 Surface tension of four different polydimethylsiloxanes with varying molecular weights in tetralin, as indicated by their viscosities (in centistokes (cst)). The figure illustrates that the surface tension drop at low polymer concentration is more pronounced at higher molecular weights and resembles the surface tension curves of surfactant solutions. Reprinted with permission from K. S. Siow and D. Patterson, *J. Phys. Chem.*, **77** (1973) 356. Copyright (1973) American Chemical Society

Measurement of Surface Tension

Equilibrium Surface Tension

A simple method to measure surface tension is by using the du Noüy ring. Here, a platinum ring is submerged in the liquid and the force that is required to pull the ring through the surface is measured (see Figure 16.11(a)). Instead of a ring a platinum sheet, or plate, can be used. The downward pull of the liquid is measured and this pull depends on the contact angle of the liquid towards the plate, as shown in Figure 16.11(b). A very simple method to estimate the surface tension is to measure the capillary rise (Figure 16.11(c)). The height of the capillary rise, h, is directly proportional to the surface tension, according to the equation $\gamma = rh\rho g/2$, where r is the capillary radius, ρ the density of the liquid and g the gravitation constant. This equation is only valid for liquids that perfectly wet the capillary surface, i.e. the contact angle of the liquid towards the capillary surface is zero.

Some systems have a very long relaxation towards equilibrium and hence surface tension measurements take a very long time to complete. This is, for example, the case with polymers in solution. Here the pendant drop (Figure 16.11(d)) is a suitable method to use. The shape of the drop is a measure of the surface tension; drops that are close to spheres are formed from liquids with a

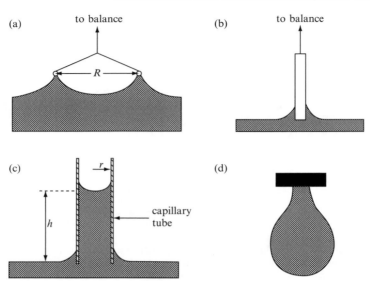

Figure 16.11 Surface tension can be measured by (a) the du Noüy method, (b) the Wilhelmy plate method, (c) the capillary rise method, or (d) the pendant drop method

high surface tension while elongated drops are formed from liquids with a low surface tension. This method is also suitable for measurements of the interfacial tension between two mutually insoluble liquids.

Dynamic Surface Tension

The dynamic surface tension can be measured by different means, e.g. by using the maximum bubble pressure method. Air is continuously blown through two capillaries, of different diameters, which are dipped into the solution. The pressure required to form a bubble is inversely proportional to the capillary diameter and directly proportional to the surface tension of the liquid. The use of two capillaries makes measurements of the immersion depth unnecessary. The methods measure dynamic surface tensions at times down to *ca.* 1–10 ms. If the dynamic surface tension at even shorter times is to be measured, then the oscillating jet method is applicable. Here, a liquid jet emerges from an elliptically formed orifice. Since a non-circular cross-section of the jet is mechanically unstable, the liquid in the jet tries to attain the circular cross-section and in doing so will oscillate between extreme values. The frequency of these oscillations is related to the dynamic surface tension.

The Surface and Interfacial Tensions can be Understood in Terms of Molecular Interactions

We will in this section illustrate the dependence of surface and interfacial tensions on the molecular interaction. For simplicity, we will ignore any entropic contributions to the surface tension, which indeed is a surface free energy quantity. Using the same nomenclature as in Chapter 10, we recognize that in pure liquid A the energy per molecule in the bulk liquid is as follows:

$$E_A^{bulk} = \frac{z}{2} w_{AA} \tag{16.4}$$

where w_{AA} is a negative quantity signifying the attraction energy between the molecules. Let the fraction of nearest neighbours at a plane parallel to the surface be l and the fraction of nearest neighbours in a plane above, or respectively below, this plane be m. Then we have the following:

$$2m + l = 1 \tag{16.5}$$

The energy per molecule present at the liquid–air surface is then:

$$E_A^{surface} = \frac{z}{2} w_{AA}(m + l) \tag{16.6}$$

since there are no neighbours in a plane above a molecule present at the liquid–air surface. The energy required to transport one molecule from the bulk to the surface is then given by the following:

$$\Delta E = E_A^{surface} - E_A^{bulk} = \frac{w_{AA} z}{2}[(m + l) - (2m + l)] = -\frac{w_{AA} z}{2} m \tag{16.7}$$

Since w_{AA} is a negative quantity, we find that there is a net positive energy required to transport a molecule from the bulk liquid to the surface. This is the reason why the liquid drop is spherical, thus minimizing the surface area and hence the total energy.

ΔE in equation (16.7), if scaled by the cross-sectional area per molecule, a_A, is equal to the surface energy, or in this treatment where entropic contributions are ignored, equal to the surface free energy, or the surface tension:

$$\gamma_A = \frac{\Delta E}{a_A} = -\frac{w_{AA} z}{2a_A} zm = -\frac{E_A^{bulk} m}{a_A} \tag{16.8}$$

once again illustrating that the surface tension of a liquid is a reflection of the cohesion energy of the liquid.

The interfacial tension not only reflects the cohesion energy of the two liquids being in contact but also the interaction between the two different liquids, or rather the interaction between the different molecules. As before, let the energy per molecule in the two bulk phases be as follows:

$$E_A^{bulk} = \frac{z}{2} w_{AA} \tag{16.9a}$$

and:

$$E_B^{bulk} = \frac{z}{2} w_{BB} \tag{16.9b}$$

respectively. The energy per molecule of type A or B at the AB interface is then:

$$E_A^{interface} = \left[\frac{w_{AA}}{2}(m+l) + w_{AB}m \right] z \tag{16.10a}$$

and:

$$E_B^{interface} = \left[\frac{w_{BB}}{2}(m+l) + w_{AB}m \right] z \tag{16.10b}$$

respectively. The energy required to create an interface becomes, expressed per unit area, equal to the interfacial tension, as follows:

$$
\begin{aligned}
\gamma_{AB} &= \frac{1}{a}(E^{interface} - E_A^{bulk} - E_B^{bulk}) \\
&= \frac{z}{a}\left[w_{AB}m + \frac{w_{AA}}{2}(m+l) + \frac{w_{BB}}{2}(m+l) - \frac{w_{AA}}{2}(2m+l) - \frac{w_{BB}}{2}(2m+l) \right] \\
&= \frac{zm}{a}\left(w_{AB} - \frac{w_{AA}}{2} - \frac{w_{BB}}{2} \right) = \frac{zwm}{a} = \frac{m}{a}kT\chi \tag{16.11}
\end{aligned}
$$

showing that the interfacial tension is directly proportional to w, or χ, i.e. the antipathy between the two liquids (cf. Chapter 10). Thus, for liquid pairs with a large antipathy the interfacial tension is large, as, for example, in the water–alkane system. If the liquid pairs are made more compatible, then the interfacial tension will decrease and eventually become zero. This occurs at the critical miscibility point. Note that the treatment given here is only approximate since it assumes no mutual solubility of the two liquids and does not take into account any entropic contribution.

The work of adhesion, W_{AB}, between two materials is defined in terms of the interfacial and surface energies as follows:

$$W_{AB} = \gamma_A + \gamma_B - \gamma_{AB} \tag{16.12}$$

which, using equations (16.8) and (16.11), can be written as:

$$W_{AB} = \frac{zm}{a}\left(-\frac{w_{AA}}{2} - \frac{w_{BB}}{2} - w\right) = \frac{zm}{a}(-w_{AB}) \tag{16.13}$$

Hence, the work of adhesion is proportional to w_{AB} and is a direct measure of the molecular interaction between the two components.

Spreading of one liquid on top of a solid, or another liquid, occurs if there is an energy gain, i.e. if the total surface energy is lower after the spreading. The energy change upon spreading one liquid, A, on top of a second liquid (or solid), B, is given by the following:

$$\begin{aligned}
W_{A/B} &= (\gamma_A + \gamma_{AB}) - \gamma_B \\
&= \frac{zm}{a}\left(-\frac{w_{AA}}{2} + w_{AB} - \frac{w_{AA}}{2} - \frac{w_{BB}}{2} + \frac{w_{BB}}{2}\right) \\
&= \frac{zm}{a}(w_{AB} - w_{AA})
\end{aligned} \tag{16.14}$$

Showing that whether or not a liquid will spread on another liquid, or solid, is a matter of balance between the cohesion forces of the spreading liquid (w_{AA}) and the molecular interaction between the two liquids (w_{AB}). Spreading is promoted by a strong attractive interaction between the two components and is counteracted by a large internal cohesion in the liquid to be spread. Thus, in order to obtain a spreading of a liquid, there should be an attraction between the two liquids and the spreading liquid should have a small cohesive energy density. This is the case, for example, for silicone oils that spread on almost any surface.

Surface Tension and Adsorption can be Understood in Terms of the Regular Solution Theory

We will illustrate here how the regular solution theory can be used to give an understanding of the relationship between surface tension and adsorption of a second component. The regular solution theory is used just for its simplicity. We can perform a similar analysis of a polymer solution by using the Flory–Huggins theory, but part of the simplicity in the following derivation would then be lost. The regular solution theory assumes that (a) the molecules are of the same size, (b) only nearest neighbours ineract, and (c) there is no volume change on mixing two components. As discussed in Chapter 10, when mixing two liquids the change in chemical potential of the liquids, when compared to the chemical potential in the pure liquids, μ_A^0 and μ_B^0, respectively, is given by the following:

$$\frac{\mu_A - \mu_A^0}{RT} = \ln x_A + \chi x_B^2 \tag{16.15a}$$

and:

$$\frac{\mu_B - \mu_B^0}{RT} - \ln x_B + \chi x_A^2 \tag{16.15b}$$

respectively. These relationships are derived in Chapter 10. The χ-parameter is related to the molecular interactions through:

$$\chi = \frac{wz}{kT}(2m+1) = \frac{wz}{kT} \tag{16.16}$$

The surface is considered as a separate phase, being in equilibrium with the bulk solution. The thickness of the surface phase is here only one lattice site. The chemical potential of the surface phase is then given by the following:

$$\frac{\mu_A^s - \mu_A^{s,0}}{RT} = \ln x_A^s + \chi^s(x_B^s)^2 - \frac{\gamma a^0}{kT} \tag{16.17a}$$

and:

$$\frac{\mu_B^s - \mu_B^{s,0}}{RT} = \ln x_B^s + \chi^s(x_A^s)^2 - \frac{\gamma a^0}{kT} \tag{16.17b}$$

where γ is the surface tension of the solution. The standard state is taken as the surface phase of the pure substances. The constant a^0 represents the molecular cross-sectional area and can be estimated from the following:

$$a^0 = \left(\frac{V}{N_A}\right)^{2/3} \tag{16.18}$$

where V is the molar volume and N_A is the Avogadro number. The χ^s parameter differs from that in the bulk since there is one lattice layer missing. Hence, at the surface, the χ^s-parameter is related to the molecular interaction through the following:

$$\chi^s = \frac{wz}{kT}(m+l) = \chi(m+l) \tag{16.19}$$

In a hexagonal lattice we have $\chi^s = 0.75\chi$.

At equilibrium between the solution and surface phases, the chemical potentials of the phases are the same, giving:

$$\frac{\gamma a^0}{kT} = \frac{\mu_A^{s,0} - \mu_A^0}{RT} + \ln \frac{x_A^s}{x_A} + \chi\left[(m+l)(x_B^s)^2 - x_B^2\right] \tag{16.20a}$$

and:

$$\frac{\gamma a^0}{kT} = \frac{\mu_B^{s,0} - \mu_B^0}{RT} + \ln \frac{x_B^s}{x_B} + \chi\left[(m+l)(x_A^s)^2 - x_A^2\right] \tag{16.20b}$$

For the pure components, equations (16.20a,b) turn into the following:

$$\mu_A^{s,0} - \mu_A^0 = \gamma_A a^0 N_A \tag{16.21a}$$

and:

$$\mu_B^{s,0} - \mu_B^0 = \gamma_B a^0 N_A \tag{16.21b}$$

where γ_A and γ_B are the surface tensions of the pure liquids. We have thus identified the surface tension of a pure liquid as the difference in chemical potential between the surface and bulk phases, normalized with the cross-sectional area of the molecules.

Insertion of equations (16.21a,b) into equations (16.20a,b) gives the following:

$$\frac{\gamma a^0}{kT} = \frac{\gamma_A a^0}{kT} + \ln \frac{x_A^s}{x_A} + \chi\left[(m+l)(x_B^s)^2 - x_B^2\right] \tag{16.22a}$$

and:

$$\frac{\gamma a^0}{kT} = \frac{\gamma_B a^0}{kT} + \ln \frac{x_B^s}{x_B} + \chi\left[(m+l)(x_A^s)^2 - x_A^2\right] \tag{16.22b}$$

Inspection of equations (16.22a,b) reveals that a change in the surface tension does not necessarily imply that the surface has changed. It could also be a reflection of a change in the bulk solution. Aqueous mixtures of inorganic electrolytes are an example of this. Here, the electrolyte has no affinity for the surface and hence is accumulated in the bulk solution, thus decreasing the concentration of the solvent, thus causing the surface tension to increase according to equations (16.22a,b). Similarly, if the surface tension is not changed upon the addition of a second component, it cannot be concluded that the solute does not adsorb at the surface. Equations (16.22a,b) show that the addition of a second component might well lead to a situation where the surface tension is not changed, with one requirement being that the solute is roughly equally partitioned between the surface and the bulk, i.e. $x_B^s/x_B \approx 1$.

The solution surface tension can easily be calculated for athermal solutions, i.e. when $\chi = 0$. Using equations (16.22a,b), we find after some algebraic exercises that:

$$\exp\left(\frac{\gamma a^0}{kT}\right) = x_A\exp\left(-\frac{\gamma_A a^0}{kT}\right) + x_B\exp\left(-\frac{\gamma_B a^0}{kT}\right) \tag{16.23a}$$

and:

$$\exp\left(\frac{\gamma a^0}{kT}\right) = x_A^s\exp\left(\frac{\gamma_A a^0}{kT}\right) + x_B^s\exp\left(\frac{\gamma_B a^0}{kT}\right) \tag{16.23b}$$

The algebraic forms of equations (16.23a,b) are identical to equations (5.5) and (5.3), respectively, for mixed micelles, in Chapter 5. Here the quantity $\exp(\gamma a^0/kT)$ corresponds to the CMC. This is not a surprising analogy since the solution surface tension is a reflection of the surface composition in an analogous way as the CMC of a mixed surfactant system is a reflection of the micellar composition. Equations (16.23a,b) reveal that it is the preferential adsorption of the component with the lower surface tension that is the reason why the surface tension of a solution is always lower than the arithmetic mean of the surface tensions of the pure components.

Combining equations (16.22a) and (16.22b), and eliminating the solution surface tension, we find an equation for the adsorption isotherm, i.e. a relationship between the surface composition and the solution composition:

$$\ln\left(\frac{x_B^s}{1-x_B^s}\right) - \ln\left(\frac{x_B}{1-x_B}\right) = \frac{a^0(\gamma_A - \gamma_B)}{kT} + \chi[(1-2x_B) - (m+l)(1-2x_B^s)] \tag{16.24}$$

If the net molecular interaction between the solvent and solute is ignored, i.e. $\chi = \chi^s = 0$, the equation can be written as follows:

$$\frac{x_B^s}{1-x_B^s} = \frac{x_B}{1-x_B}\exp\left[\frac{a^0(\gamma_A - \gamma_B)}{kT}\right] \tag{16.25}$$

which at low solute concentrations ($x_B \ll 1$), is identified as the Langmuir equation (equation (17.5) in Chapter 17), namely:

$$\frac{x_B^s}{1 - x_B^s} = x_B \exp\left[\frac{a^0(\gamma_A - \gamma_B)}{kT}\right] \qquad (16.26)$$

where the Langmuir constant, K, is identified through the following:

$$K = \exp\left[\frac{a^0(\gamma_A - \gamma_B)}{kT}\right] \qquad (16.27)$$

Bibliography

Adamson, A. W., *Physical Chemistry of Surfaces*, 5th Edn, Wiley, New York, 1990.

Aveyard, R. and D. A. Haydon, *An Introduction to the Principles of Surface Chemistry*, Cambridge University Press, Cambridge, UK, 1973.

Defay, R., I. Prigogine, A. Bellemans and D. H. Everett, *Surface Tension and Adsorption*, Longmans, London, 1966.

17 ADSORPTION OF SURFACTANTS AT SOLID SURFACES

Surfactant Adsorption is Governed both by the Nature of the Surfactant and the Surface

Adsorption of surfactants at solid surfaces is important in many industrial applications, e.g. the dispersion of solids in aqueous solution, detergency, solubilization of chemicals, etc. Many of these applications involve the dispersion of solids in water. It is illuminating to calculate that in one litre of water-borne paint there is *ca.* 15.000 m^2 surface area available (from latex and pigment particles) for the surfactants to adsorb at, as already mentioned in Chapter 1. Hence, surfactant adsorption is one very important parameter that determines the properties of such systems.

Surfactant adsorption is determined by two main factors. The first is the interaction of the surfactant with the surface and the second is the hydrophobicity of the surfactant, giving rise to what is called the hydrophobic effect. This latter driving force is of course closely related to the surfactant structure and hence the solubility characteristics of the surfactants in water. It has been found to be the dominating force in surfactant adsorption in many instances. For example, the interaction with the surface only plays a minor role on hydrophobic surfaces. The reason is that on hydrophobic surfaces the surfactants adsorb with their hydrophobic hydrocarbon moiety in contact with the surface and their hydrophilic moiety in contact with the solution, as shown in Figure 17.1(a). This configuration resembles that of micelles in the sense that the hydrophobic moiety is transferred from the aqueous environment to a hydrophobic environment upon adsorption. Indeed, it is found that the adsorption free energy of surfactants at a hydrophobic surface is very close to the micellization free energy of the surfactant.

At very polar surfaces, on the other hand, the surfactants (at low solution concentrations) adsorb with their polar moiety in contact with the surface due to the interaction between the surfactant head group and the surface, as shown in Figure 17.1(b). At higher surfactant concentrations, two different structures

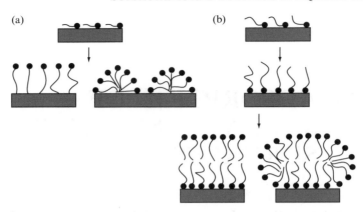

Figure 17.1 Surfactants adsorb (a) on non-polar surfaces with their hydrocarbon parts in contact with the surface, as well as (b) on polar surfaces with their polar parts in contact with the surface

at the surface are possible. If there is a strong attraction between the surfactant head group and the surface, a monolayer is formed, where the surfactant head groups are in contact with the surface and the hydrocarbon moieties are in contact with the solution. This adsorption structure will create a hydrophobic surface, which in turn will adsorb further surfactants with the same configuration as described above for hydrophobic surfaces, i.e. a surfactant bilayer is formed at higher surfactant concentrations. This occurs, for example, in the adsorption of charged surfactants at oppositely charged surfaces. In addition, a chemical reaction between the surfactant head group and the surface can create a hydrophobic monolayer. An example of a non-surfactant system is the adsorption of alkylxantogenates on sulfide minerals, which is used technically in the flotation of such minerals.

If, however, the attraction between the surfactant head groups and the surface is intermediate in strength, then micelles, or other surfactant aggregates, will form at the surface at higher surfactant concentrations. This is because the attraction between the hydrophobic hydrocarbon moieties is stronger than the interaction of the surfactant head groups with the surface. Surfactant aggregation at surfaces is thus a matter of a balance between the interaction of the surfactant head group with the surface and the interaction between the hydrophobic moieties of the surfactants.

Since the critical packing parameter (CPP) is a reflection of the balance of the interactions between the hydrophobic moieties and the polar part we realize that the critical packing parameter plays an important role in surfactant adsorption, regardless of whether the surface is hydrophobic or hydrophilic. It will be illustrated in this chapter that the adsorption increases as the CPP increases, i.e. as the surfactants are able to pack tighter at the surface with a larger energy gain. Since surfactants in aqueous solution normally are used under conditions where the critical packing parameter is small we will list below those conditions which increase the CPP and hence also the adsorption.

Increase of the critical packing parameter of a system with a single straight-chain ionic surfactant can be accomplished in the following ways:

(1) Change of a surfactant to one with a longer hydrocarbon chain.

(2) Change of a surfactant to one with a branched hydrocarbon chain.

(3) Use of surfactants with two hydrocarbon chains.

(4) Addition of a long-chain alcohol, amine or other hydrophobic amphiphilic molecule.

(5) Addition of a hydrophobic non-ionic surfactant.

(6) Addition of a small amount of surfactant with opposite charge.

(7) Addition of salt.

Point (1) above is known as Traube's rule, which states that 'the adsorption of organic substances from aqueous solutions increases strongly and regularly as we ascend the homologous series'.

Increase of the critical packing parameter of a system with a single straight-chain non-ionic surfactant can be achieved in the following ways:

(1) Change of a surfactant to one with a longer hydrocarbon chain.

(2) Change of a surfactant to one with a branched hydrocarbon chain.

(3) Use of surfactants with two hydrocarbon chains.

(4) Change of a surfactant to one with a shorter polyoxyethylene chain.

(5) Increase of temperature.

(6) Addition of salt.

Furthermore, a decrease in the critical packing parameter of non-ionic surfactants can be accomplished by adding ionic surfactants to the system.

Model Surfaces and Methods to Determine Adsorption

Dispersed Systems

Latex particles are very common as model surfaces for studying surfactant adsorption. Most often, lattices of polystyrene are used, but other polymers, such as poly(methyl methacrylate) (PMMA) have been used as well. The lattices are synthesized without any surfactant and the colloidal stability is achieved by the presence of charged initiator residues, located at the ends of the polymer chains. These charges give sufficient colloidal stability provided that the solids content is kept low, preferably in the range 5–10%. Under certain conditions, monodisperse lattices are obtained. The size of the latex particles is normally in the range 0.1–0.4 μm, giving a specific surface area of *ca.* 15–60 m²/ g. The monodispersity and the large specific surface area make these latex dispersions ideal for adsorption studies.

In dispersed systems, the method of determining surfactant adsorption is almost exclusively by adding surfactant, leaving time for the system to come to equilibrium, separating the solids and finally determining the surfactant concentration in the solution. The concentration depletion gives the adsorption (mg/m²) from the following equation:

$$\Gamma = \frac{(C_0 - C)V}{ma_{sp}} \tag{17.1}$$

where C and C_0 are the equilibrium concentration (mg/ml) and the concentration before adsorption, respectively, V is the solution volume (ml), m is the particle amount (g) and a_{sp} is the particle specific surface area (m²/g). The concentration determination can be accomplished by, for example, ion-selective electrodes, UV/vis spectroscopy, refractive index, titration, chromatography or surface tension. In fact, if the latter is used the solids need not be separated since the particles do not contribute to the surface tension, provided that they are completely wetted by the liquid. Hence, the dispersed system can be titrated with the surfactant to be investigated. This is called soap titration. We will just note that the accuracy in adsorption isotherm determination depends on the method of surfactant analysis. Low accuracy is, for example, obtained using ion-selective electrodes since the determined potential is proportional to the logarithm of the surfactant concentration, i.e. small changes in the potential correspond to large changes in surface concentration.

Some systems are suitable for the serum replacement technique. Here, the dispersion is confined in a flow cell with stirring. The confinement is achieved by a filter. The surfactant solution is allowed to flow slowly through the cell and its concentration is determined at the outlet. A schematic of such a cell is shown in Figure 17.2. Figures 17.3(a) and 17.3(b) show respectively, a surfactant concentration profile and the corresponding isotherm. Important parameters for such a set-up are the flow-through speed and stirring rate (too high a stirring rate might cause coagulation, while too low a stirring rate will cause the filter to clog).

The residence time of a surfactant in the cell should be of the order of at least one hour in order to ensure that equilibrium is attained. The adsorption isotherm is calculated from a mass balance, i.e. for the *n*th batch we have the following:

surfactant solution

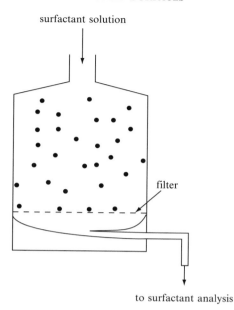

filter

to surfactant analysis

Figure 17.2 Schematic of the serum replacement cell for determination of surfactant adsorption isotherms on suspended particles

$$\Gamma_n = \frac{1}{A}\left(\sum_{i=1}^{n}\Delta V_i(C_0 - C_i) - C_{n,c}V_c\right) \qquad (17.2)$$

where A is the total surface area of the particles in the cell, ΔV_i is the volume of batch i, C_0 is the surfactant concentration in the inlet, C_i is the concentration in batch i, while $C_{n,c}$ is the surfactant concentration in the cell that has a volume V_c. The concentration in the cell is obtained from the plot of Figure 17.3(a).

Macroscopic Surfaces

In recent years, ellipsometry has proved to be a very powerful method to determine the adsorption of surfactants from solution on macroscopic surfaces. In brief, elliptically polarized light is reflected on a surface and the change in polarization is measured. This change is dependent on the presence of adsorbed molecules at the surface and hence the presence of these is detected and quantified. The detection limit is about $0.1\,\text{mg/m}^2$, which means that adsorption well below monolayer coverage can be detected. The method also allows the determination of the adsorbed layer thickness as well as the refractive index of the adsorbed layer. A great advantage is that detection is performed *in situ*,

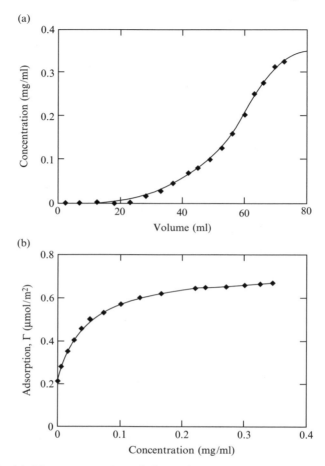

Figure 17.3 (a) The concentration of the surfactant solution in the effluent stream versus the elution volume from the serum replacement cell, and (b) the corresponding adsorption isotherm. Reproduced by permission of Academic Press from K. Steinby, R Silveston and B. Kronberg, *J. Colloid Interface Sci.*, **155** (1993) 70

i.e. when the surface is in contact with the solution, so allowing the adsorption/desorption kinetics to be determined.

Analysis of Surfactant Adsorption is Frequently Carried out in Terms of the Langmuir Equation

It is often desirable to analyse surfactant adsorption in terms of a theoretical model in order to obtain a molecular interpretation. The parameters from such an analysis can later be used to compare the adsorption behaviour of different surfactants and also to predict the adsorption in new systems. An overwhelming majority of the analysis of surfactant adsorption is performed in terms of the Langmuir equation. This equation is interpreted under the following assumptions:

(1) The surface is homogeneous.

(2) The surfactants adsorb in only one monolayer.

(3) There are no surfactant–solvent or surfactant–surfactant interactions.

(4) The surfactant and solvent molecules have equal cross-sectional surface areas.

The first two assumptions are quite reasonable but not the last two. It has been shown, however, that taking both the interaction and the cross-sectional surface area into account gives deviations, from the Langmuir equation, that are opposite to each other. Thus, the good fit using the Langmuir equation for the adsorption isotherms of surfactants is fortuitous.

The Langmuir equation can be derived in the following simple way. Let the adsorption rate be equal to:

$$\text{Rate of } adsorption = k_a C(1 - \Theta) \tag{17.3}$$

where C is the equilibrium surfactant concentration in solution, Θ is the fraction of the surface that is covered with surfactants, and k_a is a rate constant. Similarly, the desorption rate is written as follows:

$$\text{Rate of } desorption = k_d \Theta \tag{17.4}$$

where k_d is a rate constant. At equilibrium, the adsorption rate and the desorption rate are the same and hence we have the following:

$$\frac{\Theta}{1 - \Theta} = KC \tag{17.5}$$

or:

$$\Theta = \frac{KC}{1 + KC} \tag{17.6}$$

where K is the equilibrium constant ($= k_a/k_d$). Equations (17.5) and (17.6) together are termed the Langmuir equation(s). The equilibrium constant, K, describes the partitioning of the surfactant between the surface phase and the solution phase. This is realized if the Langmuir equation is applied at the limit of infinitely low solution concentrations where:

$$K = \left(\frac{\Theta}{C}\right)_{c\to 0} \tag{17.7}$$

Hence, a low K-value indicates a weak adsorption, while a large K-value is indicative of a strong adsorption. The adsorption free energy, ΔG_{ads}, is related to the K-value through the following:

$$\Delta G_{ads} = 2RT \ln (K) \tag{17.8}$$

The fraction of the surface that is covered with surfactant, Θ, is obtained by assuming that the surfactants adsorb in a monolayer and that a full coverage is obtained at high equilibrium concentrations (above the CMC) in the bulk liquid. Thus, the fraction of covered surface can be written as follows:

$$\Theta = \frac{\Gamma}{\Gamma_{max}} \tag{17.9}$$

where Γ is the adsorbed amount and Γ_{max} is the adsorption at full coverage. This equation can now be combined with equation (17.5) to give the following:

$$\frac{1}{\Gamma} = \frac{1}{\Gamma_{max}} + \left(\frac{1}{K\Gamma_{max}}\right)\frac{1}{C} \tag{17.10}$$

A plot of $1/\Gamma$ versus $1/\Theta$ gives us both the Γ_{max} and the K-value from the intercept and the slope, respectively. Thus, both the Γ_{max}-value and the K-value, and hence the adsorption free energy, ΔG_{ads}, can easily be obtained from an adsorption isotherm.

An alternative to the derivation above is to consider the adsorption of surfactants at a solid surface as a partitioning between two phases, i.e. the surface phase and the bulk phase. We thus have equilibrium of both surfactant and water molecules between the two phases. Hence:

$$\text{Surfactant in solution } + \text{ water at surface}$$
$$= \text{ surfactant at surface } + \text{ water in solution} \tag{17.11}$$

For the above equilibrium we have the following expression for the equilibrium constant, $K^{\bullet}$:

$$K^\bullet = \frac{[\text{Surfactant}]_{\text{surface}} (\text{Water})_{\text{solution}}}{(\text{Surfactant})_{\text{solution}} (\text{Water})_{\text{surface}}} \qquad (17.12)$$

Let the surfactant and water concentration at the surface be Θ and $1 - \Theta$, respectively, where Θ is the surface coverage, i.e. $\Theta = \Gamma/\Gamma_m$. The surfactant concentration in the bulk is C, while the water concentration in the bulk is taken to be constant in view of the fact that the surfactant concentration is very low.

Hence we have:

$$K = \frac{\Theta}{C(1 - \Theta)} \qquad (17.13)$$

or:

$$\Theta = \frac{KC}{(1 + KC)} \qquad (17.14)$$

We note that in this derivation we have only assumed that the surface and bulk phases are homogeneous. This is, of course, an approximation for the surface phase and upon a closer consideration this assumption leads to the assumptions 1–4 listed above.

The assumption that there is no surfactant–solvent interaction can be relaxed to assuming that the surfactant solvent interaction is the same in the surface phase as in the bulk phase. This assumption is implicit in the above derivation by the use of concentrations instead of activities.

Surfactants Adsorb on Hydrophobic Surfaces

Ionic Surfactants

Illustrations of some of the points above regarding the CPP are given in the following figures. Figure 17.4 shows the adsorption of a homologous series of fatty acids on carbon from aqueous solution, illustrating that the adsorption increases as the CPP of a surfactant increases.

Figure 17.5(a) shows the adsorption of sodium dodecyl sulfate (SDS) on polystyrene latex, illustrating that the adsorption increases as the CPP is raised through salt addition. This addition of salt reduces the electrostatic repulsion between the surfactant molecules at the surface, thus resulting in closer packing and hence a smaller effective cross-sectional area per surfactant head group. The cross-sectional area per surfactant, A, is shown as a function of salt concentration in Figure 17.5(b). The A-value, expressed in Å^2/molecule, is calculated from the maximum adsorbed amount, $\Gamma_{\max}$, through the following relationship:

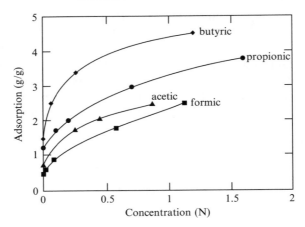

Figure 17.4 Adsorption of fatty acids on carbon from aqueous solutions. Reproduced by permission of John Wiley & Sons, Inc. from A. W. Adamson, *Physical Chemistry of Surfaces*, 5th edn, Wiley, New York 1990

$$A(\overset{\circ}{A}^2/\text{molecule}) = \frac{M}{\Gamma_{max}6.02} \qquad (17.15)$$

where M is the molecular weight of the surfactant and Γ_{max} is expressed in mg/m^2.

We noted earlier that the surfactant adsorption is relatively insensitive to the polarity of the substrate surface. This is illustrated in Figure 17.6 where the area per surfactant molecule at maximum adsorption is plotted versus the polarity of the latex polymer. This figure reveals that in going from the most non-polar latex, PVC, to the most polar, PVAc, the cross-sectional area per surfactant, and hence the adsorbed amount, changes only by a factor of *ca.* 2.

Non-Ionic Surfactants

In addition, for non-ionic surfactants, the adsorption depends on the surfactant solution properties, and hence the CPP. Figure 17.7(a) displays the adsorption data of three non-ionics, viz. nonylphenols with 10, 20 and 50 oxyethylene units, on two lattices, i.e. polystyrene and poly(methyl methacrylate).

First, we note that there is a considerable difference in adsorption between the three surfactants, with the one with the shortest polyoxyethylene chain adsorbing the most. Secondly, we note that there is only a slight difference in adsorption between the two latex surfaces, with the more polar PMMA latex adsorbing somewhat less surfactant. The increase in adsorption of surfactants with shorter polyoxyethylene chain lengths is in line with the increase in the CPP as the polyoxyethylene chains are shortened. Hence, when the CPP is

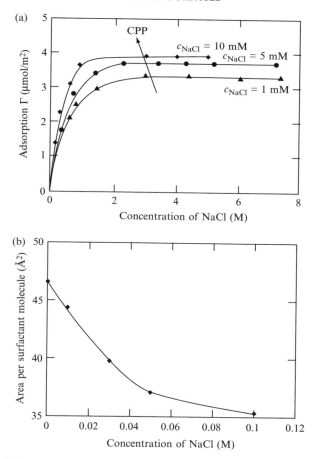

Figure 17.5 Adsorption of SDS at three different salt (NaCl) concentrations illustrating that (a) the adsorption increases with the CPP, and (b) the cross-sectional area per surfactant head group decreases with the CPP. (a) From K. Tajima, *Bull. Chem. Soc. Jpn*, **44** (1971) 1767 Copyright the Chemical Society of Japan. (b) Reproduced by permission of Academic Press from 1. Pirma and S. Chen, *J. Colloid Interface Sci.*, **74** (1980) 90

increased, the surfactant molecules pack closer at a surface. The adsorption reaches a plateau at solution concentrations slightly above the CMC of the surfactants. The small arrows in this figure indicate the CMCs.

Table 17.1 shows the results obtained from a Langmuir analysis of the isotherms, including sodium dodecyl sulfate, (SDS) as a reference. The table reveals that the cross-sectional area per molecule increases as the polyoxyethylene chain length is increased. The Gibbs free energy of adsorption decreases with polyoxyethylene chain length, although all are stronger than that of SDS.

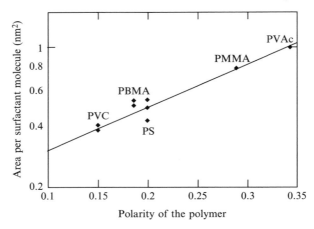

Figure 17.6 The area per surfactant molecule at maximum adsorption versus the polarity of the latex polymer. Reproduced from *Polymer Colloids II*, 1980, R. Fitch (Ed.), 1980, 'Effect of polymer polarity on the adsorption of sodium lauryl sulfate at latex/water interfaces', B. Vijayendran, pp. 209–224, Figure 6, with kind permission of Kluwer Academic Publishers

Table 17.1 Results obtained from a Langmuir analysis[a]

	$A_2(\text{Å}^2/\text{molecule})$		ΔG_{ads} (kJ/mole)		ΔG_{mic} (kJ/mole)
Surfactant	PMMA	PS	PMMA	PS	
SDS	—	52	—	−28.0	−22.6
NP-E10	60	54	−38.6	−38.2	−33.8
NP-E20	134	106	−37.3	−37.0	−31.9
NP-E50	280	200	−34.3	−34.7	−30.0

[a]SDS values taken from B. Kronberg, M. Lindström and P. Stenius, in *Phenomena in Mixed Surfactant Systems*, J. F. Scamehorn (Ed.), ACS Symposium Series 311, 1986, non-ionics on PS from B. Kronberg, L. Käll and P. Stenius, *J. Disp. Sci. Technol.*, **2** (1981) 215, and non-ionics on PMMA from B. Kronberg, P. Stenius and G. Igeborn, *J. Colloid Interface Sci.*, **102** (1984) 418.

This table also includes the Gibbs free energy of micellization, $\Delta\Gamma_{mic}$, as a comparison. The Gibbs free energy of micellization is obtained from the following relationship:

$$\Delta G_{mic} = RT\ln X_{mic} \qquad (17.16)$$

where X_{mic} is the critical micelle concentration expressed in terms of mole fractions. From the table and the figure we draw the following conclusions. There is a large difference in Gibbs free energy of adsorption between NP-E$_n$ surfactants differing in their polyoxyethylene chain lengths. Thus, in surfactant mixtures there will be a chromatographic effect in the sense that the surfactants with

(a)

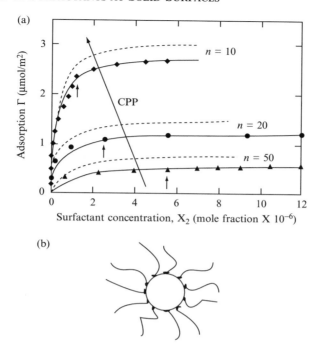

(b)

Figure 17.7 (a) The adsorption isotherms of NP-E$_n$ on polystyrene latex (continuous curves) and poly(methyl methacrylate) latex (dashed curves), and (b) a simple adsorption model of non-ionic surfactants on latexes where the hydrocarbon chains are in contact with the latex surface and the polyoxyethylene chains protrude into the aqueous solution. Reproduced by permission of Academic Press from B. Kronberg, P. Stenius and G. Igeborn, *J. Colloid Interface Sci.*, **102** (1984) 418

the short polyoxyethylene chains will be preferentially adsorbed on hydrophobic surfaces. Hence, when a technical surfactant batch is used, the long polyoxyethylene chain analogues will stay in solution and the short ones will be found at the surfaces in the system. Figure 17.8 shows a calculated example for a 50/50 mixture of NP-E$_{10}$ and NP-E$_{20}$ where both surfactant batches follow a Poisson distribution. This figure shows that the surface is enriched in the short-chain polyoxyethylene species while the solution is enriched in the long-chain species.

Table 17.1 also reveals that the non-ionic surfactants adsorb more strongly than the anionic SDS. This is due to the lack of electrostatic repulsion between the molecules at the surface. As a comparison, the CMCs of non-ionic surfactants are, in general, of the order of 10–100 times smaller than those of ionic surfactants with the same hydrocarbon chain lengths. Later in this chapter, the implications of this difference in adsorption free energies will be discussed, viz. the competitive adsorption between SDS and NP-E$_{10}$.

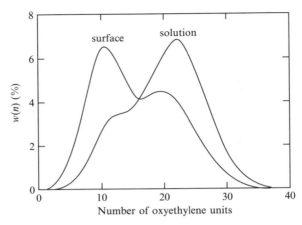

Figure 17.8 Calculated normalized distribution curves of a surfactant mixture (50% NP-E$_{10}$ + 50% NP-E$_{20}$) adsorbed on a latex and in the equilibrium solution

From inspection of the adsorption model depicted in Figure 17.7(b), one can falsely draw the conclusion that the adsorption free energy should be constant, irrespective of polyoxyethylene chain length in view of the fact that the adsorption free energy is due only to the contacts formed between the hydrocarbon moieties of the surfactant and the surface. The fact that the Gibbs free energy of adsorption decreases as the polyoxyethylene chain is increased (Table 17.1) is due to the fact that the surfactants increase in size as the polyoxyethylene chain is increased. This is not taken into account in the analysis when using the Langmuir equation. Taking into account the fact that the surfactant molecules have different sizes, and also taking the molecular interaction into account, it is possible to analyse the adsorption isotherms with a more realistic model. Table 17.2 shows the results obtained from such an analysis for the adsorption of the non-ionic surfactants on polystyrene latex. This table shows two driving forces that determine the adsorption, viz. (a) the interaction of the surfactant

Table 17.2 The two driving forces for the adsorption of non-ionic surfactants on a polystyrene latex (in units of kT). Reproduced by permission of Academic Press from B. Kronberg, *J. Colloid Interface Sci.*, **96** (1983) 55

Surfactant	Surface–surfactant interaction	Surfactant–surfactant interaction
NP-E$_{10}$	1.6	5 ± 2
NP-E$_{20}$	1.6	7 ± 2
NP-E$_{50}$	1.6	6 ± 2

hydrocarbon moiety with the surface, and (b) the interaction of the surfactant hydrocarbon moieties with one another. Strictly speaking, it is the change in interaction that determines the adsorption, i.e. the first contribution involves the exchange of surface–water molecular contacts to surface–hydrocarbon moiety molecular contacts. The second contribution involves the exchange of contacts between water and surfactant hydrocarbon moiety to contacts between surfactant hydrocarbon moieties. The data in table 17.2 clearly show that it is the latter contribution that dominates the adsorption of non-ionic surfactants. This contribution is akin to the micellization process and hence it is not surprising that the Gibbs free energies of adsorption (see Table 17.1) are very close to the micellization free energy.

Non-ionic surfactants are known to have strong temperature dependencies with respect to their solution properties in aqueous systems (see Chapter 4). Hence, one expects also a strong temperature dependence of the adsorption on surfaces. Figure 17.9(a) shows the temperature dependence of the adsorption of NP-E$_{20}$ on a PMMA latex. Here, it is shown that the adsorption increases strongly with temperature. This is in accordance with the fact that the polyoxyethylene chain coils up, i.e. decreases in size, with temperature, thus giving a larger CPP value. Figure 17.9(b) shows that the cross-sectional area per surfactant molecule decreases continuously with temperature.

Further insight into the adsorption mechanisms of non-ionic surfactants on latex surfaces can be achieved through colloidal stability measurements. It is known that non-ionic surfactants give rise to a steric stabilization through the polyoxyethylene chains that protrude into the aqueous solution when adsorbed on a surface. Colloidal stability can be measured through freezing the sample with subsequent thawing (a freeze–thaw cycle) or by mechanically shearing the suspension. Alternatively, the stability upon adding salts to the suspension can be evaluated. The amount of coagulate in any of these treatments is a measure of the colloidal instability. Figure 17.10 displays the mechanical, freeze–thaw and salt stability of a latex to which has been added either SDS or NP-E$_{10}$.

This figure shows that SDS provides good mechanical stability while it does not enhance the stability with respect to the addition of salts or subjecting the suspension to a freeze–thaw cycle. The non-ionic surfactant NP-E$_{10}$, on the other hand, gives good stability if present at high enough concentrations in all three treatments. Mechanical stability, for example, is achieved at surface concentrations above ca. 0.6–0.8 mg/m^2. This value corresponds to a surfactant monolayer where the surfactant molecules are lying down flat on the surface. Above this surfactant concentration, the polyoxyethylene chains start to protrude into the aqueous solution. This will produce stability against coagulation by mechanical stirring. Freeze–thaw stability is not obtained until the surface concentration is ca. 2–2.5 mg/m^2, corresponding to a complete surfactant monolayer on the surface where the surfactants are standing up with

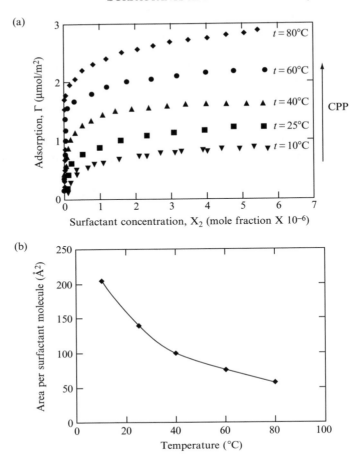

Figure 17.9 The temperature dependence of (a) the adsorption of NP-E$_{20}$ on a PMMA latex, and (b) the cross-sectional area per surfactant molecule. Reproduced by permission of Academic Press from K. Steinby, R. Silveston and B. Kronberg, *J. Colloid Interface Sci.*, **155** (1993) 70

their hydrocarbon moietys in contact with the surface and the polyoxyethylene chains are protruding into the solution.

Surfactants Adsorb on Hydrophilic Surfaces

Ionic Surfactants

At very low surfactant concentrations, ionic surfactants adsorb on charged surfaces almost exclusively by an ion-exchange mechanism. Thus, the counter-ions in the diffuse double layer just outside the surface are exchanged for

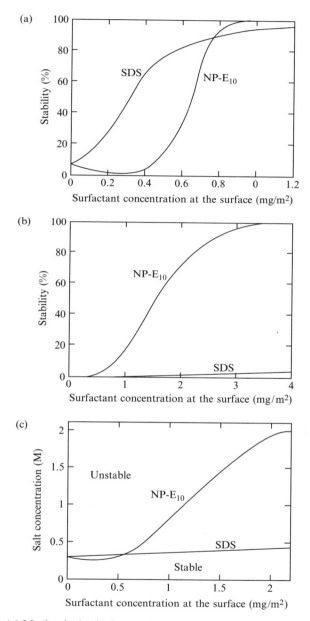

Figure 17.10 (a) Mechanical-, (b) freeze–thaw, and (c) salt stability data obtained for a latex to which has been added either SDS or NP-E$_{10}$

surfactants with the same charge. This ion exchange leads to a higher surfactant concentration close to the surface when compared to the surfactant concentration in the bulk, which in turn induces a micellization process at the surface at solution concentrations far below the bulk critical micelle concentration. Whether or not these surface micelles have a spherical or a semi-spherical shape (hemi-micelles) is still being debated. There is, however, no doubt that there is a surface aggregation at concentrations well below the CMC of the surfactant. At higher concentrations, the surfactants will form a double layer at the surface. The lower part of the double layer need not be a tightly packed layer. This double layer is completed at the CMC of the surfactant. The adsorption as a function of surfactant concentration is shown schematically in Figures 17.11(a) and 17.11(b), with an example being given in Figure 17.11(c). Note that both the x- and y-axies have a logarithmic scale in order to enhance the features at low surfactant concentrations.

Since the aggregation on the surface is similar to micellization one would expect the adsorption to be very dependent on the surfactant alkyl chain length. This is indeed found to be the case, as shown in Figure 17.12. Here, those compounds with the longest alkyl chain adsorb the most. They also show the greatest co-operativity, as revealed by the distinct step in the adsorption. This is again an example of the fact that the adsorption is dependent mainly on the surfactant structure and only marginally on the interaction with the surface. The only requirement for adsorption is that the surfactant in question should have a slight affinity for the surface such that the surfactant concentration at the surface is enhanced. This will further promote aggregate formation, or micellization, at the surface.

In this chapter, as well as in Chapter 19, the adsorption behaviours of surfactants and polymers on kaolin are used as illustrative examples of adsorption. Kaolin is a clay which consists of plate-like particles and where the two faces differ chemically from each other (one is an aluminium oxide surface, while the other is a silica surface), as well as from the edges. At neutral pH, the faces are negatively charged, while the edges are positively charged. Lowering the pH decreases the net negative charge of the clay particles and at a pH of $ca.$ 2 the kaolin particles carry no net charge (point of zero charge).

Figure 17.13(a) shows the adsorption of dodecylpyridinium chloride on kaolin as a function of surfactant concentration at three salt concentrations. This figure reveals that at concentrations below the point at which charge reversal of the particle surface is reached, the adsorption decreases with salt concentration, while at concentrations above charge reversal the adsorption increases with salt concentration. The origin of the latter is the shielding effect of the salt, which reduces the repulsion between the surfactant molecules on the surface, therefore facilitating adsorption. The origin of a decrease in adsorption with increasing salt concentration at concentrations below the point of zero charge is to be found in the competition between the surfactant ions and the salt

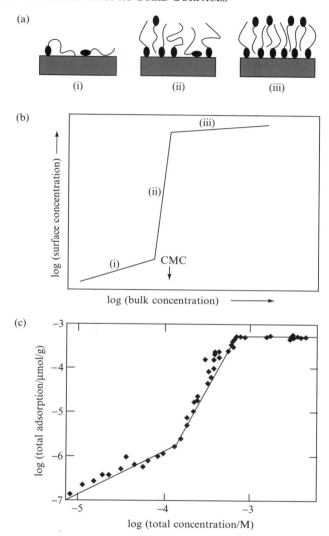

Figure 17.11 (a, b) Schematic representations of the adsorption of an ionic surfactant on a charged surface as a function of surfactant concentration, and (c) an example showing the adsorption of sodium decylbenzene sulfate on alumina. Reproduced by permission of Academic Press from J. Scamehorn, R. Schechter and W. Wade, *J. Colloid Interface Sci.*, **85** (1982) 463

ions for the sites at the surface. Another point of view is that the driving force is the release of bound counterions when the surfactants adsorb. This driving force is quenched as salt is added. This mechanism is also discussed in Chapter 19 for the adsorption of polyelectrolytes at oppositely charged surfaces.

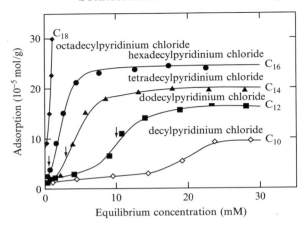

Figure 17.12 Adsorption of a number of cationic substances on silica, at pH 4, illustrating that those compounds with the longest alkyl chains show the greatest co-operativity

The adsorption of ionic surfactants on hydrophilic surfaces is almost completely independent of temperature, as exemplified in Figure 17.13(b). Here the adsorption changes only by 10–20% over a temperature range of 50°C. The maximum in adsorption at around 25°C coincides with a minimum in the CMC of the surfactant at the same temperature.

Non-ionic Surfactants

The adsorption of non-ionic surfactants on hydrophilic surfaces is dictated by the interaction between the surface and the polyoxyethylene chain. If there is an interaction present, the adsorption behaviour will be similar to the adsorption of ionic surfactants on hydrophilic surfaces. Figure 17.14 displays schematically the adsorption of a non-ionic surfactant on a silica surface. In addition, here there is a surface aggregation at concentrations well below the CMC of the surfactant. The concentration at which this surface aggregation starts to appear is called the critical surface aggregation concentration (CSAC) and is of the order of one tenth of the CMC of the surfactant. Note that the y-axis is linear and hence the isotherm is of a different shape when compared to that in Figure 17.11.

Adsorption isotherms with a shape such as that in Figure 17.14 indicate a high degree of co-operativity above the CSAC. The adsorption can thus be described as a surface-induced self-assembly. We note that a very small surfactant–surface interaction is sufficient to induce a surface self-assembly. This is completely analogous to polymer-induced self-assembly in solution, as discussed earlier in Chapter 13.

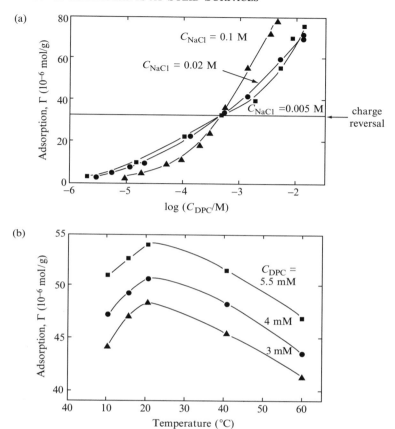

Figure 17.13 (a) Adsorption of dodecylpyridinium chloride (DPC) on kaolin at three different ionic strengths, and (b) the temperature dependence of the adsorption at three concentrations beyond the isoelectric point. Reprinted with permission from T. Mehrian, A. de Keizer and J. Lyklema, *Langmuir*, **7** (1991) 3094. Copyright (1991) American Chemical Society

Figures 17.15(a) and 17.15(b) show the adsorption behaviours of some non-ionic surfactants of type C_mE_n on silica. We first note that the adsorbed amount abruptly increases well below the CMCs, which are indicated by small arrows. We also note that plateau adsorption is reached at concentrations of around the CMCs of the surfactants. Finally, we note that the step-like isotherms are especially pronounced for those surfactants with a low polyoxyethylene content, thus indicating a high degree of co-operativity. Analysis of the results shown in Figure 17.15 clearly indicates that the surfactants form discrete structures, like micelles, at the interface. The size of these aggregates increases with increasing hydrocarbon chain length of the surfactant.

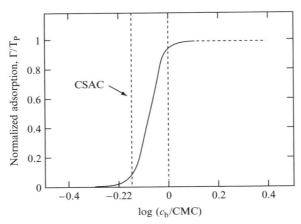

Figure 17.14 Schematic illustration of the adsorption of non-ionic surfactants on a silica surface, where c_b is the surfactant concentration in solution and the left-hand dashed vertical line indicates the critical surface aggregation concentration (CSAC). From F. Tiberg, Ph.D. Thesis, Lund University, Sweden, 1994

Figure 17.16 shows the adsorption of two alkylphenol-based surfactants on silica, i.e. OP-E$_{9.5}$ and NP-E$_{9.5}$. Both surfactants follow the same curve when plotted on a reduced scale. Note that the features are the same as those shown above in Figures 17.14 and 17.15. It has been demonstrated by fluorescence measurements that the aggregates formed at the surface are separate, i.e. a monolayer is not formed at the surface at low concentrations. Rather, the surface is covered with separate aggregates with a slow exchange rate of the fluorescent probe between the aggregates.

Figure 17.17 shows the adsorption of non-ionic surfactants (nonylphenol ethoxylates) on kaolin. This figure shows that the adsorption decreases with increasing chain length of the polyoxyethylene chain, i.e. as the CPP decreases. This is an illustration of the fact that surfactant aggregation at surfaces is more pronounced for small polyoxyethylene chain surfactants, again showing that the adsorption of surfactants is dependent on the solution properties, i.e. the adsorption increases as the CPP of the surfactant is increased.

Finally, Figure 17.18 illustrates the fact that the interaction between the surfactant and the surface is important in order for adsorption to take place. This figure shows that for the adsorption of two non-ionic surfactants on silica the adsorption decreases to zero at pH levels of *ca.* 10 and above. Note that for the more hydrophobic surfactant, i.e. C$_{12}$EO$_5$, the pH dependence is very strong at pH levels above pH 9. This indicates that only a small interaction between the surfactant and the surface is sufficient to adsorb hydrophobic surfactants. Once a few molecules are adsorbed on the surface, they form the

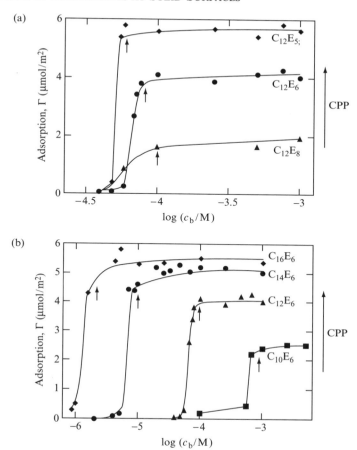

Figure 17.15 Adsorption of various poly(ethylene glycol) monoalkyl ethers, C_mE_n, on silica at 25°C showing an increased adsorption as the CPP increases; c_b is the surfactant concentration in solution. Reprinted from *Thin Solid Films*, **234**, F. Tiberg, B. Lindman and M. Landgren, 'Interfacial behaviour of non-ionic surfactants at the silica–water interface revealed by ellipsometry', 478–481, Copyright (1993), with permission from Elsevier Science

nuclei for further adsorption and surface aggregation. This effect is less pronounced for the more hydrophilic surfactant, i.e. $C_{12}EO_8$, as expected.

The results shown in Figure 17.18 can be interpreted in two ways. First, in these alkaline solutions the surface hydroxyl groups are ionized and cannot form hydrogen bonds with the surfactant and hence no adsorption takes place. Secondly, hydroxyl ions from solution compete for the adsorption sites and at pH levels above *ca.* 10 all surfactants are displaced from the surface.

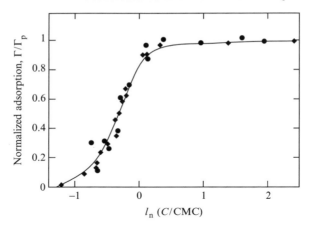

Figure 17.16   Adsorption of an octylphenol ethoxylate, OP-EO$_{9.5}$ (●) and a nonylphenol ethoxylate, NP-EO$_{9.5}$ (♦) on silica. Reprinted with permission from P. Levitz, H. Van Damme and D. Keravis, *J. Phys. Chem.*, **88** (1984) 2228. Copyright (1984) American Chemical Society

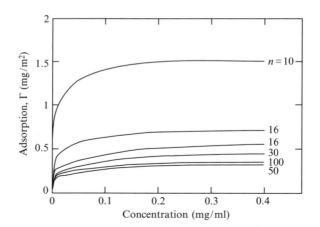

Figure 17.17   Adsorption of non-ionic surfactants (nonylphenol ethoxylates) on kaolin showing that the adsorption decreases with increasing chain length of the polyoxyethylene chain, n, i.e. as the CPP increases. Reproduced by permission of TAPPi Press from P. Stenius, J. Kuortti and B. Kronberg, *Tappi J.*, **67** (1984) 56

Competitive Adsorption is a Common Phenomenon

Technical use of surfactants often involves more than one surfactant species and hence a knowledge of the competitive adsorption between two surfactants is of interest. We will give here two illustrative examples of competitive adsorption.

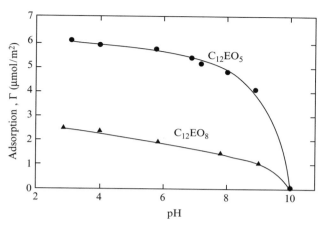

Figure 17.18 Adsorption of two non-ionic surfactants on silica as a function of pH. From F. Tiberg, Ph.D. Thesis, Lund University. Sweden, 1994

Anionic plus Cationic Surfactants

Intuitively, one would predict that a mixture of an anionic and a cationic surfactant would precipitate due to the electrostatic attraction between the species, thus forming an insoluble complex. This assumption is correct, but only at higher surfactant concentrations. At lower surfactant concentrations, the two surfactants can be present simultaneously in a system, especially if the surfactants have shorter hydrocarbon chains. The adsorption of a mixture of an anionic and a cationic surfactant will naturally be driven in such a way that the surface will contain an equivalent amount of each surfactant. We will illustrate that the relative hydrophobicity of the surfactants is also of great importance.

Figure 17.19 shows the zeta potential of a latex in the presence of mixtures of anionic, C_n^- and cationic, C_n^+, surfactants, differing in their chain length, n. The zeta potentials are obtained at the CMC of the surfactant mixture, thus ensuring that the surface is saturated with the surfactant mixture. The mixture of C_8^- and C_8^+ surfactants shows the expected behaviour, i.e. a zeta potential of zero at a surfactant ratio of 50/50. Excess of one or the other of the surfactants gives a net charge and hence a net zeta potential.

The mixtures of C_8^- and C_{12}^+ or C_{12}^- and C_8^+ surfactants show a different behaviour, however. Here, the surfactant composition leading to a zero zeta potential is in the order of 90/10, where the more hydrophilic surfactant is in excess. This is an illustration of the fact that the hydrophobicity is a very strong driving force for surfactant adsorption, irrespective of the fact that there is an electrostatic attraction between the head groups of the surfactants trying to pull the surface composition into a 50/50 situation. Hence, the hydrophobicity

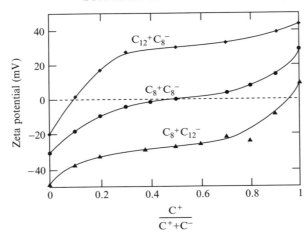

Figure 17.19 The zeta potential of a latex in the presence of mixtures of anionic, C_n^-, and cationic, C_n^+, surfactants, differing in chain length, n. From H. Rendall and A. Smith, *Proceedings of the Surface Active Agents Symposium*, Society of Chemical Industry, London, 1979

dominates over the electrostatic forces and thus the solution properties of the surfactants determine the adsorption.

Anionic and Non-Ionic Surfactants

Mixtures of anionic and non-ionic surfactants are very common in technical applications since such systems give rise to both electrostatic repulsion and steric repulsion between particles at which the surfactants are adsorbed. The composition on a surface will be very much dependent on the solution properties of the single surfactants, i.e. their CMCs, as will be shown below.

Figure 17.20 shows the simultaneous adsorption of a non-ionic, NP-E_{10}, and an anionic, SDS, surfactant on a polystyrene latex surface. The surfactant mixture is added in a NP-E_{10}/SDS ratio of 30:70. This figure shows that the non-ionic surfactant is in abundance at the surface in spite of the fact that it is at a lower concentration in the bulk surfactant mixture. Thus, the more hydrophobic surfactant, i.e. NP-E_{10}, adsorbs preferentially on the hydrophobic latex surface.

Figure 17.21 shows the surfactant composition at the surface as a function of the surfactant composition in the bulk solution. The data points in this figure are determined at the CMC of the solution, i.e. when the surface is saturated with the surfactant mixture. It can be seen that the non-ionic surfactant adsorbs preferentially on the surface at all solution compositions. Thus, for example, a solution with a NP-E_{10}/SDS ratio of 10:90 is in equilibrium with a surface with

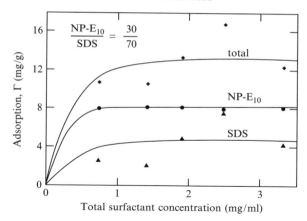

Figure 17.20 The simultaneous adsorption of a non-ionic, NP-E$_{10}$, and an anionic, SDS, surfactant on a polystyrene latex surface. Reprinted with permission from B. Kronberg, M. Lindsröm and P. Stenius, in Phenomena in Mixed Surfactant Systems, J. Scamehorn (Ed.), American Chemical Society, Washington, DC, 1986, p. 225. Copyright (1986) American Chemical Society

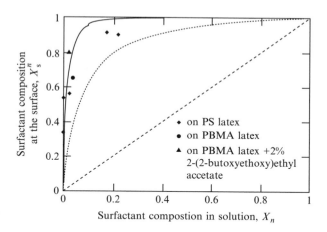

Figure 17.21 The surfactant composition on the latex surface as a function of the surfactant composition in the bulk solution for the adsorption of NP-E$_{10}$ and SDS on polystyrene (PS) or poly(butyl methacrylate) (PBMA) latexes. The dashed line shows equal distribution between the surface and bulk, while the dotted line shows the calculated distribution assuming no interaction between the surfactants (see equation (17.17)). The continuous line includes an interaction

a $NP-E_{10}/SDS$ ratio of 90:10. Hence, analysing the surfactant composition in the solution of a system with more than one surfactant component will give erroneous results if one assumes that the solution composition is the same as the total surfactant composition. We also note that the results displayed in Figure 17.21 are independent of the nature of the surface.

The surfactant composition of a hydrophobic surface can be estimated from the CMCs of the single surfactants. This is a result of the fact that the solution properties of surfactants dominate the adsorption. The surface composition, X_s^n, is related to the solution composition, X_n, through the following expression:

$$X_s^n = \frac{X_n CMC_a}{X_n CMC_a + (1 - X_n)CMC_n} \tag{17.17}$$

where CMC_a and CMC_n are the CMCs for the anionic and non-ionic surfactants, respectively. We note that the surface is not a parameter in this equation and hence it is valid for any hydrophobic surface. Equation (17.17) is akin to the expression for the surfactant composition in a mixed micelle (see equation (5.6) in Chapter 5).

Two Non-Ionic Surfactants

We now turn to the case where two non-ionic surfactants compete for the surface. Of course, also here the more hydrophobic surfactant will dominate the adsorption due to its lower solubility in the aqueous phase. We have already discussed the simultaneous adsorption of two non-ionic surfactants at a hydrophobic surface (illustrated in Figure 17.8). Here it was noted that the more hydrophobic surfactant prefers the surface, while the more hydrophilic surfactant prefers the aqueous solution.

Another example of simultaneous adsorption of two non-ionic surfactants is illustrated in Figure 17.22, which shows the simultaneous adsorption, at a silica surface, of $C_{14}E_6$ and $C_{10}E_6$ at a molar ratio of 0.5 and a total surfactant concentration of 2mM, which is far above the CMC of the surfactant mixture. This figure shows that the single surfactants, $C_{10}E_6$ and $C_{14}E_6$, adsorb with *ca.* 3 and $5\,mmol/m^2$, respectively, while for the mixture an adsorption of $4.5\,mmol/m^2$ is obtained. This indicates that the surface has an excess of $C_{14}E_6$ at the cost of $C_{10}E_6$. More interesting though is the behaviour when the system is diluted.

In order to understand the behaviour on dilution it is helpful to consider two extreme points. At very high surfactant concentrations, the surfactant composition at the surface will approach that of the added composition. On the other hand, in dilute solutions approaching the CMC of the mixture, the surfactant composition at the surface will approach that described by equation (17.17).

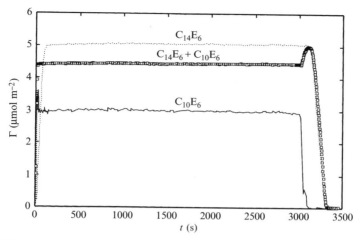

Figure 17.22 The simultaneous adsorption of $C_{14}E_6$ and $C_{10}E_6$, for a molar ratio of 0.5, at a silica surface, illustrating the competitive adsorption as well as the deposition of the hydrophobic surfactant on dilution (see text for details). Reprinted with permission from J. Brinck, B. Jönsson and F. Tiberg, *Langmuir*, **14** (1998) 5863. Copyright (1998) American Chemical Society

Hence, upon dilution the surfactant composition at the surface, X_s^n will increase for the more hydrophobic surfactant while it will decrease for the more hydrophilic surfactant. Upon dilution of the mixture, the chemical potential of the more hydrophobic component increases and hence the adsorption increases at the same time as the hydrophilic surfactant desorbs from the surface, as shown in the figure. We note in passing that this is an important phenomenon in cleaning, such as dishwashing, where rinsing could cause redeposition of the more hydrophobic species (dirt) on the surfaces.

Anionic Surfactant and Polymer

Many technical systems contain both surfactants and water-soluble polymers. We will here illustrate the importance of surfactant–polymer interactions for the adsorption at surfaces. We consider the adsorption on a polar surface where the surfactants adsorb with their head groups interacting with the surface. The surfactant hydrocarbon chains will create a nucleus for further hydrophobic adsorption. In Chapter 13, we have seen that water-soluble polymers facilitate the formation of micelles for ionic surfactants by shielding the repulsion between the head groups of the surfactants in the micelle, as depicted in Figure 17.23. This will also occur on micelle aggregates adsorbed on the surface, as will now be illustrated.

Figure 17.24 shows the simultaneous adsorption of SDS and poly(vinyl pyrrolidone) (PVP) on an Fe_2O_3 surface. Pure PVP, in the absence of SDS,

Figure 17.23 Complex formation between ionic surfactants and water-soluble polymers

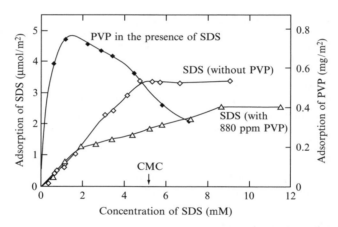

Figure 17.24 The simultaneous adsorption of SDS and poly(vinyl pyrrolidone) (PVP) on a Fe_2O_3 surface. Reproduced by permission of Academic Press from C. Ma and C. Li, *J. Colloid Interface Sci.*, **131** (1989) 485

adsorbs at levels in the order of only $0.02 \, mg/m^2$. This figure shows that PVP adsorption is strongly enhanced in the presence of SDS. However, PVP desorbs from the surface as the SDS concentration is further increased. The figure also shows SDS adsorption with and without the presence of PVP. This rather complicated pattern is due to a complex formation between the PVP and SDS, both in the solution and at the interface.

At low concentrations, the surfactant forms aggregates at the surface, as was discussed previously. PVP in turn forms a complex with the SDS micelles and hence the surface aggregates of SDS will complex with the PVP molecules. This is the reason for the enhanced PVP adsorption. At higher bulk SDS concentrations, PVP/SDS complexes also start to form in the bulk solution, with the result that the surface is depleted from PVP, thus leaving room for extra SDS to adsorb. All PVP is used up for complex formation at even higher SDS solution concentrations, at which free micelles start to form in the solution. We thus see that an enhanced adsorption of PVP on the Fe_2O_3 surface occurs with the help of SDS surfactants. This complex formation is very sensitive to the SDS concentration since complexes will be formed in solution at higher SDS concentrations.

Bibliography

Parfitt, G. D. C. H. and Rochester, *Adsorption from Solution at the Solid/Liquid Interface*, Academic Press, New York, 1983.

Rosen, M. J., *Surfactants and Interfacial Phenomena*, 2nd Edn, Wiley, New York, 1989.

18 WETTING AND WETTING AGENTS, HYDROPHOBIZATION AND HYDROPHOBIZING AGENTS

Liquids Spread at Interfaces

A drop of liquid placed on a solid surface may spread so as to increase the liquid–solid and liquid–gas interfacial areas. Simultaneously, the solid–gas interfacial area decreases and the angle of contact, θ, between the drop and the solid is reduced (Figure 18.1). The value of θ can be seen as a compromise between the tendency of the drop to spread so as to cover the solid surface and to contract in order to minimize its own surface area.

Spreading continues until the system has reached equilibrium. The degree of spreading is governed by the surface tension of the liquid, γ_{LG}, the surface tension (usually referred to as the surface free energy) of the solid, γ_{SG}, and the interfacial tension of the liquid and the solid, γ_{SL}. The forces involved are illustrated in Figure 18.2.

The surface free energy of the solid, γ_{SG}, tends to spread the drop, i.e. to shift the three-phase point outward. Thus, spreading is favoured on high-energy surfaces. The interfacial tension, γ_{SL}, and the horizontal component of the surface tension, γ_{LG}, of the liquid (i.e. $\gamma_{LG} \cos \theta$) act in the opposite direction. At equilibrium, the resultant force is zero:

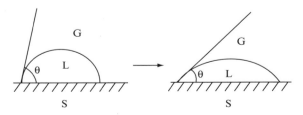

Figure 18.1   Spreading of a drop of liquid on a solid surface

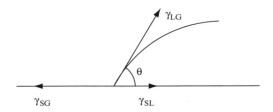

Figure 18.2   Surface forces involved in spreading

$$\gamma_{SG} = \gamma_{SL} + \gamma_{LG} \cos\theta \tag{18.1}$$

This expression is generally known as Young's equation. It has become the basis for the understanding of the phenomenon of spreading on *solid surfaces*.

Spreading of *one liquid on top of another* (assuming immiscible liquids) can be treated in an analogous way. Since the liquid–liquid interface is not planar, however, the angle of contact, θ_2, must also be taken into consideration (Figure 18.3). At equilibrium, we have:

$$\gamma_{L_2G} = \gamma_{L_1L_2} \cos\theta_2 + \gamma_{L_1G} \cos\theta_1 \tag{18.2}$$

It can easily be seen that if the interface between the liquids becomes planar—as it will do if L_2 is a high viscous material, then $\cos\theta_2$ is equal to 1. Equation (18.2) is then identical to Young's equation (Equation 18.1).

A spreading coefficient, S, defined as:

$$S = \gamma_{SG} - \gamma_{SL} - \gamma_{LG} \tag{18.3}$$

is often used. Spreading will then occur spontaneously as long as $S > 0$. For spreading of one liquid on another the spreading coefficient can be written analogously, as follows:

$$S = \gamma_{L_2G} - \gamma_{L_1L_2} - \gamma_{L_1G} \tag{18.4}$$

For the situation of one liquid drop on top of another liquid, the value of S can easily be obtained since the surface tensions γ_{L_1G} and γ_{L_2G}, as well as the

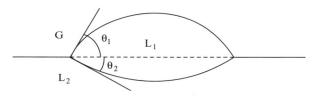

Figure 18.3   A drop of liquid L_1 on liquid L_2

interfacial tension $\gamma_{L_1L_2}$, can either be found in tables or measured by standard techniques. Thus, it is possible to estimate whether or not one liquid will spread on another.

However, it must be noted that considerations of spreading behavior based on literature values of surface and interfacial tensions may not always be accurate. First, the theory is based on total *mutual insolubility*. In addition, immiscible liquids always dissolve in each other to some extent, however, and this affects the γ-values greatly. The spreading coefficient will therefore vary with time. In some cases, this may result in a reversal of spreading. As, under the initial spreading process, the liquid phases become mutually saturated, the spreading coefficient may be reduced to below zero. This will result in contraction of the film with the formation of flat lenses on the surface.

Secondly, *impurities* in the liquid phases will have a dramatic effect on the spreading behavior. In a liquid–liquid system, $\gamma_{L_1L_2}$ will normally be reduced by impurities in either the oil or the aqueous phase. Impurities in the water phase will also have a large effect on γ_{LG}, i.e. the surface tension will be reduced. Impurities in the oil phase normally do not influence γ_{LG} to the same extent. In addition, impurities in one phase will normally partition between the phases, and thus influence all γ-values. Prediction of the net influence of contaminants on spreading is therefore difficult.

A good example of the effect of impurities on spreading behavior is the important issue of whether or not oil, for instance from an oil spill, will spread on water. If the surface and interfacial tension values are known, S can be calculated from equation (18.4). One should then keep in mind that the surface tension of water is higher for salt water than for fresh water, i.e. the value of S is larger for salt water, meaning that oil spreads easier on salt water than on fresh water. Impurities in the oil phase may reduce the surface tension of the oil considerably, which also increases S. Thus, contaminated oil, or oil containing large amount of resin or other surface active components, will spread more easily than pure hydrocarbon oil on a water surface.

The Critical Surface Tension of a Solid is a Useful Concept

Whether or not a given liquid spreads on a given substrate depends on both the liquid and the solid. In a scientific approach to wetting it is important to be able to determine the wetting characteristics of the substrate. However, determination of surface tension or surface free energy for a solid surface is not as straightforward as for a liquid. The most common approach is that devised by Zisman who introduced a useful scheme for classifying low-energy surfaces with respect to their wettability. For many series of liquids on solids—including plastics, metals and metal oxides—it was shown that the contact angle decreases with decreasing surface tension of the liquid. For a homologous series of non-polar liquids, the increase in cos θ with decreasing liquid surface tension is linear for a given solid, as illustrated in Figure 18.4.

A critical liquid surface tension, γ_c, is defined as the point where the plotted line intersects the zero contact angle line, i.e. the line representing complete wetting. In the plot shown in Figure 18.4 γ_c is 18.5 mN/m. In theory, all liquids with a γ_{LG} equal to or lower than γ_C will spread on that surface. In practice, however, γ_C is not a constant for any given solid, but varies somewhat with liquid type.

These regularities in the wetting properties of low-energy surfaces, such as polymers and adsorbed oriented monolayers of organic materials on high-energy surfaces, are significant. Even for non-homologous liquids, a plot of γ_{LG} against cos θ shows points lying in a narrow rectilinear band. However, the line may exhibit curvature if hydrogen bonding can take place between the liquid molecules and the molecules in the solid surface.

The use of a 'Zisman plot' to determine the critical surface tension is relatively straightforward and has become a widely used method to characterize a low-energy solid with respect to surface free energy. Table 18.1 gives the critical surface tension values for some common polymers, while Table 18.2 shows the values for various functional groups. The latter values have been calculated from a large collection of experimentally determined γ_c-values for polymers.

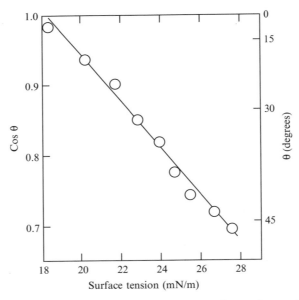

Figure 18.4   A Zisman plot for *n*-alkanes on polytetrafluoroethylene

Table 18.1 Critical surface tension (γ_c) values for various polymers at 20°C

Polymer	γ_c(mN/m)
Poly(1,1-dihydroperfluorooctyl methacrylate)	10.6
Polyhexafluoropropylene	16.2
Polytetrafluoroethylene	18.5
Polytrifluoroethylene	22
Poly(vinylidene fluoride)	25
Poly(vinyl fluoride)	28
Polyethylene	31
Polytrifluorochloroethylene	31
Polystyrene	33
Poly(vinyl alcohol)	37
Poly(methyl methacrylate)	39
Poly(vinyl chloride)	39
Poly(vinylidene chloride)	40
Poly(ethylene terephthalate)	43
Cellulose	45
Poly(hexamethylene adipamide)	46

Table 18.2 Critical surface tension (γ_c) values in relation to surface constitution at 20°C

Surface groups	γ_c(mN/m)
Fluorocarbon surfaces	
$-CF_3$	6
$-CF_2H$	15
$-CF_3$ and $-CF_2-$	17
$-CF_2-CF_2-$	18
$-CF_2-CFH-$	22
$-CF_2-CH_2-$	25
$-CFH-CH_2-$	28
Hydrocarbon surfaces	
$-CH_3$ (crystal)	20–22
$-CH_3$ (monolayer)	22–24
$-CH_2-CH_2-$	31
$-CH-$ (phenyl ring edge)	35
Chlorocarbon surfaces	
$-CClH-CH_2-$	39
$-CCl_2-CH_2-$	40
$-CCl_2-$	43

As can be seen from Table 18.2, the value of γ_c for a solid is indicative of the molecules making up its surface. The surfaces having the lowest value of γ_C, and hence the lowest surface energy, consist of closely packed CF_3 groups. Replacing one fluorine atom by hydrogen considerably raises the value of γ_c. This low value of γ_c indicates the generally low adhesion between liquids and surfaces containing trifluoromethyl groups. (Sometimes, however, the introduction of a terminal CF_3 group does not decrease the wettability very much since the introduction of the dipole associated with the CF_3-CH_2 group gives an effect in the other direction.) Another observation that can be made from Table 18.2 is that CH_3 groups have very low values when compared with CH_2 groups. This implies that surfaces rich in methyl groups should have low surface tensions, which indeed is true. The most common type of silicone oil, polydimethylsiloxane (see Figure 18.10 below), is probably the best example of such a methyl-rich surface.

The Critical Surface Tension can be Applied to Coatings

The concept of critical surface tension of a solid is useful for many practical applications, e.g. surface coatings. In order for a coating to spread on a substrate, the surface tension of the liquid coating must be lower than the critical surface tension of the substrate. (In addition, since a liquid is easier to break up and atomize when the surface tension is low, lower surface tension means better sprayability of the coating.) The well-known coatings defect, 'cratering', is also surface tension related. Contaminants on the surface, such as fingerprints and oil spots, usually have lower critical surface tensions than the surrounding areas, thus putting extra demand on the coating with regard to low surface tension.

The polymer and the solvent largely determine the surface tension of the coating. The polymers usually have relatively high surface tensions, with values between 35 and 45 mN/m being typical. The organic solvents have surface tension values between 20 and 30 mN/m, as can be seen from Table 18.3.

Table 18.3 Surface tension (γ) values for selected solvents at 20°C

Solvent type	γ(mN/m)
Alcohols	21–35
Esters	21–29
Ketones	23–27
Glycol ethers	27–35
Glycol ether esters	28–32
Aliphatic hydrocarbons	18–28
Aromatic hydrocarbons	28–30
Water	73

In general, when comparing solvents within the same solvent class it has been found that faster evaporating solvents usually have lower surface tension values than their slower evaporating counterparts. Furthermore, increased branching of the solvent molecule leads to a lowering of the surface tension. Isopropyl alcohol, which fulfils the two above-mentioned requirements, has an extremely low surface tension, i.e. 21.4 mN/m. Evidently, the higher the solvent content of the coating, then the lower the surface tension. Conventional coatings normally lie in the range of 25–32 mN/m. This is low enough to give proper wetting on most surfaces, i.e. to get below the values of the critical surface tension of the substrates (see Table 18.1).

As the resin becomes the major component of the coating, problems related to surface tension are frequently encountered. So called 'high-solids coatings' are, therefore, extremely sensitive to dirt and fingerprints (the surface tension of which is around 24 mN/m). Paint defects, such as 'cratering' and 'picture framing' occur more frequently with high-solids systems than with conventional systems.

Water is a liquid of high surface tension and obviously not suitable for the wetting of surfaces. Use of water-borne paints would have been very limited had it not been possible to use surfactants in their formulations. A good surfactant reduces the surface tension of water down to 28–30 mN/m, i.e. to the same range as that of the organic solvents used in paints and lacquers.

Surface Active Agents can Promote or Prevent Wetting and Spreading

From equation (18.3) it can be seen that spreading of a liquid on a solid surface can be promoted in two different ways. Addition of a surface active agent will lead to a reduction of both the surface tension, γ_{LG}, and the interfacial tension between the solid and the drop, γ_{SL}. The situation is illustrated in Figure 18.5.

An alternative way of promoting wetting is shown in Figure 18.6. By hydrophilizing the surface, γ_{SL} is reduced and γ_{SG} is increased. Both changes work to promote wetting, according to equation (18.3).

If, on the other hand, a hydrophilic material, e.g. paper or textile, needs to be made water-resistant one should aim at minimizing wetting. This is normally done by hydrophobization of the surface. This will lead to an increase in γ_{SL}

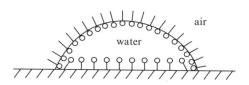

Figure 18.5 Addition of a surfactant to a drop of water on a hydrophobic solid surface will reduce both γ_{LG} and γ_{SL}

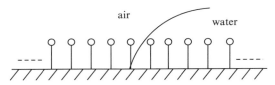

Figure 18.6 Hydrophilization of the surface will reduce γ_{SL} and increase γ_{SG}. The hydrophilization is shown schematically here as adsorption of a normal surfactant. In practice, adsorption of a surface active polymers is more useful

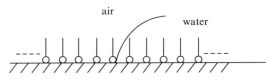

Figure 18.7   Hydrophobization of the surface will increase γ_{SL} and decrease γ_{SG}

and a decrease in γ_{SG}, with both effects contributing to reducing the spreading coefficient, S, as shown in Figure 18.7.

Wetting Agents

Wetting depends upon the effective reduction of the surface tension under dynamic conditions, i.e. as the wetting liquid spreads over the substrate, the surface active molecules must diffuse rapidly to the moving boundary between the liquid and substrate. Therefore, a good wetting agent must (i) have a strong driving force to go the solid–liquid interface, (ii) effectively reduce the surface tension, (iii) have a reasonable concentration of free, i.e. non-micelle bound surfactants, and (iv) move rapidly to the newly created surface. In order to fulfill these requirements, a good wetting agent is a surfactant that is fairly small and fairly hydrophobic (often on the borderline of water solubility) but still not having a very low critical micelle concentration. Wetting agents are often branched surfactants because these do not form micelles as readily as their straight-chain counterparts. Both non-ionic and anionic surfactants are used in commercial formulations. Figure 18.8 shows the structure of two common wetting agents. Recently, the concept of 'superspreading' has been introduced. A 'superspreader' is a surfactant which, when added in small amounts to an aqueous solution, enables the liquid to spread spontaneously and rapidly on very hydrophobic surfaces, such as a solid paraffin wax surface (Parafilm). For example, it has been demonstrated that the ethoxylated silicone-based surfactant shown in Figure 18.9 has a surface-covering rate on Parafilm of about $10 \, \text{cm}^2/\text{min}$ for a $0.008 \, \text{g}$ aqueous drop containing $0.1 \, \text{wt\%}$

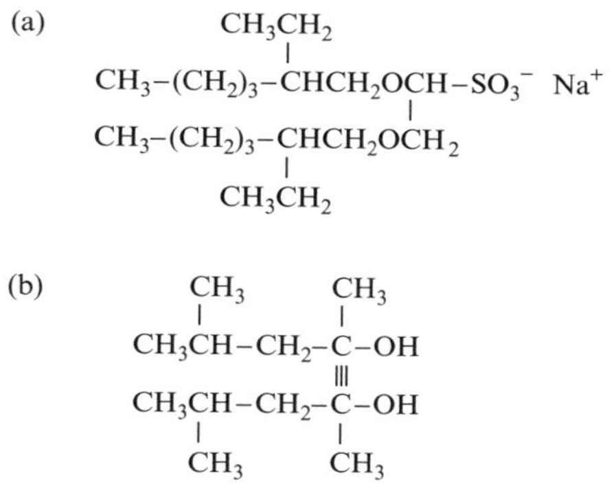

(a)
$$CH_3CH_2$$
$$|$$
$$CH_3-(CH_2)_3-CHCH_2OCH-SO_3^- \ Na^+$$
$$|$$
$$CH_3-(CH_2)_3-CHCH_2OCH_2$$
$$|$$
$$CH_3CH_2$$

(b)
$$CH_3 \qquad CH_3$$
$$| \qquad \quad |$$
$$CH_3CH-CH_2-C-OH$$
$$|||$$
$$CH_3CH-CH_2-C-OH$$
$$| \qquad \quad |$$
$$CH_3 \qquad CH_3$$

Figure 18.8 Two surfactants that are commonly used as wetting agents, (a) sodium bis(2-ethylhexyl)sulfosuccinate (AOT) and (b) an acetylene glycol (Surfynol 104, from Air Products)

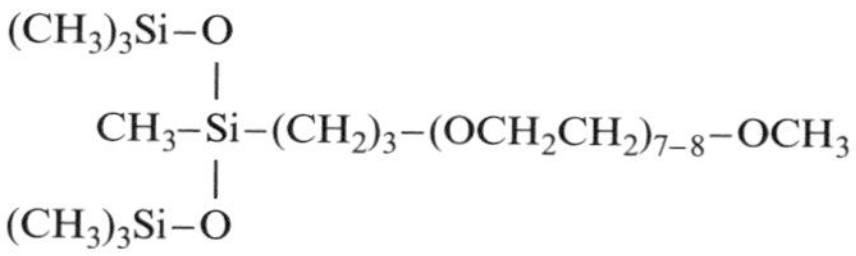

$$(CH_3)_3Si-O$$
$$|$$
$$CH_3-Si-(CH_2)_3-(OCH_2CH_2)_{7-8}-OCH_3$$
$$|$$
$$(CH_3)_3Si-O$$

Figure 18.9 Structure of a silicone-based superspreader

surfactant. Such rapid spreading is not very common and the mechanism of action of superspreaders is not completely clear. It was found that the covered area in the early stage of spreading increases monotonically with increasing wt% surfactant and time. In addition, the total area covered when the spreading stops is proportional to the concentration of surfactant in the dispersion. (The superspreader has a very low water solubility, and is best applied as a water-continuous dispersion.) A droplet of an ultrasonicated dispersion spreads faster than a hand-shaken dispersion, but the final covered area, is the same in both cases. From this and from the size of the final covered area, it was concluded that spreading stops when all of the surfactant is deposited as a bilayer on the substrate.

In some respects, such as the need for rapid movement to a newly created interface and high concentration of non-micellized surfactants, the requirements for wetting agents resemble those required for foaming agents.

Hydrophobizing Agents

Paraffin wax, silicones, silanes and fluorinated hydrocarbons are examples of efficient hydrophobizing agents (with efficiency increasing in that order). In addition, cationic surfactants are often used for this purpose. Figure 18.10 shows the structure of the most common silicone oil, i.e. polydimethylsiloxane. The conformation of the silicone on the surface is such that the siloxane backbone interacts with the surface, with the methyl groups being oriented away from it, as also illustrated in this figure. Thus, as discussed above, the silicone treatment is effectively a surface methylation. As can be seen from Table 18.2, methyl groups render a surface extremely hydrophobic.

Figure 18.11 shows the structure of a glass (or silica) surface hydrophobized with dichorodimethylsilane. The latter and other silanes are common hydrophobizing agents for mineral surfaces containing silanol groups. As with the silicones, the treatment is effectively a surface methylation.

Since both paper and textile fibres normally have a negative net charge, long-chain cationic surfactants are commonly used to impart water repellency. The surfactants also give a debonding effect, i.e. the fibre–fibre attractive interactions are considerably reduced. The most common textile softeners are quaternary ammonium compounds containing two C16–C18 hydrocarbon chains and two methyl substituents on the nitrogen atom. Due to environmental demands (increased rate of biodegradation), the traditional hydrolytically stable surfactants of this type, 'dialkyl quats', have to a large extent been replaced by ester-containing surfactants of similar structure, i.e. 'dialkyl esterquats' (for further discussion of this, see Chapters 1 and Chapter 11). The structures of both a traditional dialkyl quat and a dialkyl esterquat are shown earlier in Figure 1.10.

One may expect that proper surface hydrophobization requires a densely packed layer of the hydrophobizing agent with good surface coverage. Recent

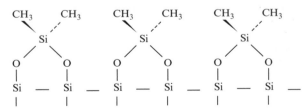

Figure 18.10   Polydimethylsiloxane, a common type of hydrophobizing agent

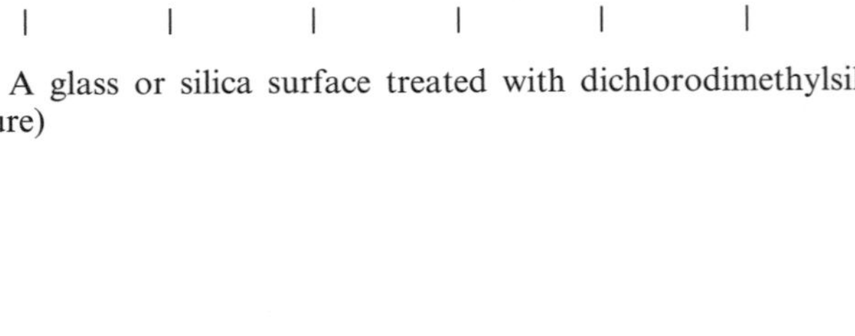

Figure 18.11   A glass or silica surface treated with dichlorodimethylsilane (idealized surface structure)

studies of the mechanism of hydrophobization has revealed that only a partial coverage by the hydrophobizing agent seems to be needed to form a non-water-wetting surface, however. Water will not spread on a surface that is partially covered with hydrophobic domains. In many cases, it seems that a coverage of 10–15% is enough. This is probably the reason why hydrophobizing agents are already effective in minute amounts.

Measuring Contact Angles

The contact angle of a drop of liquid on a solid, planar surface is usually measured directly on either a drop resting on a horizontal plane (a sessile drop) or an adhering gas bubble captured at a solid–liquid interface, in both cases using a goniometer (Figure 18.12). The angle of contact is read with the help of a microscope objective to view the angle directly. For larger drops, the shape is no longer spherical because gravitational forces come into play. This can be taken advantage of as a method to determine surface tension (see Chapter 16).

Contact angle measurements using either sessile drops or adhering bubbles are today usually automated and computerized which enables values of contact angles to be determined with a high degree of reproducibility. With modern, fast and computerized instruments, contact angles can also be determined on penetrating substrates, such as paper. By using instruments that record data several hundred times per second, the apparent contact angle of a drop on a highly absorbing paper can be measured as a function of time. By extrapolating to time zero, the contact angle of the paper before penetration has started can be obtained. This is a useful way to determine the effect on the surface free energies of various types of additives used in paper production. An example of such an experiment is given in Figure 18.13.

A very simple way of measuring contact angles, suitable as a screening method, is to measure the diameter of a standardized drop put on the surface. The contact angle can then be obtained from the following equation:

$$V = \pi a^3/6[3 \tan \theta/2 + (\tan \theta/2)^3] \qquad (18.5)$$

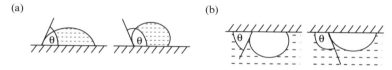

Figure 18.12 Measuring the contact angle by using (a) a sessile drop, or (b) an adhering bubble

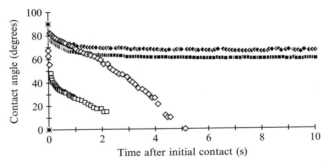

Figure 18.13 Measuring the contact angle on four different materials with roughly the same surface free energy but with varying penetrating properties

where V is the known volume of the drop and a is the radius measured. A table to correlate θ-values with values for $2a$ (the drop diameter) is given in the literature for 10 μl drops (*J. Immunol. Methods*, 40 (1981) 171–179). As in all determinations of contact angles, it is important to conduct the measurement immediately after application of the drop. Evaporation of water leads to drop contraction and a change in the contact angle (or, more specifically, to a change from advancing to receding contact angle).

An indirect method of determining the contact angle is by measuring the capillary rise of a liquid at a vertical plate (Figure 18.14). A solid surface is aligned vertically and brought into contact with the liquid and measurement of the height, h, of the meniscus gives the contact angle, θ, from the following equation:

$$\sin \theta = 1 - (\Delta P g h^2 / 2\gamma_{LG}) \qquad (18.6)$$

where ΔP is the density difference between the aqueous and the gas phase and h is the measured capillary rise. This method has been found particularly effective

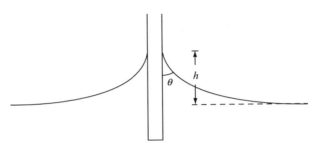

Figure 18.14   The capillary rise method used to measure the contact angle

for measuring contact angles as a function of rate of advance and retreat, and for determining the temperature coefficient of θ.

Contact angle measurement is, however, not as straightforward as one would imagine. Regardless of the method used, a number of complications may arise, namely:

(1) Few solid surfaces are effectively flat at the scale of observation. The angles that a fluid–fluid interface makes with a solid at a singularity are directly dependent on the macroscopic geometry of the solid. This is illustrated in Figure 18.15. Roughness on the scale that affects contact angle measurements can conveniently be observed by using scanning electron microscopy.

(2) Incorporation of low-energy domains into a high-energy surface will strongly influence the contact angle. Grease contaminants on a glass surface are an example of this.

(3) The phases must be in equilibrium in order to give accurate θ-values. This is not always the case due to:
   —penetration of the liquid into the surface region of the solid;
   —reorientation of surface groups of the solid, in particular the exposure of polar groups on treatment with water.

In practice, the conditions under which contact angles are measured are far from ideal. In most cases, the observed contact angle will depend greatly on whether the liquid is advancing over a dry surface or receding from a wet surface. This is referred to as *contact angle hysteresis* and is noticeable on all rough or dirty surfaces.

Many surfaces exhibit both roughness and heterogeneity and on such surfaces both advancing and receding contact angles should be measured. The Wilhelmy plate technique (see Figure 16.11 above) is a commonly used method for making contact angle hysteresis measurements.

In general, it can be stated that on a heterogeneous surface the advancing angle is a measure of the wettability of the low-energy part of the surface while the receding angle is more characteristic of the high-energy part of the surface.

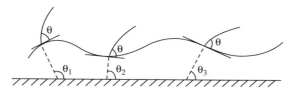

Figure 18.15 Different θ-values may be measured on a rough surface

Bibliography

Adamson, A. W. and A. P. Gast, *Physical Chemistry of Surfaces*, Wiley, New York, 1997.

Grundke, K., Wetting, spreading and penetration, in *Handbook of Applied Surface and Colloid Chemistry*, K. Holmberg (Ed.), Wiley, Chichester, UK, 2001.

Jarvis, N. L. and W. A. Zisman, *Kirk-Othmer Encyclopedia of Chemical Technology*, 2nd Edn, Vol. 9, VCH, New York, 1966, pp. 707–738.

Neumann, W. A., C. Lam and J. Lu, Measuring contact angle, in *Handbook of Applied Surface and Colloid Chemistry*, K. Holmberg (Ed.), Wiley, Chichester, UK, 2001.

19 INTERACTION OF POLYMERS WITH SURFACES

Technically, the adsorption of polymers is utilized in many applications, such as in the dispersion of particles, flocculation processes, treatment of surfaces, etc. In these processes, the purpose of polymer adsorption is to modify interactions between surfaces. From a thermodynamic point of view, the main driving force is an effective interaction between polymer segments and the surface, such as the adsorption of a cationic polymer at an anionic surface. If the solvent is a poor solvent (i.e. when the interaction energy between the polymer segments and the solvent molecules is unfavourable when compared to the polymer segment–segment and solvent–solvent molecular interaction energy) then the effective polymer–surface interaction can be strongly attractive and the polymer will seek any opportunity to escape contact with the solvent and hence it adsorbs on almost any surface, even the liquid–air surface. Adsorption is therefore always accentuated before precipitation occurs. The surfaces in the system will simply act as nucleation sites for polymer precipitation. The general conclusion is that the poorer the solvent–polymer interaction, then the better the adsorption and vice versa.

High molecular weight species are more prone to adsorb than low molecular weight polymers. This fact has many implications. The first implication can be seen in Figure 19.1, which is a typical adsorption isotherm of a polymer. Characteristic for polymer adsorption isotherms is the sharp increase in adsorption at very low polymer concentrations, followed by a plateau that does not change with polymer concentration. The very strong adsorption at low concentrations is a reflection of the 'unhappiness' of the polymer in the solution, for reasons stated above, i.e. there is poor solvent–polymer interaction in the bulk solution. This sharp rise in adsorption at low concentrations becomes even more pronounced as the polymer molecular weight is increased. The constant adsorption at higher polymer concentrations is due to the fact that the surface is saturated with the polymer.

In measuring adsorption isotherms of polymer solutions it is important to realize that the polymer needs time to reach equilibrium. Normally an

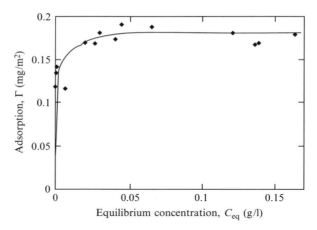

Figure 19.1 Adsorption of poly(acrylic acid) on kaolin from aqueous solution showing the typical features of polymer adsorption, namely the steep rise in adsorption at low polymer concentrations and a constant adsorption at high concentrations. Reprinted from *Colloid Surf.*, **50**, L. Järnström and P. Stenius, 'Adsorption of polyacrylate and carboxy methyl cellulose on kaolinite: salt effects and competitive adsorption', 47–73, Copyright (1990), with permission from Elsevier Science

equilibration time of *ca.* 24 h under gentle stirring is sufficient, but sometimes even longer times are needed, especially when a porous substrate is used and the polymer needs to diffuse into the pores in order to find a free surface to adsorb at.

The word polyelectrolyte is sometimes used for all types of aggregates that carry a high net charge. Some of the literature reserves it for charged polymers and it is the latter type we will consider now. A flexible polymer often achieves a net charge from carboxylate or sulfate groups ($-COO^-$, $-SO4^-$) or from amino groups ($-NH^+$) or analogous nitrogen compounds. Polyelectrolytes can be classified as strong (quenched) or weak; the net charge of the latter varies with pH. The charge dependence of the pH can be quite complex and will affect the adsorption isotherm. The electrostatic interactions will have an influence on the adsorption process, although non-electrostatic effects often dominate the process.

The Adsorbed Amount Depends on Polymer Molecular Weight

Analysing the molecular weight dependence of the adsorbed amount gives an idea of how the polymer is adsorbed on the surface. If the polymers lie flat on the surface there will not be any molecular weight dependence. On the other hand, if the polymer adsorbs 'head on', i.e. with only the polymer chain end at the surface, the adsorbed amount will be directly proportional to the molecular weight. In practice, however, most polymer systems are found to adsorb in a coil configuration and the molecular weight dependence of the adsorbed amount is proportional to M^α, where M is the polymer molecular weight and α is a constant with a value in the range 0.3–0.5. Figure 19.2 shows a typical molecular weight dependence of an adsorption, in this case adsorption of poly(vinyl alcohol) in aqueous solution at a polystyrene latex surface. Notice how the adsorption becomes progressively steeper at low polymer concentrations as the polymer molecular weight increases. Contrary to these results is the adsorption of polyoxyethylene on a silica surface from water. In this system, there is no molecular weight dependence of the adsorption. The reason is probably the high affinity between the ether groups of the polymer and the hydroxyl groups on the surface, thus resulting in a flat configuration of the polymer at the surface.

We have seen that a polymer with a high molecular weight shows a larger adsorption than polymers with lower molecular weights (see Figure 19.2, for example). Thus, in adsorbing a polydisperse polymer sample the high molecular weight species will be preferentially adsorbed at the expense of the low molecular weight species. As an illustration, Figure 19.3(a) shows the adsorption isotherms of a mixture of two polymers with different molecular weights, displaying a common isotherm when plotted with Γ_b as the concentration variable. This figure also shows the adsorption of the two single polymers. The figure reveals that the adsorption of a binary polymer mixture is divided into three different regions. In the first region, at very low polymer concentrations, there is sufficient surface available for all polymer to adsorb, thus resulting in a very strong adsorption. Eventually, at higher polymer concentrations, the binary polymer mixture covers the surface and an exchange process

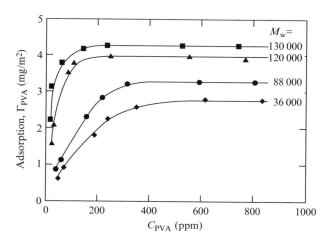

Figure 19.2 Adsorption of poly(vinyl alcohol) (PVA) on a polystyrene surface, illustrating that the adsorption increases with the polymer molecular weight. Reproduced by permission of Academic Press from S. Chibowski, *J. Colloid Interface Sci.*, **143** (1990) 174

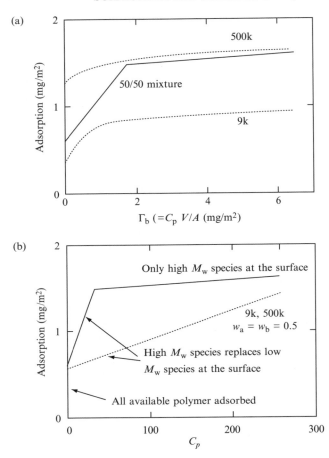

Figure 19.3 (a) Adsorption of two dextrans, with molecular weights of 9000 and 500 000 g/mol, respectively, and a 50/50 mixture, illustrating that all polymers adsorb at low concentrations, followed by a region where the low M_w polymer is replaced by the high M_w polymer at the surface. (b) The effect of the surface/volume ratio on the 50/50 mixture. The continuous and dashed curves represent low and high surface/volume ratios, respectively. Reproduced by permission of Academic Press from V. Hlady, J. Lyklema and G. Fleer, *J. Colloid Interface Sci.*, **87** (1982) 395

takes place where the high molecular weight polymer adsorbs at the expense of the low molecular weight material. Finally, at even higher polymer concentrations, the high molecular weight polymer covers the whole surface and a nearly constant adsorption is found.

Since the adsorbed amount is dependent on the molecular weight, we will have a system where the measured adsorption depends on the ratio of the polymer solution volume to the available surface area. A small available surface area will adsorb only high molecular weight species and hence a large adsorption is detected. A high available surface area will adsorb even the lower molecular weight species and hence a lower adsorption is detected. This problem is circumvented by changing the polymer concentration variable, C_p, according to the following:

$$\Gamma_b = C_p V / A \qquad (19.1)$$

where V is the total liquid volume and A is the total available surface area. Figure 19.3(b) illustrates that when plotting the adsorption versus the polymer concentration, the second concentration region, where the low molecular weight species are replaced by the high molecular weight polymer, differs depending on the surface/volume ratio. If the ratio is high, i.e. there is a lot of surface available in the system, this region will be extended to high polymer concentrations. However, when exchanging the polymer concentration variable for Γ_b, there will be a single curve (not shown in the figure).

The Solvent has a Profound Influence on the Adsorption

The solvent–polymer interaction, or the solvent quality, influences the polymer adsorption in two ways. First, the solvent–polymer interaction influences the polymer configuration. As was discussed earlier in Chapter 8, polymers expand in good solvents and contract in poor ones. Thus, a polymer already adsorbed at a surface will occupy a larger surface area in a good solvent and a smaller surface area in a poor solvent. Hence, a larger adsorption is found from poor solvents. Secondly, a solvent influences the adsorption by the stability in solution, as was discussed above, i.e. if the polymer is not 'happy' in the solution it will seek any opportunity to escape the solvent, e.g. by adsorbing at a surface. Table 19.1 shows the adsorption of polystyrene from methyl ethyl ketone with the addition of methanol, which is a non-solvent. This table reveals

Table 19.1 Adsorption of polystyrene on carbon

Concentration of methanol in methyl ethyl ketone (%)	Adsorbed amount (mg/g)
0	34
0.5	35
1.0	48
1.5	46
2.5	81
5.0	144
10.0	144
12.3	Precipitation

that the adsorbed amount increases as the point of precipitation is approached, again illustrating that the surface acts as a nucleus for the precipitate and that adsorption is accentuated before precipitation occurs.

The adsorption of water-soluble polymers at surfaces is indeed a reflection of the solution properties of the polymer. This is illustrated in Figure 19.4 for the adsorption of ethyl hydroxy ethyl cellulose, (EHEC) on silica. In Figure 19.4(a), the cloud point dependence on salt addition for various salts is shown, while in Figure 19.4(b) the corresponding adsorption at a silica surface is shown. It is clearly demonstrated that those salts that decrease the cloud point, i.e. those salts that decrease the solubility, increase the adsorption of the polymer, thus illustrating the fact that adsorption is a measure of the escaping tendency of the polymer from the solution. Equivalently for those salts which increase the cloud point, i.e. increase the solubility, the adsorption decreases, as illustrated in Figure 19.4 for NaI and NaSCN. We stress that this is a general phenomenon and is not specific to salt additions. Figures 19.5(a) and 19.5(b) show the corresponding results for the adsorption of EHEC on silica when various alcohols are added. Here, we can also conclude that those additives that decrease the solubility, i.e. decrease the cloud point, also increase the adsorption, and vice versa.

Electrostatic Interactions Affect the Adsorption

The adsorption of polyelectrolytes at a surface can be categorized as described in Figure 19.6. Here, the pH dependences of both the surface and the polyelectrolyte are shown. The general trend is that the charge is positive at low pH values and negative at high pH values. Of course, the curves could also level off towards zero charge at either low or high pH values. In this figure, we have indicated the different regions where one has either oppositely charged polyelectrolyte and surface (case I) or where they both have the same charge (case II).

In the first case, i.e., when the polymer and the surface have opposite charges, as illustrated in Figure 19.7(a), the driving force of adsorption could at first sight seem obvious, i.e. the Columbic attraction between the polyelectrolyte and the surface. However, in view of the fact that the counterions are also attracted to the surface and the polymer it might not be so obvious. The driving force indeed stems from the presence of the counterions. When a polyelectrolyte adsorbs at an oppositely charged surface, the counterions from both the polymer and the surface are released into the bulk solution. This increases the entropy in the system, which brings the system into a lower free energy state. This mechanism is the driving force for the adsorption of a polyelectrolyte at an oppositely charged surface (case I). From the above, we predict that the adsorption of a polyelectrolyte will decrease on addition of salt, since the entropy gain of the released counterions is now less when compared to a salt-free system. Of course, there is the alternative view that the added salt will

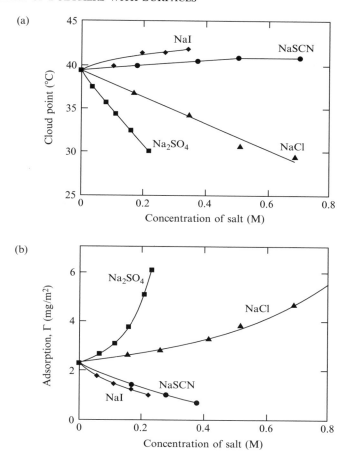

Figure 19.4 (a) The cloud point variation of aqueous ethyl hydroxy ethyl cellulose (EHEC) salt solutions, and (b) the adsorption of EHEC from the same solutions on a silica surface, illustrating that the adsorption increases as the EHEC solubility decreases. Reprinted with permission from M. Malmsten and B. Lindman, *Langmuir*, **6** (1990) 357. Copyright (1990) American Chemical Society

shield the attractive electrostatic forces between the polymer and the surface. Another viewpoint is that the added salt will compete with the polymer for the adsorption sites at the solid surface, (see Figure 19.7(b)).

In the second case, i.e. when the polymer and the surface have the same charge, as illustrated in Figure 19.7(c), the driving force of adsorption originates from attractive van der Waals forces between the polymer and the surface. Interestingly, here the addition of salt has the opposite effect when compared to the first case, i.e. it increases the adsorption (see Figure 19.7(d)). One view of

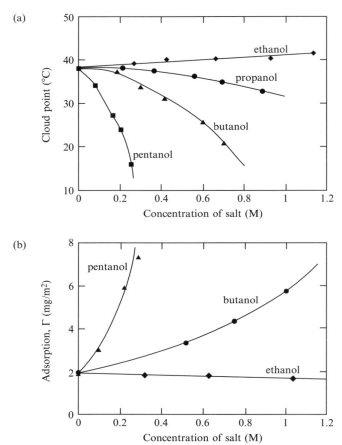

Figure 19.5 (a) The cloud point variation of aqueous EHEC – alcohol solutions, and (b) the adsorption of EHEC from the same solutions on a silica surface, illustrating that the adsorption increases as the EHEC solubility decreases. Reprinted with permission from M. Malmsten and B. Lindman, *Langmuir*, **6** (1990) 357. Copyright (1990) American Chemical Society

the mechanism for this effect is that in the salt-free solution the local concentration of counterions will increase upon adsorption (the presence of counterions, around the adsorbed polymer and in the vicinity of the surface, is required for the maintaining of electrical neutrality). This local increase of counterions represents a sort of 'ordering' in the system and hence corresponds to a decrease in the entropy, and thus the free energy increases. Thus, the presence of counterions counteracts the adsorption of a polyelectrolyte at a similarly charged surface. Upon addition of salt, this effect is diminished, thus increasing the adsorption of the polyelectrolyte. The reason is that the added salt swamps

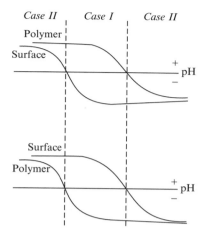

Figure 19.6 Schematic illustration of the variation of charge of a polyelectrolyte and a surface as a function of pH for the two cases when the pK of the polymer is larger and smaller than the pK of the surface

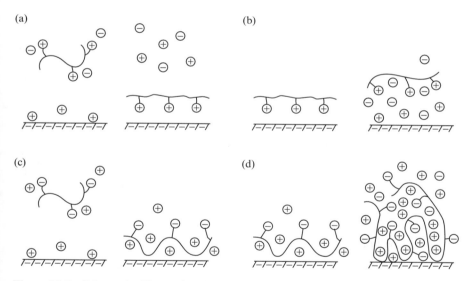

Figure 19.7 Schematics illustrating the two cases shown in Figure 19.6 and the effects of addition of salt

the counterion effect. An alternative point of view for the mechanism is that the added salt causes a shielding of the repulsive forces between the polymer and the surface and also between different polymers. Therefore, added salt causes a higher adsorption of the polymer.

We will now discuss some illustrative examples for the adsorption of poly-electrolytes on charged surfaces. Table 19.2 shows how the adsorbed amount of poly(acrylic acid) on kaolin varies with addition of salt. One observation is that the adsorption increases with ionic strength, as illustrated by the addition of NaCl. This is therefore clearly a case-II adsorption and we thus conclude that the adsorption occurs at one of the negative faces of kaolin, since poly(acrylic acid) is negatively charged.

When adding a salt with a divalent cation, such as $CaCl_2$, we find that this salt is far more effective in increasing the adsorption when compared to NaCl. This is due to the higher efficiency of divalent cations in shielding the repulsive charges on the polymer/kaolin, thus resulting in a large adsorption. Another way to describe the same phenomenon is in terms of the counterion entropy as outlined above. With increasing salt concentration, the entropic gain in dissolving a polyelectrolyte decreases and hence the adsorption is favoured. The entropy effect is larger with divalent cations like Ca^{2+}, simply because the number of entities is less, i.e. the presence of salts diminishes the unfavourable local concentrations of ions upon polymer adsorption. With trivalent or other forms of multivalent ions, the solubility would decrease even more efficiently and the adsorption would increase. Another point of view is that the addition of salts drives the polyelectrolyte towards phase separation, and hence enhanced adsorption.

The initial adsorption of a polyelectrolyte is a fast process, but the approach to a true equilibrium situation could be very slow. This is due to rearrangement and/or adsorption/desorption of different molecular weight species. The process can, for example, be followed in surface force measurements and can proceed for several days or more.

Figure 19.8 shows the adsorption of a cationic polymer at montmorillonite. This figure shows that once the surface is neutralized, there will be no further adsorption due to chain–chain repulsion and the maximum adsorption will be found for the neutral chain. This is an illustration of how the electrostatic interactions, although the surface and polymer have different charges, will act to reduce the adsorption, i.e. case I above.

Table 19.2 Adsorption of poly(acrylic acid) on kaolin from aqueous salt solutions of varying ionic strengths

Electrolyte	Electrolyte concentration (M)	Ionic strength (M)	Adsorption (mg/m^2)
NaCl	0.20	0.20	0.18
NaCl	0.01	0.01	0.08
$CaCl_2$	0.0033	0.01	0.30

Quite naturally, the adsorption will also show a pH dependence. This could come from a charge dependence of the surface, or of the polyelectrolyte, on the pH. Figure 19.9 shows an example of the latter situation, where the protein net charge becomes increasingly more negative at high pH levels.

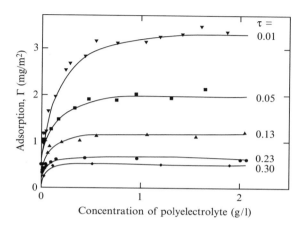

Figure 19.8 Adsorption of a cationic polymer on negatively charged montmorillonite. The different curves demonstrate how the adsorption varies with the degree of ionization (τ). From R. Audebert, personal communication

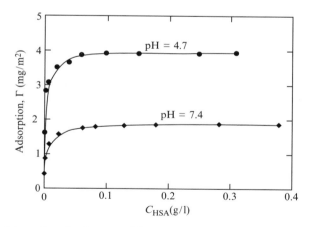

Figure 19.9 Adsorption isotherms of human serum albumin (HSA) on negatively charged AgI sols (the isoelectric point of HSA is 4.9); electrolyte concentration is 10 mM; $T = 295$ K. Reproduced by permission of Academic Press from B. Matuszewska et al., *J. Colloid Interface Sci.*, **84** (1981) 403

Increasing the charge density of the surface normally leads to an increased adsorption, as demonstrated in Figure 19.10. This is, of course, a consequence of the fact that the system prefers to maintain charge neutrality. In many cases, however, non-electrostatic forces govern the adsorption process. This manifests itself in the adsorption of a negatively charged chain at negatively charged surfaces (see Figure 19.11), i.e. a case-II situation. The adsorption in Figure 19.11

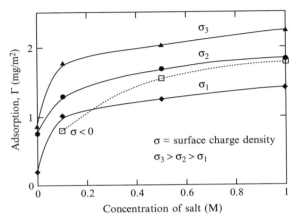

Figure 19.10   Adsorbed amount of poly(styrene sulfonate), Γ, on positively charged surfaces with varying charge density and as a function of salt concentration. Reproduced by permission of Academic Press from T. Cosgrove *et al.*, *J. Colloid Interface Sci.*, **111** (1986) 409

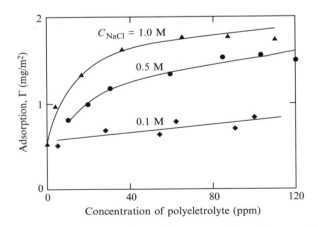

Figure 19.11   Adsorption isotherms for poly(styrene sulfonate) on negatively charged surfaces. Reproduced by permission of Academic Press from T. Cosgrove *et al.*, *J. Colloid Interface Sci.*, **111** (1986) 409

increases with added salt as the interchain repulsion becomes more screened at high salt concentrations. It is also occasionally possible to obtain a charge reversal when an anionic polymer adsorbs on a positively charged surface, or vice versa. Figure 19.12 shows such a situation for polyglutamate adsorbed on charged polystyrene particles.

We will here briefly just caution the experimentalist when designing an experiment where the adsorption of polyelectrolytes in salt solution is determined. In these systems, it is important to control the surface-to-volume ratio. The reason is that the salt in the system is co-adsorbed with the polymer. If the adsorption is determined in a system where there is a small surface area available, there will be plenty of salt that can co-adsorb with the polymer. If, however, the adsorption is determined in a system where there is a large surface area available, there is a risk that all of the salt in the system is used up in the co-adsorption with the polymer and hence the factor determining the adsorption is not the polymer solubility but rather the availability of salt in the system. This is illustrated in Figure 19.13 for the adsorption of carboxymethyl cellulose on kaolin from aqueous solutions containing 0.0033 M $CaCl_2$ at two different solid contents. Here, one finds a lower adsorption at a higher solid content, which is due to the fact that the Ca^{2+} ions are used up in the adsorption process, thus leaving a low Ca^{2+} ion concentration in solution.

We conclude that for meaningful measurements of polymer adsorption one should have a control over the following parameters: the polymer molecular weight and molecular weight distribution, pH, ionic strength, the presence of

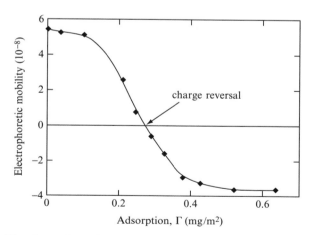

Figure 19.12   The electrophoretic mobility of positively charged polystyrene particles as a function of the adsorbed amount of poly-L-glutamate (1 mM KBr and pH 6.5). Reproduced by permission of Academic Press from B. C. Bonekamp *et al.*, *J. Colloid Interface Sci.*, **118** (1987) 366

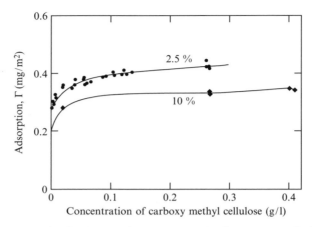

Figure 19.13   Adsorption isotherms of carboxymethyl cellulose on kaolin from aqueous solutions containing 0.0033 M CaCl₂ at two different solid contents. Reproduced by permission of TAPPI Press from L. Järnström, G. Ström and P. Stenius, *Tappi J.*, **70** (1987) 101

multivalent ions, the presence of small amount of additives (such as displacers, see Figure 19.29 below), the surface-to-volume ratio and the temperature.

Concerning the polymer configuration on a surface, the following values for the adsorbed amounts are useful as guidelines. At surface concentrations below $1 \, mg/m^2$ the polymers lie flat on the surface, at concentrations around $3 \, mg/m^2$ the polymers adsorb in a monolayer in a coil configuration, and finally at surface concentrations above $5 \, mg/m^2$ the polymers adsorb as aggregates or multilayers at the surface. These values are, of course, only approximate and it is assumed that all of the surface is available for polymer adsorption.

Polyelectrolyte Adsorption can be Modelled Theoretically

A wide range of different polyelectrolytes is commercially available with very different structures and topologies. One simple example is poly(ethylene imine), a positively charged polymer with $- CH_2CH_2 -$ units between the nitrogen atoms which are protonated at normal pH levels (see Figure 19.14). The commercial polymer is highly branched and it is only the dimers and tri-mers that are truly linear. Amino acids such as glutamic acid and lysine can polymerize to produce anionic and cationic polyelectrolytes, respectively. The charges will, in these cases, reside on side chains of the polymer. In a first simple approach, one may assume that these details should be of minor import-ance and that the main effect is that we have connected charges in the poly-electrolyte. With this assumption, one can model the chain as a 'necklace' of

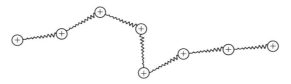

Figure 19.14 A schematic illustration of a simple polyelectrolyte model consisting of charged monomers connected with harmonic springs. Such springs do not model single chemical bonds, but rather a number of bonds connecting the charged atoms

charges connected either by stiff bonds or by harmonic springs. The theories discussed below are all based on this assumption.

Mean Field Approximations

All theories available in the literature for polyelectrolytes at surfaces are based on various mean field approximations (cf. the Poisson–Boltzmann approximation of Chapter 8). Some also include a lattice approximation as in the Bragg–Williams theory (cf. Chapter 10) and some treat continuous chains. Below we give a short description of a few approaches to the problem of polyelectrolyte adsorption and how it affects the forces between particles. This is an active area of research and the exposé below is by no means complete.

Scheutjens–Fleer   This is the most common theory for polymer and polyelectrolyte adsorption and has proved to be a quite successful model (see Figure 19.15). The theory gives good agreement with experiment, although it involves many parameters—citing the authors, 'all data and trends are explainable with our theory, using reasonable parameter values'. One advantage is that quite long chains can be handled in the formalism. However, the Scheutjens–Fleer theory has not been used very much for force calculations.

Miklavic–Marcelja   This theory treats only a single chain and essentially neglects the effect of all other adsorbed chains. It is limited to short chains, but on the other hand it treats the excluded volume term within the chain exactly.

Polyelectrolyte Poisson–Boltzmann   This is a direct extension of the Poisson–Boltzmann equation to also incorporate the chain connectivity. It is not a theory for adsorption problems, but it turns out to be very accurate for force predictions, as comparisons with Monte Carlo simulations have shown. The theory is limited to rather short chains and the resulting equations can only be solved numerically.

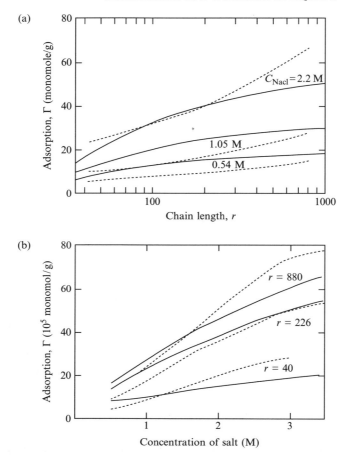

Figure 19.15 Polyelectrolyte adsorption as functions of (a) chain length (r), and (b) ionic strength (C_s). Continuous curves represent the Scheutjens–Fleer theory, while dashed curves are experimental results

Podgornik and Opheusden   This theory has treated an infinite continuous chain between two charged walls. The relevance of this type of model remains to be shown, although qualitatively it seems to capture some important physical features.

Distribution Profiles

The distribution and configuration of polyelectrolytes outside a charged surface can in principle be determined via scattering experiments, although these turn out to be difficult to interpret. Indirect information can be obtained from other

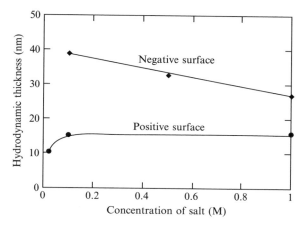

Figure 19.16 The hydrodynamic thickness of poly(styrene sulfonate), adsorbed on polystyrene latex particles, as a function of ionic strength. Reproduced by permission of Academic Press from T. Cosgrove *et al.*, *J. Colloid Interface Sci.*, **111** (1986) 409

sources, such as measurements of the hydrodynamic radius. Figure 19.16 and 19.18 show that polyelectrolytes adsorb in a flat configuration at oppositely charged surfaces, while they take on a more extended structure when adsorbed on an equally charged or neutral surface. This can be understood from the attractive electrostatic surface–chain interactions in the former case and from the chain–chain repulsions in the latter case. The addition of salt has only a weak influence on the distribution profile, as can be seen in Figure 19.17.

Polyelectrolytes Change the Double-Layer Repulsion

The addition of flexible polyelectrolytes to a charged colloid has a strong impact on the colloidal force and could cause both a stabilization and destabilization of the colloid. This is used in many technical applications, e.g. in pulp production, food industry and water refinement (purification) to mention just a few. Biological systems also contain flexible polyelectrolytes, such as spermine/spermidine for 'packing' of DNA and appear in the blood coagulation process and as glycolipids at membrane surfaces.

The force behaviour in the presence of polyelectrolytes is diverse and we cannot expect the Poisson–Boltzmann (PB) equation to explain experimental observations. In fact, a large variety of 'non-DLVO' behaviour is seen in such systems. We will discuss here a few idealized situations using a simplified polymer model, which however still retains two important properties—connectivity and flexibility. These are of special importance when polyelectrolytes interact with charged surfaces and modulate their interactions.

(a)

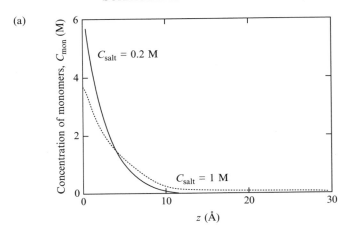

(b)

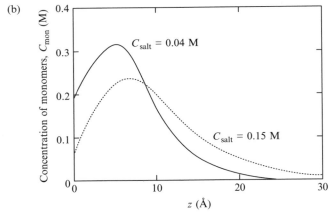

Figure 19.17 Distribution profiles from Monte Carlo simulations for charged monomers of terminally anchored polyelectrolytes to two oppositely charged walls 60 Å apart. Only one half of the system is shown with one of the walls placed at $z = 0$. (a) High surface charge density, and (b) low surface charge density

Bridging Attraction by Polyelectrolytes

Consider a simple double-layer system where the interaction between two charged surfaces is strongly repulsive. Now convert the ions into polyelectrolyte chains by connecting them with harmonic bonds as depicted in Figure 19.19, i.e. flexible polycounterions. One can imagine three distinct situations with:

(1) A fraction of the counterions is connected, giving an 'undercompensated' system.

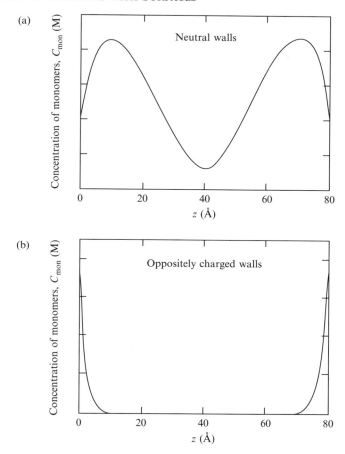

Figure 19.18 Distribution profiles for charged monomers in a terminally anchored polyelectrolyte to two (a) neutral walls and (b) oppositely charged walls. Data obtained from Monte Carlo simulations

(2) All counterions are connected, giving a 'perfectly matched' system.

(3) Extra salt is added and some of these ions are connected as well, giving an 'overcompensated' system.

In the perfectly matched system, one finds that the original double-layer repulsion has completely disappeared and a very strong attraction appears at short separation (Figure 19.20). The extra attraction comes from a bridging of the polyelectrolyte chains at short separations. The driving force for the bridging is a gain in chain entropy. At large surface–surface separations, the chains are confined to their respective surfaces (Figure 19.21(a)) with comparatively low

(a)   (b)   (c)

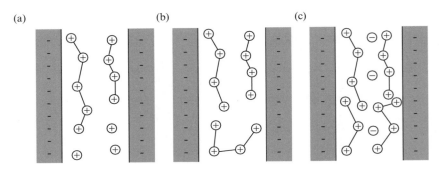

Figure 19.19 Two planar charged walls neutralized by polyelectrolytes and simple small ions: (a) an 'undercompensated' system where a fraction of counterions are connected; (b) a 'perfectly matched' system with all counterions connected; (c) an overcompensated system, which contains additional salt ions that are connected

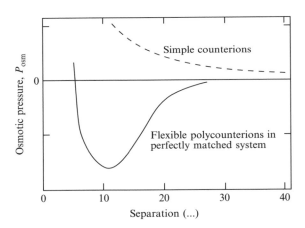

Figure 19.20 The osmotic pressure as a function of separation in a 'perfectly matched' system (continuous line), and in a double layer with unconnected small ions (dashed line)

chain entropy. When the separation decreases and becomes of the order of the monomer–monomer distance, then the chains can, with a small electrostatic energy cost, bridge over to the other surface and thereby substantially increase the chain entropy (cf. Figure 19.21(b)).

The osmotic pressure can be determined via the so-called contact theorem (cf. equation (8.5)). For a polyelectrolyte system, the contact relationship should be modified to also include the bridging term, as follows:

$$p_{\text{osm}} = kTc(0) + p^{\text{bridge}} + p^{\text{corr}} \tag{19.2}$$

(a)   (b)

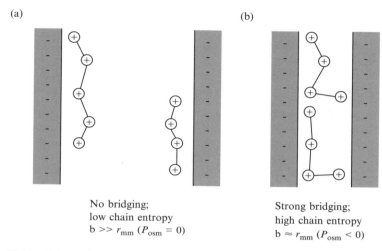

No bridging;
low chain entropy
$b \gg r_{mm}$ ($P_{osm} = 0$)

Strong bridging;
high chain entropy
$b \approx r_{mm}$ ($P_{osm} < 0$)

Figure 19.21 Schematic representations of the bridging mechanism; r_{mm} is the average distance between *charged* monomers

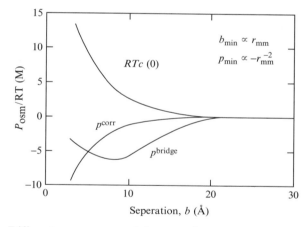

Seperation, b (Å)

Figure 19.22 Different components of the osmotic pressure in a 'perfectly matched' polyelectrolyte double layer as function of separation. The variation of the force minimum with the monomer–monomer distance is also indicated

The third term in equation (19.2) becomes more important in systems with divalent counterions. Figure 19.22 shows a comparison of the entropic term, $kTc(0)$, and the bridging term, p^{bridge}, as a function of the separation between the surfaces. The position of the force minimum is found at a distance approximately equal to the monomer–monomer separation in the polyelectrolyte and

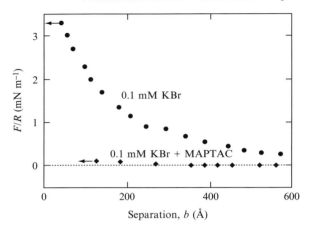

Figure 19.23   The interaction of two mica surfaces in 10^{-4} M KBr (●) and 10^{-4} M KBr plus cationic polyelectrolyte (MAPTAC) (♦). Data are taken from a surface force experiment. Reprinted with permission from M. Dahlgren *et al.*, *J. Phys. Chem.*, **97** (1993) 11769. Copyright (1993) American Chemical Society

the magnitude varies approximately as the inverse square of the monomer–monomer separation.

Up until now, the focus has been on theoretical results and it seems legitimate to ask whether they are directly reproduced experimentally. Figure 19.23 shows a surface force experiment with 10^{-4} M KBr in one experiment and the addition of a polyelectrolyte, i.e. poly((3-(methacrylamido)propyl)trimethylammonium chloride) (MAPTAC), in another. MAPTAC has the following chemical structure:

$$[-CH_2-CCH_3-CO-NH-(CH_2)_3-N^+(CH_3)_3]_n$$

The addition of MAPTAC completely wipes out the double-layer repulsion and an attraction appears at about 100 Å. Increasing the salt concentration to 10^{-2} M makes the double-layer repulsion appear again. These results can be interpreted as being due to an increased adsorption at high salt concentrations. The salt screens the repulsion between the polyelectrolyte charges and the charged walls adsorb more chains than necessary for neutralization and a charge reversal appears. This acting as an 'overcompensated' system. Similarly, repulsion will also appear if the amount of polymer adsorbed is less than what is required for neutralization. In both of these cases we will find ordinary double-layer repulsion. In fact, it is only when we have an approximately perfectly matched system that the double-layer repulsion disappears, as in Figure 19.24.

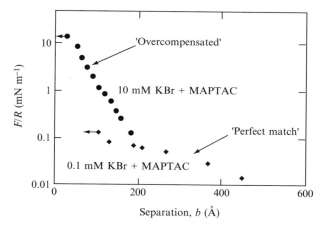

Figure 19.24 The interaction of two mica surfaces in a MAPTAC solution containing 10^{-4} M KBr (♦) and 10^{-2} M KBr (●). Reproduced with permission from M. Dahlgren *et al.*, *J. Phys. Chem.*, **97** (1993) 11769. Copyright (1993) American Chemical Society

In an ordinary double-layer system with small ions, the addition of salt will, in almost every situation, lead to a decreased repulsion. In a polyelectrolyte system, the opposite is possible, since the salt concentration will directly affect the amount of adsorbed polyelectrolyte as well as its configuration (see Figures 19.23 and 19.24). The addition of salt to a polyelectrolyte system also has a more subtle effect, since it affects the Donnan equilibrium. The salt balance in an 'ordinary' double layer and in one with polyelectrolytes is qualitatively different, as can be seen in Figure 19.24.

Asymmetric Systems

An additional attraction has been observed, both experimentally and theoretically, in asymmetric systems, e.g. when one wall carries adsorbed polyelectrolytes and the other is a bare charged surface (e.g. mica). To understand why an attraction takes place, one has to once more consider the contact theorem. In the previous formulations, it was assumed that the two halves are, on the average, electroneutral. If this is not the case another term must be added, as follows:

$$p_{osm}^{\bullet} = kTC(0) + p^{bridge} + p^{corr} - \frac{\sigma_0^2}{2\varepsilon_0\varepsilon_r} \qquad (19.3)$$

where $|\sigma_0|$ is the net charge in each half. The charge asymmetry could give rise to a very long ranged attraction. The driving force for the creation of a charge asymmetry is usually the mixing entropy of the counterions. Whether the force is asymptotically attractive or repulsive depends on the salt concentration and on the details of the charged surfaces.

Neutral Walls

The 'ordinary' double layer referred to above is usually created by charged surface groups, as in mica, or adsorbed ions, such as AgI sols. One can also imagine charging up a surface by adsorbing polyelectrolytes at a neutral wall (as a matter of fact, such a situation is not very difficult from the overcompensated system discussed earlier). Polyelectrolytes adsorbed at a neutral surface will, of course, be neutralized by counterions. The counterions will for entropic reasons distribute into the solution and 'drag' the chains with them. The final situation will be one with extended chain configurations, with a magnitude largely dependent on the salt concentration. This is yet another example of the importance of the entropic forces and the fact that the behaviour of a charged polymer is determined by its counterions.

The long-range interaction between two such polyelectrolyte 'brushes' will be the same as for an ordinary double layer, although the separation to use in the PB equation is not the wall–wall separation, but has to be modified by the polymer extension. Approximately, one can use the following:

$$b^* = b - 2 \text{ (length of the chain)} \tag{19.4}$$

At short separation, the interpenetration of chains will give rise to additional repulsive forces.

Energetic Bridging

Another type of bridging will occur between finite charged aggregates like micelles or sols. The situation that can appear is depicted in Figure 19.25. In this case, a polyelectrolyte chain neutralizes two or several particles. If the aggregates are far apart (the separation is longer than the contour length of the chain), the chain will be captured by one of the aggregates and the interaction

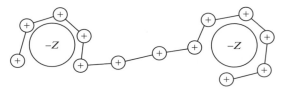

Figure 19.25 Schematic representation of a polyelectrolyte chain interacting with two oppositely charged spheres

will be weakly attractive. At intermediate separations, the chain may 'bridge' both particles, which is, from an energetical point of view, very favourable and a strong attraction is found. The main driving force for the bridging is the neutralization of the aggregates. Thus, one may speak of an 'energetic' type of bridging in contrast to the 'entropic' bridging discussed in the section on the bridging attraction by polyelectrolytes at planar interfaces.

In summary, we find that the addition of polyelectrolytes to charged colloids can induce attractive interactions, but it can also cause repulsive forces. Extra salt will screen the electrostatic interactions between the colloidal particles, as well as between the polyelectrolyte chains. The net result can be either an increased repulsion or an increased attraction. Thus, electrostatics alone is capable of producing a large variety of effects in polyelectrolyte systems. Specific polymer–surface interactions will, of course, further complicate the picture.

Polymer Adsorption is Practically Irreversible

The question of whether or not a polymer is reversibly adsorbed has been the subject of many studies. The conclusion is that although a polymer is reversibly adsorbed, it can practically be considered as irreversibly adsorbed. This apparent contradiction is explained by the slow dynamics in polymer systems. In order for a polymer to desorb from a surface, all polymer segments that are attached to the surface must detach at approximately the same time. If only a few segments detach, there is a high probability that other segments will adsorb before the whole polymer desorbs. Thus, the polymer chain is grounded on the surface by its own inertia. In analogy, this is the same inertia that causes a 'busload of people to move more slowly than a single person'. Thus, over a limited time-scale, polymers can be considered to be irreversibly adsorbed, while on very long time-scales the adsorption is indeed reversible.

The fact that polymers adsorb almost irreversibly at surfaces can be used for surface modification. Hydrophilization of surfaces with surface active polymers was discussed in Chapter 12. An example where a polyelectrolyte is used to change the charge of a surface is shown in Figure 19.26. Here, poly(ethylene imine), which is a cationic polymer, is used to recharge an anionic surface. The cationic polymer is fully charged at pH 4 and hence it will adsorb strongly to the anionic surface, as shown in Figure 19.26(a), giving a perfectly matched system. This configuration in the adsorbed state does not lead to a cationic charged surface, however. If the polymer is adsorbed at a higher pH (pH 8) where only a fraction of the cationic sites are charged, the polymer will adsorb in a more coiled configuration, as shown in Figure 19.26(b). The pH of the system can subsequently easily be changed to pH 4 and the already adsorbed polymer will be fully ionized, thus resulting in a charged cationic surface, i.e. giving an overcompensated system, as shown in Figure 19.26(c). Here, the

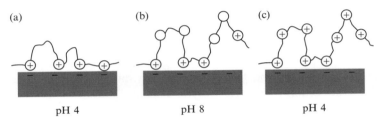

Figure 19.26 Schematics for the adsorption of poly(ethylene imine) on an anionic surface, illustrating that the apparent irreversibility can be utilized to reverse the charge in order to obtain a cationic surface

inertia of a polymer is used since sufficient time is not given for the polymer to desorb and reach the equilibrium state at pH 4, as depicted in Figure 19.26(a). Given sufficient time, however, the system will relax into the configuration shown in Figure 19.26(a).

The Acid–Base Concept can be Applied to Polymer Adsorption

The operational concept of Lewis acids and bases has been fruitfully applied to the adsorption of polymers from solution. A Lewis acid is an acceptor of electron pairs, while a Lewis base is a donor of electron pairs. Examples of the two groups are chloroform and tetrahydrofuran, respectively. In the adsorption process there is always a competition between the three components involved, i.e. the polymer, the solvent and the surface, and here the acid–base concept can be utilized in order to rationalize the adsorption behaviour. Figure 19.27(a) shows the adsorption of poly(methyl methacrylate) (PMMA), which is a Lewis base, at a silica surface, which is a Lewis acid. The adsorption is measured from different solvents of varying acidity or basisity. The solvents are categorized with respect to their acidity or basisity along the x-axis in Figure 19.27. To the left, in the very basic solvents there is no polymer adsorption. This is due to competition for the surface between the basic polymer and the basic solvent; the more basic the solvent, then the lower the polymer adsorption. On the right-hand side, in acidic solvents, we have the opposite situation. Here, there is a low polymer adsorption, which is due to a competition for the basic polymer between the acidic surface and the acidic solvent. The more acidic the solvent, then the more the basic polymer wants to stay in the solution, so resulting in a low adsorption at the less acidic surface. If neutral solvents are used, however, there is a large polymer adsorption. These principles have technical implications, as indicated in Figure 19.28, where the modulus of a PMMA plastic is plotted as a function of the volume fraction of silica filler. The plastic was cast from two different solvents, i.e. tetrahydrofuran (THF), which is a very basic solvent, and methylene chloride, which is a slightly acidic solvent. The figure shows that the modulus increases with filler

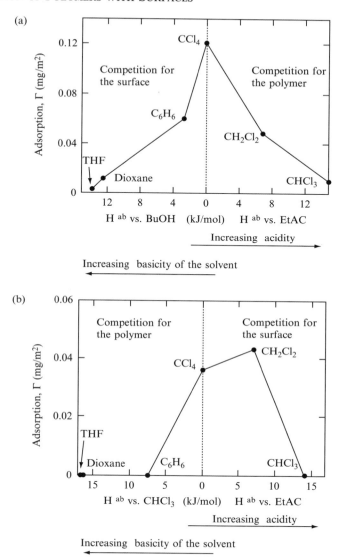

Figure 19.27 (a) Adsorption of PMMA on silica, and (b) adsorption of chlorinated PVC on calcium carbonate from a variety of solvents that differ with respect to their acid/base properties. Reprinted with permission from F. Fowkes and M. Mostafa, *Ind. Eng. Chem. Prod. Dev.*, **17** (1978) 3. Copyright (1978) American Chemical Society

content for the plastic cast from methylene chloride, while it is more or less constant for the plastic cast from tetrahydrofuran. The poor result for the latter system is due to an adsorption of basic THF on the acidic silica filler

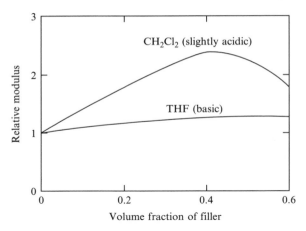

Figure 19.28   Relative modulus of PMMA as a function of silica content for samples cast from methylene chloride and THF, respectively. From F. Fowkes, in *Corrosion Control by Coatings*, H. Leidheiser (Ed.), Scientific Press, New Jersey, 1979, p. 483

surface, preventing polymer-filler molecular contacts, and hence no mechanical strengthening is obtained. Thus, in order to obtain a high-modulus plastic cast, a solvent that has no affinity for the filler surface should be chosen.

The same Lewis acid–base treatment is, of course, valid for the adsorption of an acidic polymer on to a basic surface, which is shown in Figure 19.27(b) for the adsorption of chlorinated PVC on to calcium carbonate.

The competition between polymer and solvent molecules for the surface can be measured accurately in the following experiment. First, the polymer is allowed to adsorb to the surface from a neutral solvent. Then a second solvent is added. This solvent has a certain affinity to the surface and thus competes with the polymer for the surface sites. At higher concentrations of this second solvent, the polymer will completely desorb and therefore this second solvent has been termed a 'displacer'.

Figure 19.29 shows how poly(vinyl pyrrolidone) (PVP) is continuously desorbed from a silica surface in water as the displacers are added in increasing concentration until the critical displacer concentration (at which all polymer is desorbed) is reached. The figure shows that *N*-methylpyrrolidone (NMP) is the most efficient displacer, being able to completely desorb the polymer at a volume fraction in water of *ca.* 1%, while dimethyl sulfoxide, (DMSO) does not completely desorb the polymer even if present as a neat solvent. These results correlate perfectly with the adsorption strength of the displacers at silica from aqueous solutions. The adsorption strength at low concentrations is proportional to the slope of the adsorption isotherm, which for the displacers have the following values: *N*-methylpyrrolidone (NMP), 4300; *N*-ethylmorpholine (NEM), 44; pyridine (PYR), 5.3; *N*-ethylpyrrolidone (NEP), 3.5; dimethyl

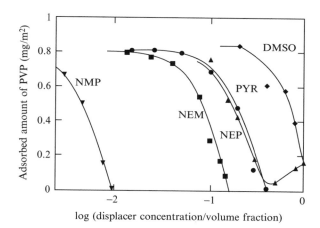

Figure 19.29 Desorption of poly(vinyl pyrrolidone) from silica in aqueous solutions with varying concentration of 'displacer', showing that total polymer desorption is obtained at a critical displacer concentration. Reproduced by permission of Academic Press from M. Cohen Suart, G.Fleer and J. Scheutjens, *J. Colloid Interface Sci.*, **97** (1984) 515 and 526

sulfoxide (DMSO), 2.8 (see Chapter 17 for an analysis of adsorption isotherms). Thus, the stronger the adsorption of the displacer, then the less displacer is needed to totally remove a polymer from the surface. We conclude that a polymer can be 'washed out' from a surface by adding a displacer that has a stronger affinity for the surface than the polymer itself.

Measurement of Polymer Adsorption

Polymer adsorption is most commonly measured through the concentration depletion of a solution after being in contact with the particle surface. Thus, to a solution of known volume and polymer concentration, the particles, e.g. latex or pigment, are added. After equilibration, the particles are separated and the polymer concentration in the solution is determined. Commonly, most problems occur in the selection of a suitable and accurate analysis method for the polymer. If only qualitative results are sufficient, there is an alternative method through the determination of the electrophoretic mobility of the particles. In electrophoresis, the mobility of the particles in an electric field is determined. This mobility is a measure of the potential at the slipping plane just outside the particle surface. The potential depends on the distance from the particle surface, as shown in Figure 19.30(b). When a non-ionic polymer is adsorbed at a particle surface the slipping plane will be extended further out from the particle surface, as shown in Figure 19.30(a). Thus, a lower absolute electrophoretic mobility, or zeta potential, will be detected, as shown in Figure 19.30(c). Such

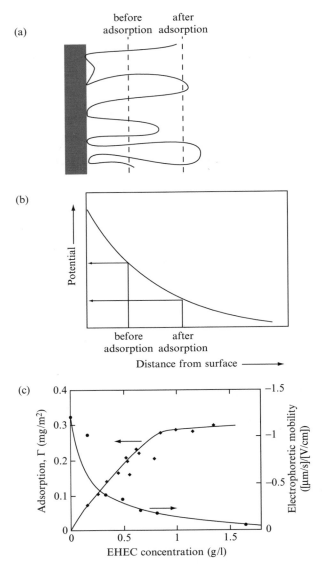

Figure 19.30 Illustrations of how the electrophoretic mobility changes upon the addition of a neutral polymer. The adsorbed polymer ethyl hydroxy ethyl cellulose (EHEC) shifts the slipping plane out from the surface, thus resulting in a lower electrophoretic mobility

measurements are rapidly performed and can serve as a guideline for the interaction between the polymer and the particle surface. The electrophoretic mobility can also be used to detect the adsorption of ionic polymers on particle surfaces, but care should be taken since the results might not always be unambiguous.

Polymer adsorption on macroscopic surfaces is conveniently measured by ellipsometry. Here, elliptically polarized light is reflected at the surface at which the polymer adsorbs. The polarization of the reflected light is sensitive to the polymer layer and the method allows both the adsorbed amount and the adsorbed layer thickness to be determined.

Polymer adsorption at the water–air surface is easily detected by surface tension measurements. In Chapter 16, it was demonstrated that the polymer might have a profound effect on the surface tension, even at very low polymer concentrations. This rapid lowering of the surface tension is due to the low solubility of the polymer in the solution.

We will finally illustrate the macromolecular nature of polymer adsorption using surface tension measurements. Consider the adsorption of poly (methacrylic acid) from aqueous solution at the liquid–air interface. As a model for the monomer unit, isobutyric acid is used. First, we stipulate that it is the non-ionized species that are surface active, i.e. which adsorb at the liquid–air interface. Thus, at the zero degree of ionization the surface tension, which is a reflection of the adsorption (see Chapter 16), has the lowest possible value for both the polymer and monomer systems. As a consequence, the surface tension has its highest possible value at very high degrees of ionization.

Figure 19.31(a) shows that the surface tension of isobutyric acid solutions increases linearly with the degree of ionization. However, for the polymer system the surface tension makes a jump at a degree of ionization of *ca.* 0.1. The reason for this jump is found in the polymeric nature of the system. When the degree of ionization is increased, some segments of the adsorbed polymer will be ionized. However, the non-ionized segments are still in sufficient amounts to keep the polymer at the surface. At this critical degree of ionization, the whole polymer will leave the surface, thus causing the jump in surface tension seen in Figure 19.31(a). An analogy is shown in Figure 19.31(b) where corks (non-dissociated monomers, or segments) float on water and where weights are hung on to the corks (dissociation process), pulling them down into the liquid. However, if the corks are connected by a string (a polymer) all corks will float until a critical weight (critical degree of dissociation) is reached, whereupon the surface will be free of any corks. A further discussion on polymers at liquid interfaces was given in Chapter 16.

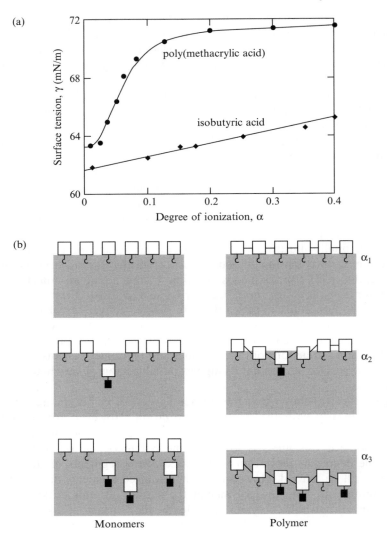

Figure 19.31 (a) Surface tension of poly(methacrylic acid) and isobutyric acid as a function of the degree of ionization. (b) An analogy using corks and weights, illustrating the adsorption of monomers and polymers at an interface From A. Katchalsky and P. Miller, *Atti del 1° Congresso Internazionale Meterie Plasiche*, Torino, Italy 19–22 September 1949

Bibliography

Fleer, J. G., M. A. Cohen Stuart, J. M. H. M. Scheutjens, T. Cosgrove, and B. Vincent, *Polymers at Interfaces*, Chapman & Hall, London, 1993.

Sato, T. and R. Ruch, *Stabilization of Colloidal Dispersions by Polymer Adsorption*, Marcel Dekker, New York, 1980.

20 FOAMING OF SURFACTANT SOLUTIONS

A foam is a dispersion of a gas in a liquid or solid. Solid foams will not be dealt with here. The ratio of gas to liquid determines the appearance of the foam, with this ratio being called the foam number. At low foam numbers, the gas bubbles are spherical and the liquid lamellae between the bubbles are very thick. A well-known example of such a foam is whipped cream. At higher foam numbers, the gas bubbles will be separated by thin and planar liquid lamellae. The region where three such lamellae meet is called the *plateau* border.

There are Transient Foams and Stable Foams

There are two types of foams and they require different methods of characterization. These are unstable foams, or so-called transient foams, and stable foams. Transient foams are commonly characterized by the Bikerman method. Here a gas is blown through the liquid in a column with a sintered glass frit at the bottom and the foam height, or foam volume, is recorded when the foam height has reached a steady state (see Figure 20.1(a)). This is repeated at different gas flows and the foam volume is normally linearly dependent on the gas flow, as illustrated in Figure 20.1(b). The slope of a plot of the foam volume versus the volumetric gas flow is interpreted as the average lifetime of a foam bubble, or a foam lamella. It is essential that there is excess liquid at the bottom of the cell at all times, since otherwise the foam volume is limited by the amount of liquid available.

Stable foams, on the other hand, are characterized by their foam ability and their foam stability. A measure of the foam ability is the foam volume immediately after the generation of the foam, while a measure of the foam stability is the lifetime of the formed foam. Some protein solutions, for example, show a very low foam ability and high foam stability, whereas some surfactant solutions show high foam ability but low foam stability.

The foam volume, formed after shaking the liquid in a test tube, can be used to characterize the foam ability of stable foams. This method has the disadvantage of being dependent on the operator, viscosity and volume of the liquid, as

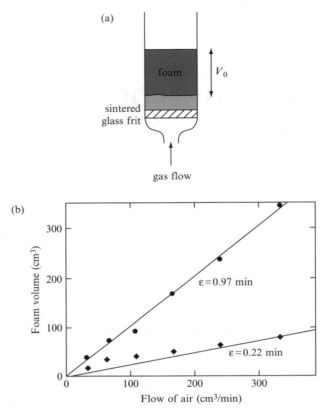

Figure 20.1   In the Bikerman method, gas is continuously blown through the liquid in a column (a), and the foam volume is recorded as a function of the air flow (b) as exemplified here for two crude oils

well as the size and shape of the container. On the other hand, the Ross–Miles characterization method is an ASTM standardized method that gives both a measure of the foaming ability and the foam stability in a reproducible manner. In this procedure, foaming liquid is poured into the foam column and the immediate foam volume is recorded, giving a measure of the foam ability. The foam volume is then again recorded after 5 or 10 min, giving a measure of the foam stability. The set-up for the Ross–Miles foam test is shown in Figure 20.2.

Two Conditions must be Fulfilled for a Foam to be Formed

Foams are always formed from mixtures, whereas pure liquids never foam. There are two important conditions that have to be fulfilled in order for a liquid mixture to foam. The first condition is that one component must be surface

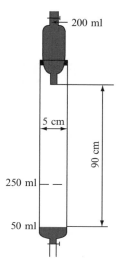

Figure 20.2 In the Ross–Miles foam test (ASTM D 1173–63), a specified amount of liquid is poured into a foam column and the immediate foam height, as well as that formed after 5 or 10 min, is recorded

active. The lowering of the surface tension upon adding the second component is a measure of the surface activity. Since most organic compounds have a relatively low surface tension when compared to that of water (see Chapter 16), it is not surprising that aqueous solutions with organic additives form foams relatively easily.

The second condition is that the foam film must show a surface elasticity, i.e. there must be a force pulling back the foam film if it has been stretched (see Figure 20.3). The surface elasticity, E, is defined as the increase in surface free energy, or the surface tension, γ, as the surface area, A, is increased, i.e. as the lamella is stretched:

$$E = A \frac{d\gamma}{dA} \tag{20.1}$$

This condition of a surface elasticity must be valid in the time period during which the lamella is stretched and restored. Thus, it is a prerequisite for foaming that the diffusion of the surface active component from the bulk solution to the newly created surface is sufficiently slow. If this is not the case, the adsorption at the surface will decrease the surface tension and the temporary stretch of the foam lamella will be made permanent with a weakening of the lamellae as the result.

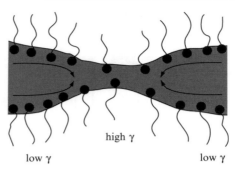

high γ

low γ   low γ

Figure 20.3 When a foam lamella is stretched, by mechanical vibrations or thermal instability, the surfactant concentration at the surface, i.e. at the stretched part, is suddenly lowered, thus resulting in an increased surface tension and hence a restoring force

Thus, in order to attain elasticity, the foaming agent, i.e. the surfactant, is not allowed to diffuse from the interior of the lamellar film to the newly created surface before the film retracts. Surfactants with a very high critical micelle concentration (CMC), i.e. surfactants that give high concentrations of unimer, will not form stable foams since the high bulk concentration will allow the newly created surface to be covered with surfactant diffusing from the interior of the lamellar film before the film retracts. In addition, aqueous solutions of solutes that do not form micelles show low foamability. An example is aqueous solutions of ethanol that do not foam, despite the fact that ethanol lowers the surface tension of water, as shown earlier in Figure 16.4.

There are Four Forces Acting on a Foam

The most obvious force acting on a foam is the gravitational force, causing drainage of the liquid between the air bubbles (see Table 20.1). The drainage can be slowed down by either increasing the viscosity of the bulk liquid or by

Table 20.1 A summary of the four forces acting on a foam

Cause	Effect
Gravitation	
Pressure difference in lamellae and plateau borders	Drainage to foam base
Pressure difference of the gas in bubbles of different size	Drainage to plateau borders
	Diffusion of gas from small to large bubbles
Overlap between the electrical double layers	Increase in foam stability

adding particles either solid particles or emulsion droplets. These particles become trapped in the plateau borders, so hindering further drainage by a local increase in the viscosity, as shown in Figure 20.4.

For stable foams, i.e. those with high foam numbers where thin lamellae form, the drainage due to the gravitational force will gradually be substituted for drainage due to capillary forces. These forces originate from the fact that the hydrostatic pressure in the plateau borders is lower than in the lamellae. This lower pressure is caused by the negative curvature of the liquid surface at the plateau borders. Thus, there is a driving force for the liquid to flow from the lamellae to the plateau borders, further decreasing the stability of the foam.

The third force acting on foams is less obvious. This stems from the fact that the gas pressure inside a bubble is inversely proportional to the size of the bubble. Thus, small bubbles have higher pressures than larger ones. There will therefore be a transport of gas from the small to the large gas bubbles through the liquid. It is thus possible for a foam to disappear without any lamellae breaking. This transport through the liquid is proportional to the solubility of the gas in the liquid. The foam stability of argon in aqueous solutions is therefore higher than of carbon dioxide in the same system, since carbon dioxide is more soluble in water than argon.

The fourth force acting on foams comes into play in very stable foams, where the lamellae become very thin. There will be an overlap between the electrical double layers originating from the surfactants adsorbed at the liquid–air interface. This overlap causes repulsion, so preventing further thinning of the lamellae. This phenomenon occurs at distances of the order of 10–100 nm, depending on the ionic strength of the system. It has recently been shown that even non-ionic surfactants give rise to a small negative charge at the interface. The origin of this charge is still being debated.

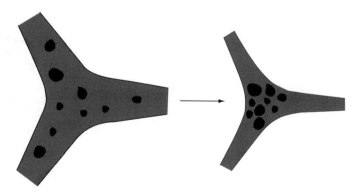

Figure 20.4 Emulsion droplets, or suspended particles, can hinder further drainage of the foam liquid, if they are trapped in the plateau borders

The addition of salt compresses the electrical double layer, whereby the stability is diminished. On the other hand, as will be discussed later in this chapter, salt addition also increases the critical packing parameter (CPP) of ionic surfactants and hence also the surface activity, in most cases causing an increased tendency of the solution to foam. Hence, there are two counteracting effects of salt and the literature can sometimes be very confusing on this matter.

Finally, a factor that has an obvious effect on foam life is the viscosity of the liquid. Naturally, very highly viscous foams are very stable, like foam mattresses, whipped cream or shaving cream. In many cases, it is not necessary that the bulk liquid is viscous as long as the air–liquid interface has a high viscosity. This is the case when lamellar liquid crystalline phases are formed in the system. These phases concentrate at the air–liquid interface, thereby locally increasing the viscosity and hence the foam stability.

The Critical Packing Parameter Concept is a Useful Tool

The foaming ability is closely coupled with the surfactant critical packing parameter (CPP). As the CPP increases, the surfactants at the air–water surface pack closer together and hence a higher cohesion is attained. This gives the liquid lamellae a good cohesion, increasing both the surface elasticity and viscosity, hence resulting in a high foaming ability and foam stability. Thus, according to this mechanism the foaming ability should continuously increase with increasing CPP.

On the other hand, the foaming ability is not only determined by the cohesion of the surfactant monolayer. Another important factor is the ease with which holes are spontaneously formed and grow in the foam lamellae. Thermal and mechanical fluctuations in the foam films lead to the formation of transient holes of molecular size. These holes are more easily formed with surfactant systems where the CPP is large. This is because the curvature of the hole is very large and the energy of forming a hole for surfactant systems with low CPPs is much larger than for surfactants with high CPPs (see Figure 20.5). Hence, according to this mechanism the foaming ability, and foam stability, should decrease as the CPP is increased.

We therefore have two counteracting phenomena acting on the foam as the CPP of the surfactant system is increased. The cohesion within the lamellae will increase with CPP and cause a higher foaming ability while the ease of hole formation will increase and cause a lower foaming ability. We therefore expect the foaming ability to display a maximum as the CPP is altered, as depicted in Figure 20.6.

At this maximum the two phenomena balance each other. At higher CPP values, the ease of hole formation dominates, while at lower CPP values the lack of cohesion in the surfactant monolayer dominates. This general behaviour will be illustrated below by a few examples.

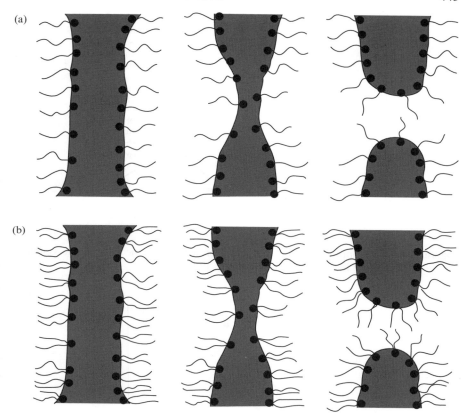

Figure 20.5 (a) Thermal and mechanical fluctuations lead to spontaneous formation of holes in the foam lamellae. (b) Such hole formation is facilitated for surfactant system with higher CPP value

The CPP of non-ionic surfactants is easily altered by changing the length of the polyoxyethylene chain. Figure 20.7 shows the foaming ability of aqueous solutions of ethoxylated nonylphenols, NP-E$_n$. This figure displays a pronounced maximum when the surfactant contains *ca.*75–85 wt% polyoxyethylene, corresponding to NP-E$_{20}$. Hence, for ethoxylated nonylphenol surfactants with a shorter polyoxyethylene chain, the ease of hole formation dominates, while for surfactants with longer polyoxyethylene chains the disability to form a highly cohesive surfactant monolayer dominates.

The CPP of non-ionic surfactants can also be altered by changing the temperature. At low temperatures, the polyoxyethylene chain is expanded, thus resulting in a large head group and hence a low CPP, while at higher temperatures the polyoxyethylene chain contracts, giving an increased CPP

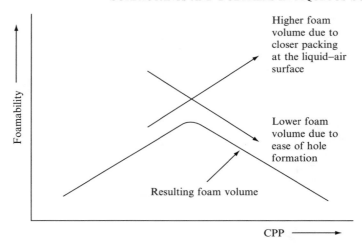

Figure 20.6 Two opposing effects are acting on a foam as the CPP of the system is altered, thus leading to a maximum in foaming ability

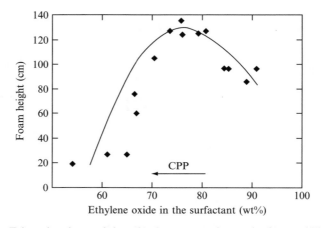

Figure 20.7 Ethoxylated nonylphenols show a maximum in foam ability when the polyoxyethylene chain constitutes *ca.* 75 wt% of the surfactant. This figure shows the immediate foam height as determined from a Ross–Miles test

value. The foaming ability shows a maximum at temperatures below the cloud point, in accordance with the mechanisms just outlined.

In ionic surfactant systems, the CPP can be changed by altering the hydro-carbon chain length, which is demonstrated in Figure 20.8. Here, the foam volume of aqueous solutions of alkyl sulfates at 60°C is plotted versus the number of carbon atoms in the alkyl chain. This figure shows a maximum in

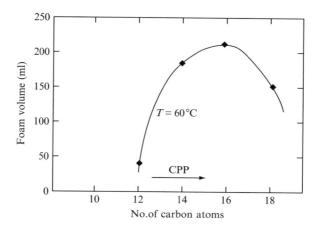

Figure 20.8 Foam volume at 60°C for a series of alkyl sulfates showing a maximum when the surfactant alkyl chain contains 16 carbons. From J. J. Bikerman, *Foams*, Springer-Verlag, Berlin, 1973, p. 114

foam volume when the surfactant alkyl chain contains 16 carbons. We thus conclude that for shorter-chain surfactants the disability to form a highly cohesive surfactant monolayer dominates, while for longer-chain surfactants the ease of hole formation dominates. We note in passing that these experiments were performed at 60°C in order to be well above the Krafft point of the systems.

The CPP of anionic surfactant systems is also susceptible to the addition of non-ionic long-chain amphiphiles, such as fatty acids or alcohols. These compounds increase the CPP of the total system and thus the foaming ability is increased if the CPP of the original surfactant is small. This is demonstrated in Figure 20.9, which shows that foams from aqueous solutions of soaps are sensitive to changes in pH. At high pHs, the fatty acids are dissociated (forming soaps with low CPPs), hence resulting in a poor foaming ability due to the disability to form a highly cohesive surfactant monolayer. On the other hand, at low pHs, where the undissociated fatty acids dominate, there is also poor foaming ability. This is due to the ease of hole formation in these systems since the CPP values are now large. Note that the foam maximum is displaced towards larger pH values for the longer alkyl chains. When lengthening the alkyl chain, the CPP is increased; thus, more charged species (higher pH) are needed in order to balance the above two forces.

We thus conclude that the formation of a foam is very sensitive to the CPP of the system, showing a maximum as the CPP is changed. There are pitfalls in this approach, however. One example is revealed when studying the effect of salts on foams. In ionic surfactant systems, the addition of salts will increase the CPP. Thus, for those systems with low CPPs the foaminess should increase with

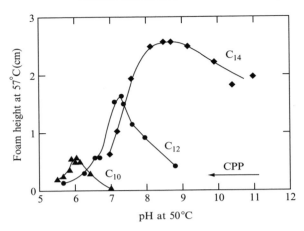

Figure 20.9 The pH dependence of the foam volume of some fatty acids, showing a maximum at intermediate pH values. From J. J. Bikerman, *Foams*, Springer-Verlag, Berlin, 1973, p. 111

salt addition. This is not always observed, however, since an increased salt content also lowers the double-layer repulsion between the two liquid–air interfaces in the lamellae, thereby decreasing both the foaminess and the foam stability.

Polymers might Increase or Decrease Foam Stability

Water-soluble polymers are present in many technical applications of surfactant solutions and hence it is of interest to know how the interaction between polymers and surfactants influence the foam stability. As an illustration we choose the system polyvinylpyrrolidone (PVP) and Sodium dodecyl sulfate (SDS) as described in Chapter 13.

In the region below T_1, where there is no surfactant association in the bulk, the polymer and the surfactant interact strongly at the surface, as can be seen from the lowering of the surface tension. The latter is most likely due to an enhanced surfactant adsorption, induced by the presence of the polymer. In this region, the foam stability increases upon addition of the polymer. The latter is likely to be situated in the vicinity of the surfactant head groups, creating both an increased surface viscosity and a steric repulsion between the two surfaces in the foam lamellae, thus enhancing the stability of the foam

In the region between T_1 and T_2, i.e. where the polymers associate with micelles that form in the bulk solution, there is a desorption of the polymer from the surface. There is now insufficient surfactant at the surface to produce the required elasticity needed to stabilize the foam films, while in the polymer-free system the surface is saturated with surfactant in this concentration region.

Hence, the foam stability is decreased upon addition of the polymer in this concentration region.

Finally, for surfactant concentrations above T_2, where all polymer in the system has been used up in the formation of polymer–micelle complexes, the foam stability increases, when compared to the polymer-free solutions. The simple reason is that the presence of polymer–micelle complexes increases both the bulk and the surface viscosity, thereby increasing the drainage and hence the foam stability.

Particles and Proteins can Stabilize Foams

Particles and surface active polymers, such as proteins, are two other groups of foam stabilizers besides surfactants in aqueous solutions. Particles stabilize foams by two different mechanisms. The first mechanism was already mentioned in connection with the drainage of foams. Fully dispersed particles become trapped in the plateau borders, thereby lowering the drainage rate. Note that this mechanism requires that the particles become completely dispersed, i.e. they have no affinity for the liquid–air surface.

In the second stabilizing mechanism, the particles themselves are surface active. For already hydrophilic particles this is achieved by adsorbing, or chemically reacting, hydrophobic moieties on the surface of the particles. Too large a hydrophobicity will, of course, precipitate the particles. If the particles are made only partly hydrophobic, however, they will show surface activity, i.e. they will have an affinity for the liquid–air interface. Such systems give very stable foams. The best foam stability is achieved when the contact angle of water is around 90°, i.e. when one half of the particle is in the liquid and the other half is in the air, as shown in Figure 20.10.

Similarly, proteins stabilize foams when the pH is close to the isoelectric point. Highly charged proteins far from the isoelectric point are in general

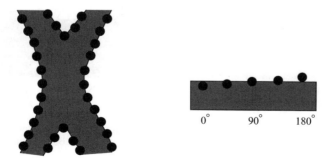

Figure 20.10   Solid particles act as foaming agents if they are made partly hydrophobic. The best foam stability is achieved when the contact angle is around 90°

easily soluble in water and do not show any surface activity. At pH values close to the isoelectric point, on the other hand, the protein is less soluble. In Chapter 19 it was demonstrated that the surface activity of polymers increases as the bulk solubility decreases. This is also the case for proteins. Hence, the proteins will be partitioned to the liquid–air surface as the pH approaches the isoelectric point. This increased surface activity is reflected in an increased foaminess. If, however, the pH is changed even closer towards the isoelectric point, the protein will, of course, precipitate.

Various Additives are Used to Break Foams

Foams in industrial processes are generally not wanted and can sometimes be the bottleneck for a high production rate. When unwanted foams appear in such systems it is common to add 'foam killers'. These are surface active compounds, or systems, that break foams by spreading on the foam lamellae. For example, a foam stabilized by an ionic surfactant can easily be broken down by spraying octanol on top of the foam. The octanol droplets have a very low surface tension and hence spread on the foam lamellae. This process carries with it a layer of liquid, below the surface, thus thinning the lamellae to the point of breakage (see Figure 20.11). Note that octanol can only be used by a spraying application. Mixing it with the system, in order to prevent foaming, might stabilize the foam since the addition of octanol might lead to the formation of lamellar liquid crystalline phases, which are incredibly efficient in stabilizing foam lamellae. Octanol should therefore be used with caution, since the final system might give even more stable foams. Another mechanism of foam breaking upon the addition of organic molecules, such as octanol, could be due to the fact that the total CPP of the system increases locally, since

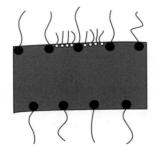 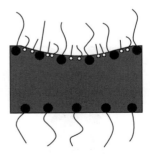

Figure 20.11 In the most common defoaming mechanism, the defoamer spreads on the surface of a foam lamella pulling with it a layer of liquid, below the surface, thus thinning the lamella to the point of breakage

the organic molecules partition in the hydrocarbon part of the surfactant monolayer at the air–water interface. The ease of hole formation is thereby increased, so leading to reduced foaming ability.

This spreading mechanism is also used in machine dishwashing detergents. Here, very hydrophobic surfactants of $EO_n-PO_m-EO_n$ or $PO_m-EO_n-PO_m$ have traditionally been used. These surfactants are only dispersible in water and act by the described mechanism.

Another and more common type of foam breaker is based on silicone oil or mineral oils. Here, the oil droplets will enter the liquid–air surface from the bulk and then spread on the surface. Since the oil itself does not show any surface elasticity, it will cause the foam film to break. The condition for an oil to spread on the aqueous surface is that the spreading coefficient should be positive, as follows:

$$S = \gamma_w - \gamma_{w/o} - \gamma_o > 0 \qquad (20.2)$$

where γ_w and γ_o are the surface tensions of the foam liquid and foam breaking oil, respectively. The $\gamma_{w/o}$ is the interfacial tension between the foam liquid and the foam breaking oil. Normally the surface tension of organic compounds is found in the range 25–40 mN/m, which is roughly that of the foam liquid if it is stabilized by surfactants. Thus, a prerequisite for spreading is that the foam breaking oil shows a very low interfacial tension towards the aqueous solution according to the inequality above. If not, the oil will just form a lens at the aqueous–air surface. Silicone oils are well suited as foam breakers since they have a low surface tension, as well as a very low interfacial tension towards water. In some applications, the use of silicone oils interferes with the performance of the product and thus substitutes have to be used. The most common of those are based on mineral oils. Such oils, however, have in general a higher interfacial tension towards water and will therefore not spread spontaneously. In order to achieve spreading, surface active compounds, such as long-chain fatty amides, are added to the mineral oil. These adsorb at the oil–water interface, thereby decreasing the interfacial tension. These additives must not be soluble in the aqueous phase since they will otherwise be depleted from the mineral oil and thereby decrease the efficiency of the foam breaker.

Particles are also added to the foam-killing oils. The purpose of the particles is to break a so-called pseudo-emulsion film that sometimes can be created between the oil droplet and the air. This pseudo-emulsion film prevents the oil droplet from entering the surface, despite the fact that energy would be gained by the process. The mechanism is most likely related to the high viscosity of this pseudo-emulsion film.

Bibliography

Akers, R. J., *Foams*, Academic Press, London, 1975.

Bikerman, J. J., *Foams*, Springer-Verlag, Berlin, 1973.

Garrett, P. R. (Ed.), *Defoaming: Theory and Industrial Applications*, Surfactant Science Series, 45, Marcel Dekker, New York, 1993.

Prud'homme, R. K. and S. A. Khan, *Foams: Theory, Measurements and Applications*, Marcel Dekker, New York, 1996.

Wilson, A. J., *Foams: Physics, Chemistry and Structure*, Springer-Verlag, Berlin, 1989.

21 EMULSIONS AND EMULSIFIERS

Emulsions are Dispersions of One Liquid in Another

An emulsion is a dispersion of one liquid in another liquid. The two liquids are obviously immiscible. Almost all emulsions contain water as one phase and an organic liquid as the other phase. The organic phase is normally referred to as the 'oil' but by no means it needs to be an oil in the normal meaning of the word. Non-aqueous emulsions also exist, such as emulsions of fluorocarbons in hydrocarbon. There are also examples of emulsions of two aqueous phases. In such systems, the two phases are normally solutions of two different polymers, such as dextran and poly(ethylene glycol).

There are two main types of emulsions, i.e. oil-in-water (o/w) and water-in-oil (w/o). Oil-in-water emulsions are by far the most important and common examples are paints, glues, bitumen emulsions, agrochemical formulations, etc. Spreads (margarines) are well-known examples of w/o emulsions. There also exist so-called double emulsions which may be w/o/w or o/w/o. Such systems are of interest for drug delivery. The droplets in an emulsions are referred to as the dispersed phase, while the surrounding liquid is the continuous phase.

If an oil is dispersed in water without any amphiphile or other kind of stabilizer added, the stability will be very poor. The oil droplets will collide and the collisions will lead to droplet fusion, i.e. coalescence. The rate with which individual droplets move due to gravity is proportional to (i) the density difference between the dispersed and the continuous phase, (ii) the square of the droplet radius, and (iii) the inverse of the viscosity of the continuous phase. For an emulsion of a typical hydrocarbon oil in water, calculations show that a 0.1 μm droplet moves 0.4 mm/day, a drop of 1 μm moves 40 mm/day and a drop of 10 μm moves 4 m/day. It is evidently important to make the droplets small.

Further calculations indicate that the half-life of a droplet of 1 μm, assuming a 1:1 oil-to-water ratio and a viscosity of the continuous phase of 1 mPa s, is 0.77 s. Six months shelf life means a half-life of 1.6×10^7 s. From these calculations, it can clearly be seen that the droplets need to be stabilized. As we shall see later, there are various ways to stabilize emulsions.

Emulsions can be Very Concentrated

Most technical emulsions have a concentration of dispersed phase in the range 25–50%. Such emulsions have approximately the same viscosity as the continuous phase. Since a high viscosity of the continuous phase is advantageous for emulsion stability, it is common practice to add a polymer or some other additive that increases the viscosity of this phase.

For emulsions of monodispersed droplets, 50% dispersed phase means rather close packing of the particles. In many cases, the particles in such emulsions may, in fact, reside in the secondary minimum (see below). The theoretical maximum for random packing of spheres is 64 vol % internal phase, while for hexagonal packing the value is 74 vol%. Nevertheless, emulsions can contain far above 90 vol% dispersed phase. For spherical droplets, this requires a broad distribution of droplet sizes so that smaller droplets can fill the spaces between the larger ones. One technically important example of such concentrated emulsions is emulsion explosives. These are w/o emulsions containing around 90% water in which ammonium nitrate is dissolved. Ammonium nitrate oxidizes the surrounding hydrocarbon oil to give carbon dioxide, water and nitrogen, all gaseous components that require a large volume.

In extreme cases, emulsions can have 99% dispersed phase. Such concentrated emulsions are 'gel-like' and their structure resembles that of concentrated foams with a polyhedral rather than a spherical structure of the droplets/air bubbles.

Emulsions can Break Down According to Different Mechanisms

Emulsion break-down can occur by various mechanisms, as illustrated in Figure 21.1. Creaming or sedimentation occurs as a result of the density difference between the phases. Creaming is more common than sedimentation because most emulsions are of the o/w type and the oils tend to be of lower density than the aqueous phase. Some oils, such as chlorinated hydrocarbons, have higher densities than water, however, and addition of a chlorinated solvent to the oil component was once used as a way to improve the storage stability of emulsions. Droplets can also flocculate, which means that they enter the so-called 'secondary minimum', an energetically stable situation where the droplets are close to each other but still retain their integrity. Creaming, sedimentation and flocculation are all reversible phenomena and the original state can often be regained by application of high shear. A much more severe phenomenon is coalescence, a process in which droplets merge into each other. Coalescence can be seen as an irreversible phenomenon. Sometimes, if the secondary minimum is weak or non-existent, there is no real distinction between flocculation and coalescence.

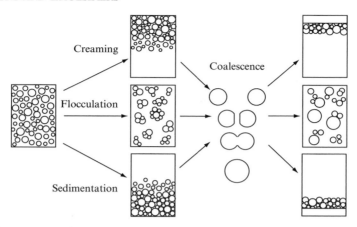

Figure 21.1 Mechanisms for the destabilization of emulsions. Note that different processes can occur simultaneously

Yet another destabilization mechanism (not illustrated in Figure 21.1) is Ostwald ripening. In this process, small drops, which have a large area per volume, lose material and finally disappear while bigger drops grow in size. The process is caused by diffusion of molecules of the dispersed phase through the continuous medium. For o/w emulsions, the rate of Ostwald ripening depends on the solubility of the oil in water and the process is much more pronounced for a water-continuous emulsion of a small hydrocarbon, such as hexane, than of a larger hydrocarbon, such as hexadecane.

When discussing emulsion stabilization, a parallel may be drawn with the stabilization of solid dispersions, i.e. suspensions. In suspensions, the dispersed phase is normally of a higher density than the solvent, and so sedimentation, not creaming, occurs. Flocculation also takes place and, as was the case with emulsions, this may lead into the secondary minimum which means that the suspension can be redispersed relatively easily. Sooner or later the solid particles will aggregate, which is an irreversible process, equivalent to the coalescence of emulsion droplets. If the solid particles have some water solubility, which is the case for most salts and oxide materials, Ostwald ripening will lead to a gradual increase of particle size, and often to a more narrow size distribution.

The Emulsion Droplets Need a Potential Energy Barrier

As mentioned above, a standard emulsion droplet with a radius of 1 μm has a half-life of less than 1 s. If a potential energy barrier is introduced at the surface, the half-life can be extended to days or even years. One can show that an energy barrier of $10kT$ (where k is the Boltzmann constant and T is the absolute temperature) reaching one radius out from the droplet gives a half-life of 2 h, a barrier of $15\,kT$ a half-life of 1.3 days, and a barrier of $20\,kT$ a half-life of 5 years. The potential energy barrier is obviously decisive in emulsion stability. How can such barriers be created?

It is common practice to distinguish between different modes of emulsion stabilization, although in reality these are often combined.

Electrostatic Stabilization

Electrostatic stabilization by ionic (in particular anionic) surfactants is very common. In addition, polyelectrolytes can be used for this purpose. Electrostatic stabilization is based on the repulsive interaction that results when the diffuse double layers around the particles start to overlap. The overlapping gives an increase in ion concentration, which results in a loss in entropy.

Steric Stabilization, sometimes called Polymer Stabilization

Steric stabilization can be achieved by non-ionic surfactants having long polyoxyethylene chains as the polar head groups. Non-ionic polymers are also frequently employed to provide steric stabilization. A necessary requirement for steric stabilization to be effective is that the continuous phase is a good solvent for the polymer chains which protrude into the surrounding medium. The solvent should be a 'better-than-theta solvent' (see Chapter 9). Steric stabilization is caused by the repulsive force that arises as a result of a decrease in entropy when chains from two droplets start to entangle.

Figure 21.2 illustrates the electrostatic and steric stabilization provided by anionic and non-ionic surfactants, respectively.

Particle Stabilization

Solid particles can be used to stabilize emulsions. The particles should be small in comparison to the emulsion droplets and they should be relatively hydrophobic. The best effect is obtained with particles that form a 90° contact angle with the oil droplets. Such particles are balanced in that they protrude equally into the two liquid phases, as shown in Figure 21.3. Hydrophobic proteins, often proteins at their isoelectric points, can function in a similar way.

Stabilization by Lamellar Liquid Crystals

Surfactants can pack in multilayers around the droplets, forming lamellar liquid crystals (see Chapter 3). Such a multilayer arrangement is stable and can provide very long-lived emulsions. Formation of lamellar liquid crystals can sometimes also be the key to making emulsions with a minimum of energy

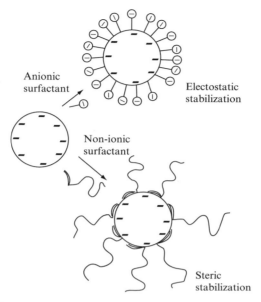

Figure 21.2 Electrostatic and steric stabilization of an emulsion by surfactants

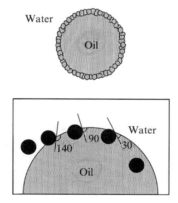

Figure 21.3 Stabilization of an emulsion by small particles

input, so-called 'spontaneous emulsification'. Figure 21.4 illustrates such lamellar liquid crystals at the surface of an oil droplet.

Combination of Stabilizing Mechanisms

As mentioned above, emulsions are often stabilized by more than one mechanism. In many systems, electrostatic and steric stabilization are combined. Such

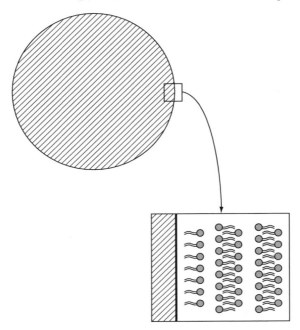

Figure 21.4 Stabilization of an emulsion by a lamellar liquid crystal. Reproduced by permission of John Wiley & Sons, Inc. From D.F. Evans and H. Wennerström. *The Colloidal Domain. Where Physics, Chemistry, Biology and Technology Meet*, 2nd edn, Wiley-VCH, New York, 1999, p. 579

a combination is sometimes called 'electrosteric stabilization'. Figure 21.5 shows one typical example of electrosteric stabilization, i.e. a mixture of fatty alcohol ethoxylate and sodium dodecyl sulfate, a common combination to stabilize technical emulsions. Spreads (margarines), which are w/o emulsions, may be stabilized by three mechanisms, namely the use of an anionic phospholipid to give electrostatic stabilization, a biopolymer to give steric stabilization and fat crystals to give particle stabilization.

The DLVO Theory is a Cornerstone in the Understanding of Emulsion Stability

The DLVO theory, named after the four scientists behind it, i.e. Derjaguin, Landau, Verwey and Overbeek, pictures the interplay between van der Waals attractive forces and double-layer repulsive forces. It is the competition between these forces that determines the stability of dispersed systems. The attractive term dominates at both small and large distances between the particles. In between, the repulsive double-layer interaction can take over provided

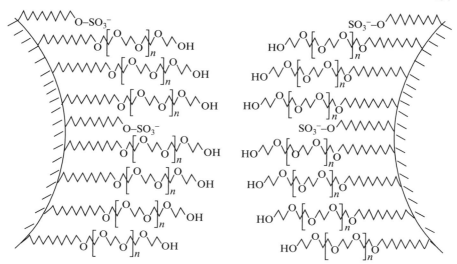

Figure 21.5 An o/w emulsion stabilized by a combination of an alcohol ethoxylate and sodium dodecyl sulfate

that the surfaces are highly charged and the electrolyte concentration in the continuous phase is not too high. (When the particles come very close together there is a strong repulsion due to overlapping of electron clouds, the 'Born repulsion'.) Salts shield the charges on the particles and are therefore detrimental to emulsion stability. The DLVO theory predicts, and experiments also show, that at a salt concentration of 0.1 M there is virtually no electrostatic repulsion left. Electrostatic stabilization can be extremely efficient in model systems although the electrolyte sensitivity is a severe drawback in many technical applications.

Figure 21.6 shows the interaction energy curves for two situations. The attractive interaction is the same in both cases but the electrostatic repulsion is strong in (1) and weak in (2). The resulting curve V(1) shows a repulsive maximum, although no such maximum is present in curve V(2).

The DLVO theory and most general concepts of emulsion stabilization and destabilization can also be applied to solid dispersions, i.e. suspensions. Suspensions usually have larger density differences between the dispersed and the continuous phases which makes stabilization more difficult. On the other hand, solid particles are often very highly charged when compared to emulsion droplets, a fact which favours stability. In actual situations, tailor-made formulations are worked out for each individual system, emulsion or suspension. However, all systems have one thing in common: they are thermodynamically unstable. Sooner or later, sometimes after years of storage, all emulsions and suspensions will phase-separate.

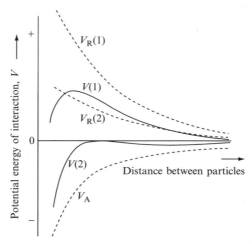

Figure 21.6 Potential energy of interaction versus interparticle distance for two situations, i.e. (1) one with a strong charge on the particles and (2) one with a weak charge. V_R is the repulsive double-layer interaction, V_A is, the attractive van der Waals interaction, and V is the resulting interaction curve. From *Introduction to Colloid and Surface Chemistry* by D. J. Shaw. Reprinted by permission of Butterworth-Heinemann

Emulsions also show many similarities with foams. Both are dispersed systems consisting of a polar phase, i.e. water, and a non-polar phase, i.e. oil or air. Low molecular weight surfactants, i.e. emulsifiers or foamers, are needed to give a fine dispersion with a large interfacial area and both types of systems can be further stabilized by the action of polymers or small particles which accumulate at the interface. In addition, the mechanisms behind the destabilization of emulsions and foams have much in common and the reagents used for this, e.g. demulsifiers or defoamers, are often similar. Foams are discussed above in Chapter 20.

Emulsifiers are Surfactants that Assist in Creating an Emulsion

As discussed above, surfactants are used to stabilize emulsions. For this purpose, other types of compounds can also be used, such as amphiphilic polymers, small particles, etc. Surfactants are also used to create the emulsion, i.e. to help in finely dispersing either oil into water or water into oil. In order to achieve this, two requirements must be fulfilled: the surfactant must reduce the oil–water interfacial tension to low values and the surfactant must rapidly diffuse to the newly created interface. The last requirement is crucial; only if the new interface is rapidly covered by a surfactant monolayer will that interface be stable against coalescence. High molecular weight polymers, hydrophobic particles, proteins at their isoelectric points and liquid crystals may be good at stabilizing an oil–water interface but these are all large species that diffuse slowly to the newly formed interface. Low molecular weight surfactants are therefore superior when it comes to creating emulsions. Other substances may then play a larger role in stabilizing the system. (Compare the earlier discussion about the role of Emulsan (see p. 262.) Surfactants that are added in order to assist in forming an emulsion are referred to as 'emulsifiers'.

The HLB Concept

A rule of thumb in emulsion technology is that water-soluble emulsifiers tend to give o/w emulsions and oil-soluble emulsifiers w/o emulsions. This concept is known as Bancroft's rule.

Bancroft's rule is entirely qualitative. In an attempt to extend it into some kind of quantitative relationship between surfactant hydrophilicity and function in solution, Griffin introduced the concept of the hydrophilic–lipophilic balance (HLB) of a surfactant. The HLB numbers for normal non-ionic surfactants were determined by simple calculations, as follows.

(1) For alcohol ethoxylates and alkylphenol ethoxylates:

$$HLB = \frac{wt\% \text{ ethylene oxide}}{5}$$

(2) For polyol ethoxylates:

$$HLB = \frac{wt\% \text{ ethylene oxide} + wt\% \text{ polyol}}{5}$$

(3) For fatty acid esters of polyols:

$$HLB = 20 \left(1 - \frac{\text{saponification number}}{\text{acid number}} \right)$$

Griffin's HLB number concept was later extended by Davies, who introduced a scheme to assign HLB group numbers to chemical groups which compose a surfactant. Davies' formula and some typical group numbers are shown in Table 21.1. From this table, it can be seen, for instance, that sulfate is a much more potent polar group than carboxylate and that a terminal hydroxyl group in a polyoxyethylene chain is more powerful in terms of hydrophilicity than a sugar hydroxyl group.

The HLB number concept, and particularly Griffin's version (which is restricted to non-ionics), has proved useful in the first selection of a surfactant for a given application. Table 21.2 shows how the appearance of aqueous

Table 21.1 Determination of HLB numbers according to Davies

Group	HLB number
Hydrophilic	
$-SO_4Na$	35.7
$-CO_2K$	21.1
$-CO_2Na$	19.1
$-N$ (tertiary amine)	9.4
Ester (sorbitan ring)	6.3
Ester (free)	2.4
$-CO_2H$	2.1
$-OH$ (free)	1.9
$-O-$	1.3
$-OH$ (sorbitan ring)	0.5
Lipophilic	
$-CF_3$	-0.870
$-CF_2-$	-0.870
$-CH_3$	-0.475
$-CH_2-$	-0.475
$-CH-$	-0.475
$-\underset{|}{C}H-$	

$$HLB = 7 + \sum \text{(hydrophilic group numbers)} + \sum \text{(lipophilic group numbers)}$$

Table 21.2 Use of Griffin's HLB number concept

HLB number range	Appearance of aqueous solution
1–4	No dispersibility
3–6	Poor dispersibility
6–8	Milky dispersion after agitation
8–10	Stable milky dispersion
10–13	From translucent to clear
13–20	Clear solution

HLB number range	Application
3–6	w/o Emulsifier
7–9	Wetting agent
8–14	o/w Emulsifier
9–13	Detergent
10–13	Solubilizer
12–17	Dispersant

surfactant solutions depends on the surfactant HLB. It also indicates some typical applications of surfactants with different HLB number intervals. It can be seen that an emulsifier for a w/o emulsion should be hydrophobic with an HLB number of 3–6, while an emulsifier for an o/w emulsion should be in the HLB number range of 8–18. This is obviously in line with Bancroft's rule.

For room-temperature operations, the HLB numbers calculated according to Griffin (or Davies) give a useful prediction for emulsifier selection, as will be demonstrated below. However, problems arise if the temperature is raised during emulsification or when the ready-made emulsion is stored at very low temperatures. Non-ionic surfactants of the polyoxyethylene type are very temperature sensitive. Many of them give o/w emulsions at ambient conditions and w/o emulsions at elevated temperatures. As discussed below, factors such as electrolyte concentration in the water, oil polarity and the water-to-oil ratio may also influence which type of emulsion forms. Evidently, the HLB number cannot be used as a universal tool to select the appropriate emulsifier or to determine which type of emulsion will form with a specific surfactant.

The HLB Method of Selecting an Emulsifier is Crude but Simple

It has been found empirically that a combination of surfactants, one of which is more hydrophilic while the other is more hydrophobic, is often superior to a single surfactant of intermediate HLB in making a stable emulsion. Most probably, the combination of two surfactants with very different CPP (critical packing parameter (see Chapter 2)) values gives better packing at the interface than one single surfactant. The advantage of the surfactant mixture may also be related to the rate of supply of surfactant molecules during the emulsification process. With both oil-soluble and water-soluble emulsifiers present, the newly created oil–water interface can be furnished with stabilizing surfactants from both sides simultaneously. Regardless of the underlying mechanism, the combination of a low-HLB and a high-HLB surfactant is a useful concept in emulsification, and is frequently used in practice. One typical example of such a combination is given in Figure 21.7. Some general guidelines for the selection of surfactants as emulsifying agents are given in the following:

(1) The surfactant should have a strong tendency to migrate to the oil–water interface.

(2) Oil-soluble surfactants preferably form w/o emulsions, and vice versa.

(3) Good emulsions are often formed by using a mixture of one hydrophilic and one hydrophobic surfactant.

(4) The more polar the oil phase, then the more hydrophilic the emulsifier should be, and vice versa.

Emulsification of a mixture of 20% paraffin oil (HLB=10) and 80% aromatic mineral oil (HLB=13) in water

HLB number of oil: $10 \times 0.20 + 13 \times 0.80 = 12.4$

A mixture of $C_{12}E_{24}$ with HLB = 17.0 and $C_{16}E_2$ with HLB = 5.3 is used. A 60:40 mixture of the two gives a surfactant HLB number as follows:

$17.0 \times 0.60 + 5.3 \times 0.40 = 12.3$

This surfactant combination is found to give excellent emulsion stability.

Figure 21.7 Illustration of the use of the HLB method in selecting an emulsifier. Reproduced by permission of John Wiley & Sons, Inc. From M.J. Rosen, *Surfactants and Interfacial Phenomena*, 2nd Edn, Wiley, New York, 1989, p. 327

HLB numbers have been assigned to various substances that are frequently emulsified, such as vegetable oils, paraffin wax, xylene, etc. These numbers are based on emulsification experience, i.e. the compounds have been emulsified with a homologous series of non-ionic surfactants and then assigned the HLB number of the optimum surfactant. Table 21.3 gives the HLB numbers for some common 'oils'.

In an emulsification process, an emulsifier, or a combination of emulsifiers, with an HLB number which equals that of the ingredient to be emulsified should be selected. If a surfactant mixture is used, the HLB number of the mixture is the weighted average of the individual HLB numbers. A typical example of use of the HLB method to select an emulsifier is given in Figure 21.7.

Although the HLB method is useful as a rough guide to emulsifier selection, it has some serious limitations. For example, it is rapidly set aside by temperature variations. Moreover, the method can also be severely affected by the following:

(1) Impurities in the oil.

(2) Electrolytes in the water.

(3) The presence of cosurfactants or other additives.

The PIT Concept

As discussed in some detail in Chapter 4, the physicochemical properties of non-ionic surfactants based on polyoxyethylene chains are very temperature dependent. The same surfactant may give a water-continuous emulsion at low temperatures and an oil-continuous one at high temperatures. The concept of

Table 21.3 HLB numbers assigned to various organic liquids

Compound	HLB number
Acetophenone	14
Acid, lauric	16
Acid, linoleic	16
Acid, oleic	17
Acid, ricinoleic	16
Acid, stearic	17
Alcohol, cetyl	15
Alcohol, decyl	14
Alcohol, lauryl	14
Alcohol, tridecyl	14
Benzene	15
Carbon tetrachloride	16
Castor oil	14
Chlorinated paraffin	8
Cyclohexane	15
Kerosene	14
Lanolin, anhydrous	12
Mineral oil, aromatic	12
Mineral oil, paraffinic	10
Mineral spirits	14
Petrolatum	7–8
Pine oil	16
Propene, tetramer	14
Toluene	15
Wax, bee	9
Wax, candelilla	14–15
Wax, carnauba	12
Wax, microcrystalline	10
Wax, paraffin	10
Xylene	14

using the phase inversion temperature (PIT) as a more quantitative approach to the evaluation of surfactants in emulsion systems has been found useful. (The PIT is sometimes referred to as the HLB temperature.) As a general procedure, emulsions of oil, aqueous phase and approximately 5% non-ionic surfactant are prepared by shaking during a rise in temperature. The temperature at which the emulsion inverts from o/w to w/o is defined as the PIT of the system. The phase inversion can easily be detected by an abrupt drop in conductivity when the emulsion transforms from a water-continuous to an oil-continuous system.

In laboratory work the PIT is often first determined with model non-ionic surfactants that are 'homologue-pure'. It is important in this context to note that the PIT of such a surfactant is different from the PIT of a technical surfactant of the same average degree of ethoxylation. The difference is particularly large for surfactants with relatively short polyoxyethylene chains. The reason for this difference is that for the surfactant with a broad homologue distribution the fraction of amphiphilic molecules with short polyoxyethylene chains, i.e. the hydrophobic fraction, will preferentially partition into the oil phase, whereas the fraction with long polyoxyethylene chains will, to a large extent, dissolve in the aqueous phase. The solubility is normally higher in the oil than in the water phase. A larger fraction will therefore be 'lost' in the oil than in the water phase and the resulting surfactant at the interface will be more hydrophilic than the average surfactant added to the system. Hence, the PIT will be higher than if no fractionation had occurred. Partitioning into the oil and water phases occurs for the homologue-pure surfactant as well, but in this case the PIT is not affected since all molecules, i.e. those in the oil phase, those in the water phase and those residing at the interface, are the same. This situation is illustrated in Figure 21.8.

Whereas the HLB number is a characteristic property of a surfactant molecule considered in isolation, the PIT is a property of an emulsion at which the hydrophilic–lipophilic property of the non-ionic surfactant used as the emulsifier just balances. There is, of course, a correlation between the PIT and the HLB number. Increasing the length of the polyoxyethylene chain in non-ionic surfactants gives higher HLB numbers and leads to an increase in the PIT. Other factors that affect the PIT are as follows:

(1) *Nature of the oil.* The more non-polar the oil, then the higher the PIT. For instance, ethoxylated nonylphenol (9.6 EO) has a PIT in benzene–water (1:1) of around 20°C. When benzene is replaced by cyclohexane, the PIT is raised to 70°C, while with hexadecane as the oil the PIT is above 100°C.

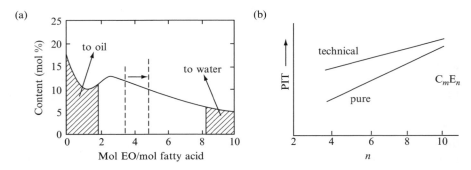

Figure 21.8 A technical alcohol ethoxylate has a higher PIT than a homologue-pure surfactant of the same average degree of ethoxylation. The difference is due to partitioning of the hydrophobic and hydrophilic fractions into oil and water, respectively. More surfactant partitions into oil than into water

(2) *Electrolyte concentration and type of salt.* The PIT decreases with addition of most, but not all, salts. Replacing distilled water by a 5% NaCl solution gives a PIT reduction of the order of 10°C. The salt dependence is the same as that discussed in Chapter 4 for the cloud point.

(3) *Additives in the oil.* Additives that make the oil more polar, such as fatty acids or alcohols, give a considerable reduction in the PIT. Very water-soluble additives, e.g. ethanol and isopropanol, have an opposite effect on the PIT.

(4) *Relative volumes of oil and water.* The PIT for homologue-pure surfactants can usually be regarded as constant for oil–water volume ratios of 0.2–0.8. For technical surfactants that contain species with varying HLB values, the oil–water ratio will affect the distribution of species between the phases, leading to an increased PIT with increasing oil–water ratios. If the PIT is measured at different oil–water ratios and the values extrapolated to an oil–water ratio of zero, the value obtained will roughly correspond to the cloud point of the surfactant.

These effects are all in harmony with Bancroft's rule of emulsifier solubility governing the emulsification process. For instance, the addition of a polar, oil-soluble organic substance leads to an increase in oil polarity, which, in turn, leads to increasing oil solubility of the surfactant. Bancroft's rule states that this will favour formation of a w/o emulsion. In order to achieve a balanced system, water solubility of the surfactant then needs to be improved. This is achieved by lowering the temperature. (Polyoxyethylene-based non-ionics become more water soluble at lower temperatures.) Thus, addition of the additive lowers the PIT. Figure 21.9 illustrates the point.

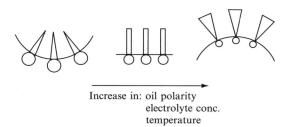

Increase in: oil polarity
electrolyte conc.
temperature

Figure 21.9 Effect of oil polarity, electrolyte concentration and temperature on the curvature of the oil–water interface

The PIT Method of Selecting an Emulsifier is often Useful

As discussed in detail in Chapter 6, the oil–water interfacial tension is at a minimum at the PIT. Emulsions made at this temperature will therefore have the finest particle size. Such an emulsion is not stable to coalescence, however. In the 'PIT method of emulsification', a non-ionic emulsifier (or better still, a surfactant mixture) is chosen which gives a PIT of around 40°C above the storage temperature of the ready-made emulsion (equal parts oil and water, 5% surfactant). Emulsification is carried out at some 2–4°C below the PIT, and the emulsion is then rapidly cooled down to the storage temperature at which the coalescence rate is low. An efficient way to bring about cooling is to carry out the emulsification with a relatively small amount of water, and subsequently add cold water.

One may also make the emulsion slightly above the PIT, in which case a w/o emulsion is formed. Subsequent addition of water (cold water is usually employed) will give a phase inversion into a o/w system. This is a useful way of emulsifying very viscous oils, such as alkyds and other resins, but the droplet size will normally not be as small as those produced by the PIT method without phase inversion.

Different Types of Non-Ionic Surfactants can be Used as Emulsifiers

Alkylphenol ethoxylates have traditionally been widely used as emulsifiers. With the increasing concern about biodegradability and aquatic toxicity, these surfactants are being substituted by alcohol ethoxylates of similar HLB numbers. Sometimes, such a replacement is not straightforward, a fact that may be attributed to the differences in the hydrophobic tail of the two surfactant types. Whereas alcohol ethoxylates are normally based on a straight-chain, aliphatic hydrocarbon, alkylphenol ethoxylates (in practice, octyl-or nonylphenol ethoxylates) have a bulky and polarizable tail. Figure 21.10 shows the structures of the two types of compounds.

For a surfactant to align properly at an interface, the molecular structure is important. The oil–water interface of an emulsion droplet is relatively planar at

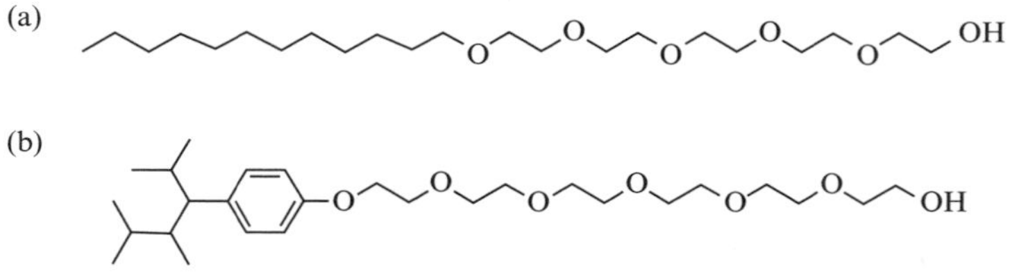

Figure 21.10 Structures of (a) a straight-chain alcohol ethoxylate, and (b) a nonylphenol ethoxylate

the surfactant dimension. Consequently, in order to obtain optimum packing at this interface, which is beneficial for emulsion stability, the surfactant should have a geometry such that the size of the polar head matches that of the hydrophobic tail. In other words, the CPP of the surfactant should preferably be close to 1 (see Chapter 2). It can easily be seen, e.g. by viewing molecular models, that the volume of the hydrocarbon tail is much smaller than that of the polar head group for linear alcohol ethoxylates of the type normally used as emulsifiers for o/w systems. For the corresponding alkylphenol ethoxylates, the volume of the hydrophobic tail is also smaller than that of the polar head but the difference in size is less pronounced. Consequently, linear fatty alcohol ethoxylates pack less efficiently at interfaces (unless these are strongly curved convex against water) than alkylphenol ethoxylates. Another way of viewing this difference is that the driving force for such alcohol ethoxylates to align at interfaces is smaller than for alkylphenol ethoxylates.

Non-ionic surfactants based on branched alcohols have a more balanced geometry than their linear counterparts. So-called 'Guerbet alcohols', namely alcohols with long side chains at the 2-carbon atom (made by the Guerbet reaction—a kind of aldol condensation) constitute an example of raw materials of interest in this respect. Guerbet alcohol ethoxylates have been found useful as replacements for alkylphenol ethoxylates in some applications. 'Oxo alcohols' represent another type of branched alcohols, but with mostly methyl side chains.

Another difference between nonylphenol ethoxylates and alcohol ethoxylates is the presence of the six π-electrons in the hydrophobic tail of the former. These are likely to affect the interactions between these surfactants and unsaturated components of the oil. Phenols are known to be able to act as donors in electron donor–acceptor (EDA) complexes, donating π-electrons to suitable acceptor molecules. This type of interaction can be of considerable magnitude although the exact nature of the bonding is not yet well understood. It is reasonable to assume that EDA complex formation will play a role in the interaction between alkylphenol ethoxylates and unsaturated bonds in the oil phase. No such contribution is, of course, present in the case of alcohol ethoxylates. Figure 21.11 illustrates the formation of an EDA complex involving nonylphenol ethoxylate.

The electronic effect is not as general as the effect of the geometrical packing. EDA complexes can only form with oils containing components that can function as π-electron acceptors. Olefins and aromatics, particularly those that contain electron-withdrawing substituents, act as acceptor molecules. Many emulsions, both in the food area and for industrial applications, are based on oils which contain unsaturated components. Formation of such EDA complexes will enhance the interactions between the emulsifier and the oil phase. This, in turn, enables the formulator to use surfactants with slightly longer polyoxyethylene chains than would otherwise be possible. Longer polyoxyethylene chains mean a higher water solubility of the surfactant. Without

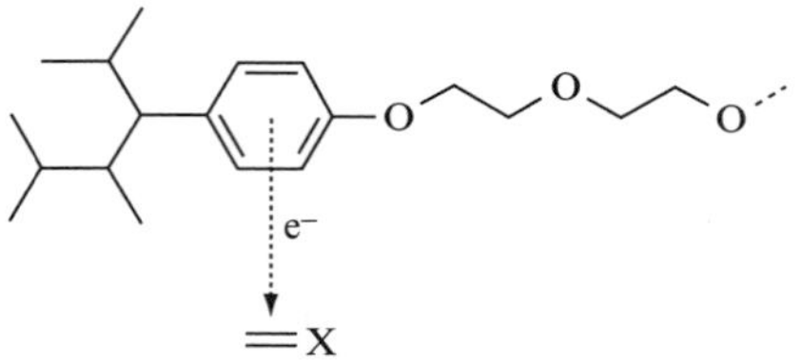

Figure 21.11 An electron donor–acceptor complex between nonylphenol ethoxylate and an unsaturated moiety. Reproduced by permission of the Oil & Colour Chemists Association from K. Holmberg, *Surf. Coatings Int.*, **76** (1993) 481

the extra contribution to the interactions involving the hydrophobic tail, such surfactants would partition too much into the water phase. The longer chains are beneficial as such since the steric repulsions between the droplets, which help to prevent coalescence, will be enhanced.

Bancroft's Rule may be Explained by Adsorption Dynamics of the Surfactant

Figure 21.12 illustrates an attempt to explain Bancroft's rule regarding what type of emulsion is formed with a specific surfactant. By applying shear to a mixture of oil and water, the oil–water phase boundary has been extended, having oil 'fingers' in water and water 'fingers' in oil, as shown in the top figure (a). This situation is unstable. If the oil fingers break up, an o/w emulsion will form and if the water fingers break, a w/o emulsion will be created. In order to stabilize the newly formed droplets—regardless of what type of structure is being formed—enough time must be available to allow the emulsifier to go to the interface and adsorb there in an amount sufficient to prevent immediate coalescence.

In the experiment of Figure 21.12, the emulsifier is oil-soluble. The discussion below is based on the fact that adsorption of surfactant at a newly formed oil–water interface is not instantaneous. The aqueous finger breaks up and immediately afterwards the situation shown in the bottom figure (b) may appear. The interfacial concentration of emulsifier is uneven. It is higher on the relatively older top and bottom sides and lower in the regions between the droplets, because these interfaces were formed more recently. The emulsifier molecules have not yet had time to reach these sites. Transient gradients in interfacial tension are created, thus leading to a flow of oil towards the gap by viscous traction, an example of the 'Marangoni effect'. The further consequence is that the droplets are transiently pushed apart, so giving them enough time to stabilize. If, instead, the oil fingers had broken up, the emulsifier molecules would have been inside the droplets and coverage of the interface would have been equal all over. No Marangoni effect would then be operative to assist in separating the droplets during the critical initial period.

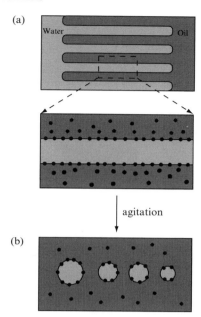

Figure 21.12 Dynamic interpretation of Bancroft's rule; the dots represent emulsifier molecules. Reprinted from *Colloid Surf., A*, **91**, J. Lyklema, 'Adsorption at solid–liquid interfaces with special relevance to emulsion systems', 25–38, Copyright (1994), with permission from Elsevier Science

Bancroft's Rule may be Related to the Surfactant Geometry

An alternative approach to explain Bancroft's rule is that of the surfactant geometry being decisive of whether o/w or w/o emulsions form. Surfactants with large CPP values need more space on the oil side; thus, they should give w/o emulsions. Surfactants with large head groups have the opposite requirement and should favour o/w emulsions. Ideas along this line were among the early attempts to explain why some surfactants gave oil-continuous emulsions while other give water-continuous systems. However, when considering and comparing the dimensions of emulsion droplets and surfactants, it is obvious that at the scale of the surfactant molecule the oil–water interface is almost planar. Thus, the energy difference between the two different orientations of the surfactant must be very small.

More recently, the surfactant geometry approach has been advanced again, but from a different viewpoint. It is argued that the spontaneous curvature of an oil–water interface governs the rates of coalescence of w/o and o/w emulsions via the values of the corresponding coalescence energy barriers. In the formation of a passage in an emulsion film, two highly curved monolayers will have to form, as illustrated in Figure 21.13. Given time, the radii will grow large

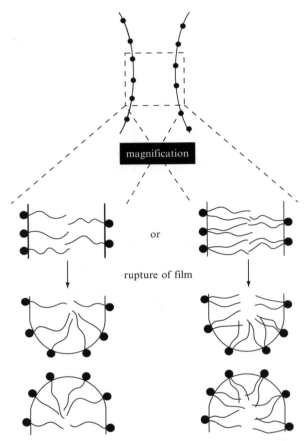

Figure 21.13 Rupture of a thin oil film in an oil–water–surfactant system creates a highly curved monolayer. The double-chain surfactant has a geometry unsuitable for stabilizing the newly raptured oil film (but would be well suited in stabilizing a raptured water film). Compare the breaking of foam films (see Figure 20.5)

but at the time of film rupture the dimensions are small. At this intermediate situation, surfactants with high CPP values, e.g. twin-tailed surfactants, will favour w/o structures, and vice versa. The monolayer energy bending penalty can be large for such highly curved surfaces.

A striking example of how the curvature concept can be used to explain the type of emulsion formed is in oil–water–surfactant systems close to the PIT where the surfactant is a polyoxyethylene-based non-ionic. As discussed in Chapter 6, such mixtures may give rise to three phases, namely a microemulsion in equilibrium with excess oil and water. If the microemulsion phase, which contains most of the surfactant, is removed and the remaining two phases are subjected to agitation, either an o/w or a w/o emulsion may form. Which type of emulsion will form depends entirely on whether the experiment is made above or below the PIT of the system. Below the PIT, the spontaneous curvature of the microemulsion is convex against water, i.e. the surfactant CPP is slightly smaller than 1, and an o/w emulsion is formed. Above the PIT, the curvature is convex against oil, i.e. the surfactant CPP is slightly larger than 1, and a w/o emulsion is formed. In this case, the relative surfactant solubility in oil and water seems not to play any role. A normal alcohol ethoxylate, such as $C_{12}E_5$, has a molecular solubility several orders of magnitude higher in hydrocarbons than in water.

Hydrodynamics may Control what Type of Emulsion will Form

At very low surfactant concentration or when a very weak amphiphile is used, the type of emulsion formed, oil-in-water or water-in-oil, is controlled by the mixing procedure rather than by the choice of surfactant. In such systems, Bancroft's rule may or may not be obeyed. For instance, addition of oil to a solution of a non-ionic surfactant in water may lead to an oil-in-water emulsion, also at temperatures above the phase inversion temperature if the surfactant concentration is very low. These hydrodynamics-controlled emulsions usually have poor stability, however, and are not of major practical importance.

Bibliography

Becher, P. and M. J. Schick, Macroemulsions, in *Nonionic Surfactants*, M. J. Schick (Ed.), Surfactant Science Series, 23, Marcel Dekker, New York, 1987.
Larsson, K. and S. E. Friberg (Eds), *Food Emulsions*, 2nd Edn, Marcel Dekker, New York, 1990.
Shinoda, K. and S. Friberg, *Emulsions and Solubilization*, Wiley, New York, 1986.
Sjöblom, J. (Ed.), *Emulsions and Emulsion Stability*, 2nd Edn, Marcel Dekker, New York, 2001.

22 MICROEMULSIONS FOR SOIL AND OIL REMOVAL

Surfactant-Based Cleaning Formulations may act by *in situ* Formation of a Microemulsion (Detergency)

A Detergent Formulation is Complex

Removal of soil from fabric, i.e. detergency, is a complex process involving interactions between surfactants, soil and the textile surface. The choice of surfactant is the key to success and surfactant-based cleaning compositions are one of the oldest forms of formulated products for technical use. The traditional surfactants are sodium and potassium salts of fatty acids, made by saponification of triglycerides. Soaps are still included in smaller amounts in many detergent formulations but the bulk surfactants in use today are synthetic non-ionics and anionics, with alcohol ethoxylates, alkylaryl sulfonates, alkyl sulfates and alkyl ether sulfates being the most prominent.

The soil present on fabrics may vary widely and different mechanisms are responsible for removal of different types of soil components. The term 'oily soil' refers to petroleum products, such as motor oil and vegetable oil, e.g. butter, but also skin sebum. Surfactants play the key role in removing oily soil but hydrolytic enzymes, which are present in most detergent formulations today, are important for hydrolysis of triglycerides, proteins and starch. In practice, oily soil is removed by the combined action of surfactants and enzymes.

Particulate soil, which may consist of clay and other minerals, is usually removed by a wetting and dispersing process in which the surfactant is assisted by highly charged anionic polyelectrolytes normally included in the formulation.

Some stains, such as those from tea and blood, may be difficult to remove even with optimized surfactant–enzyme combinations. Bleaching agents are often effective in eliminating such spots. These normally act by oxidizing the stain chromophores into non-coloured products which may, or may not, be removed by the action of surfactant.

A detergent formulation also contains a large amount of so-called 'builders', such as zeolite or phosphate, which function as sequestering agents for divalent ions, thus preventing surfactant precipitation as Ca or Mg salts in areas of hard water. Other, less vital, additives are anti-redeposition agents, fluorescing agents, perfumes, etc.

In the following, the discussion will be limited to the action of surfactants on oily soil. The need for builders, action of enzymes on triglyceride soil, etc., is outside the scope of present chapter.

Various Mechanisms have been Proposed for Oily Soil Removal

The three most important mechanisms for the removal of oily soil are as follows:

(1) *Roll-up* (Figure 22.1). This mechanism is related to fabric wetting, i.e. the surfactant–fabric interaction is decisive. Good soil release is usually obtained when the contact angle is larger than 90°. This is typically the case for oily soil on polar textiles such as cotton. On polyester and other more non-polar fabrics, a contact angle of less than 90° is usually obtained.

(2) *Emulsification* (Figure 22.2). A low interfacial tension between oil and the surfactant solution is required. The mechanism involves surfactant–oil interaction and the process is independent of the nature of the fabric.

(3) *Solubilization* (Figure 22.3). The oily soil is solubilized into an *in situ* formed microemulsion. Similar to the emulsification mechanism, solubilization

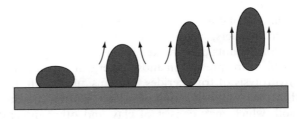

Figure 22.1   The roll-up mechanism of removal of oily soil from a solid surface

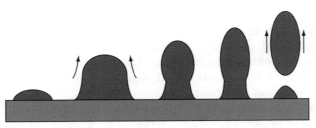

Figure 22.2   The emulsification mechanism of removal of oily soil from a solid surface

Figure 22.3 The solubilization mechanism of removal of oily soil from a solid surface

occurs independently of the underlying surface. The process requires ultra-low interfacial tension between oil and the surfactant solution.

These three mechanisms for the removal of oily soil obviously do not operate independently. It is probably true to say, though, that whereas roll-up is the most important mechanism on polar textiles, emulsification and solubilization govern the results on non-polar materials. It is also true to say that whereas roll-up is relatively simple to achieve—most surfactants are good enough wetting agents—emulsification and, in particular, solubilization need fine-tuning of the surfactant composition. Effective solubilization is only obtained with surfactants that bring the interfacial tension of oil and water down to ultra-low values. The processes of solubilization and formation of a micro-emulsion in which the soil constitutes the oil component are exactly the same as those which take place in enhanced oil recovery, which will be discussed later in this chapter.

How is Ultra-low Interfacial Tension Achieved?

The oil–water interfacial tension is at minimum at the phase inversion tempera-ture (PIT) of an oil–water–surfactant system. When applied to oily soil re-moval, the soil constitutes the oil component of the ternary system.

Non-ionic surfactants are often characterized by their cloud points (see Chapter 4). However, the cloud point value which is given does not take into account the presence of the oil. It is important to realize that it is the PIT of the oil–water–surfactant system, and *not* the cloud point of the aqueous surfactant solution, that is the temperature of relevance for solubilization. What correl-ation is there then between the cloud point and the PIT?

(1) For an oil that penetrates the surfactant hydrocarbon layer, the effect of adding oil to a micellar solution of surfactant will be that the hydrocarbon volume, v in the expression $CPP = v/(l_{max}\ a)$ (see Chapter 2), increases. The CPP will increase and the change in geometry will be manifested as a shift towards more elongated micelles, as seen in Figure 22.4.

(2) Oils that do not penetrate the surfactant hydrocarbon layer will form an oil core in the micelle. As shown in Figure 22.5, this will lead to an increased area per surfactant head group, i.e. to an increase in a, which, in turn,

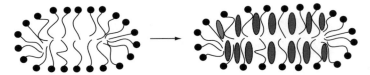

Figure 22.4 A penetrating oil gives more elongated micelles

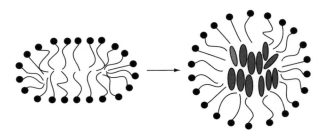

Figure 22.5 A non-penetrating oil gives more spherical micelles

means a decrease in the CPP. The effect on micelle geometry will be that of a shift towards a more spherical aggregate.

Small oil molecules generally penetrate and mix with the surfactant hydro-carbon chains while large oil molecules do not. For straight-chain aliphatic oils and straight-chain surfactants, oil molecules of chain lengths smaller or equal to that of the surfactant hydrocarbon tail tend to mix well. Longer chain length oils form a separate core.

Figure 22.6 illustrates the effect on the cloud point of the addition of a low molecular weight aliphatic hydrocarbon. As can be seen, there is a substantial drop in the cloud point which is consistent with the view that the added hydrocarbon penetrates into the surfactant tail region, so leading to an increase in the CPP value. At high amounts of added hydrocarbon, there is a cloud point increase that indicates saturation of the surfactant tail region, forcing excess oil to form a micelle core (see Figure 22.5).

The effect on the cloud point of addition of a high molecular weight aliphatic hydrocarbon is shown in Figure 22.7. The hydrocarbon does not mix with the surfactant tail region and the almost linear increase in the cloud point correlates with the increase in a and a decrease in the CPP value discussed above.

To summarize, in order to obtain maximum solubilization, washing should be carried out at a temperature corresponding to the PIT of the oil–water–surfactant system, in which the oily soil constitutes the oil. When the oil molecules are larger than the surfactant tail (which is the normal situation), the PIT is higher than the cloud point. As an example, the best result in

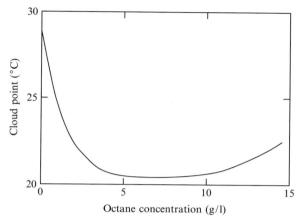

Figure 22.6   Effect on the cloud point of the addition of a low molecular weight oil

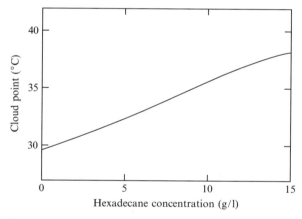

Figure 22.7   Effect on the cloud point of the addition of a high molecular weight oil

removing motor oil stains (with cleaning formulations based on non-ionic surfactant) is often obtained at some 20°C above the surfactant cloud point. At this temperature, the surfactant–water mixture present in the initial washing bath is typically a dilute dispersion of a lamellar liquid crystalline phase in water. For removal of oils of a chain length smaller than that of the surfactant tail, washing should be carried out at a temperature below the surfactant cloud point.

   In the discussion above, temperature has been used as the variable to obtain maximum oil solubilization. The other parameter that can be used to optimize cleaning efficiency is the choice of surfactant. Particularly in the case of alcohol ethoxylates, the surfactant can be fine-tuned to meet a specific need. The two parameters, temperature and choice of surfactant, are, of course, interrelated. Washing at 40 and 60°C requires different surfactants for optimum results.

In real situations, the oil, i.e. the dirt, is not a variable (unless the cleaning formulation contains hydrocarbon, in which case we have a microemulsion cleaning formulation, see p. 484). In a model experiment one may, however, vary the oil while keeping the temperature and choice of surfactant constant. Figure 22.8 shows that a distinct minimum in the oil–water interfacial tension is obtained for a specific oil molecular weight. For that specific oil, the temperature used in the experiments corresponds to the PIT of the system (compare with Figure 22.19 below).

Oil Solubilization around the PIT Leads to the Formation of a Microemulsion

Solubilization of an oil at a temperature close to the PIT of an oil–water–surfactant mixture usually gives a three-phase system in which a microemulsion is in equilibrium with excess oil and water. (If enough surfactant is present, the whole mixture will be transformed into a microemulsion, but such high surfactant concentrations are not used in normal textile cleaning.) The middle-phase microemulsion, which invariably has a bicontinuous structure (see Chapter 6), exhibits an extremely low interfacial tension towards both oil and water. Once formed through oil solubilization, such a system is therefore very efficient in solubilizing more oil. Formation of a middle-phase microemulsion in oily soil removal can be directly observed by video microscopy. As is often the case with

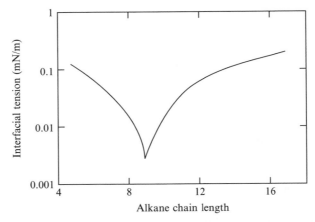

Figure 22.8 Effect of the oil chain length on the oil–water interfacial tension for a system containing non-ionic surfactant

microemulsions, a lamellar liquid crystalline phase may coexist with the micro-emulsion. In addition, this four-phase system—excess oil, microemulsion, liquid crystalline phase and water—can be observed by using the video technique. It can also be seen how the two intermediate phases grow rapidly by solubilization of the oil.

If the washing temperature is considerably below the PIT, no three-phase system with a middle-phase microemulsion will be formed. Instead, an oil-in-water microemulsion in equilibrium with excess oil will be obtained. Such a microemulsion will have a higher interfacial tension towards oil and the rate and capacity of oil solubilization will not be as good as for the bicontinuous middle-phase microemulsion.

If, on the other hand, washing is performed well above the PIT, a water-in-oil microemulsion in equilibrium with excess water will form. Such conditions often give poor detergency since both surfactant and water may be dissolved into the oil without removing it from the fabric. In the video microscopy experiments, this can be seen as a swelling of the oil phase. Moreover, when washing is carried out far above the PIT, redeposition of the dirt, in the form of a water-in-oil emulsion or microemulsion, is often a problem.

A more detailed discussion about microemulsion phase behaviour is given in Chapter 6, and Figure 6.3 can be used to illustrate the phase behaviour at temperatures below, at and above the PIT. Figure 22.9 illustrates the importance of a well-balanced system for soil removal. For the two non-ionic surfactants $C_{12}E_4$ and $C_{12}E_5$, optimum detergency occurs at the temperature at which the ternary system surfactant–hydrocarbon soil–water forms a three-phase system.

In determining the PIT for the surfactant–oily soil–water system, one should use a very high surfactant-to-oil ratio in order to mimic the washing situation. One should also take into account the fact that salts, including the detergent builders, normally decrease the PIT.

Polar Components in the Soil Affect Phase Behaviour

In the discussion above, only pure hydrocarbon soil has been considered. In real situations, the soil is seldom entirely non-polar, however. Polar, but still water-insoluble, components of various types, e.g. fatty acids, fatty alcohols and fatty amines, are common ingredients in many oily soils. Such polar substances will increase the CPP value of the system and will lead to a reduction in the PIT. The effect on the PIT can be considerable if the amount of polar component is large, as can be seen in Figure 22.10.

In practice, the soil is seldom uniform. Different spots of oily soil may have different PITs. In such a situation, the best result is usually obtained by starting the washing process at a temperature corresponding to the PIT of the soil with the highest PIT and then decreasing the washing temperature. The high-PIT soil will then be removed first, while soils with a lower PIT may solubilize water

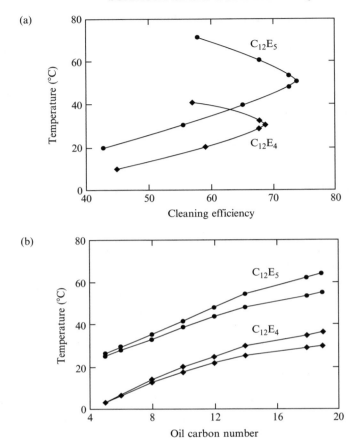

Figure 22.9 (a) Efficiency in hexadecane removal corresponds well with (b) the extension of the middle-phase microemulsion for the surfactant–hexadecane–water system for two different alcohol ethoxylates. The area between the lines in part (b) is the temperature interval at which the middle–phase microemulsion exists. Reproduced by permission of Carl Hanser Verlag from K. Stickdorn, M.J. Schwuger and R. Schomäcker, *Tenside Surf. Det.*, **31** (1994) 4

since washing is carried out above the PIT of those soils. On decreasing the temperature, the PITs of the other soils will eventually be reached and complete soil removal can be achieved. This process is illustrated in Figure 22.11.

Removal of Triglyceride Soils is Difficult

Triglycerides are generally much more difficult than hydrocarbons to remove by surfactant action. The reason for the poor detergency with triglyceride soils

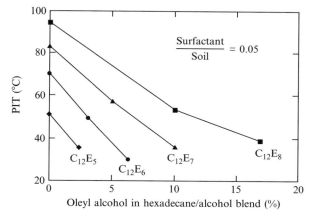

Figure 22.10 Effect of the addition of oleyl alcohol to the non-ionic surfactant–hexadecane–water system on the phase inversion temperature (PIT). Reproduced by permission of the American Oil Chemists' Society from K.H. Raney and H.L. Benson, *J. Am. Oil Chem. Soc.*, **67** (1990) 722

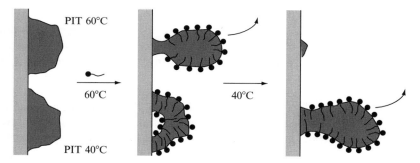

Figure 22.11 Removal of two soil spots which give different phase inversion temperatures (PITs)

may partly be related to specific interactions between the slightly polar oil and the fabric surface, but the main reason is undoubtedly related to the fact that triglycerides are not readily solubilized. With normal non-ionic surfactants, i.e. straight-chain alcohol ethoxylates, a systematic screening of different compositions reveals essentially no microemulsion regions. Instead, lamellar and sponge phases appear (see Chapter 4), both of which solubilize only small amounts of oil. The interfacial tensions between these phases and either oil or water are therefore not very low. A typical phase diagram of the alcohol ethoxylate–triglyceride–water system is shown in Figure 22.12.

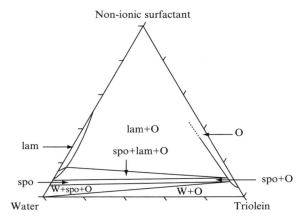

Figure 22.12   Phase diagram for the non-ionic surfactant–triolein–water system: 'O' denotes a triolein-rich phase, while 'spo' represents a sponge phase which resembles the L3 phase of binary surfactant–water systems. Reprinted from *Colloid Surf.*, **40**, F. Mori, J. C. Lim, O. G. Raney, C. M. Elsik and C. A. Miller, 'Phase behaviour, dynamic contacting and detergency in systems containing triolein and nonionic surfactants', 323–345, Copyright (1989), with permission from Elsevier Science

One way to facilitate the formation of microemulsions of triglyceride oils is to add a medium-chain alcohol as a cosurfactant. With such formulations, microemulsions can be obtained provided that the concentration of both the surfactant and cosurfactant is relatively high. In practice, however, such high loading of a volatile alcohol to a washing formulation is not feasible.

Addition of a cosurfactant improves solubilization partly by increasing the CPP value and partly by disturbing the ordering of the surfactant hydrocarbon tails at the oil–water interface. Another way to attain this dual effect is to use a branched-chain surfactant instead of the usual straight-chain compound. With branched-tail non-ionics, such as secondary alcohol ethoxylates, a moderately large microemulsion region can be obtained. This leads to effective triglyceride soil removal, most certainly by an emulsification–solubilization mechanism, provided that a composition is chosen so that washing occurs close to the PIT of the system. Triglycerides generally give PITs far above the cloud points of non-ionic surfactants. As an example, the PIT for the system $C_{12}E_5$–triolein–water is around 65°C, while the cloud point for $C_{12}E_5$ is around 30°C.

Fortunately, under real washing conditions triglyceride soil removal is facilitated by the action of other detergent components. The triglycerides are hydrolysed during the washing process, either by saponification due to the high pH of the washing solution, or by the action of lipases, which are becoming an increasingly common detergent ingredient. The mono- and diglycerides formed are polar lipids which are easily removed and which may assist in solubilizing the remaining triglycerides.

Monitoring Soil Removal

The methods used to study soil removal under controlled conditions include reflectance measurements ('whiteness') by the use of a launderometer, radio-active determinations of remaining soil and direct optical probing of the amount of surface-adsorbed soil. By ellipsometry it is possible to measure *in situ* the amount of soil at a surface with a sub-molecular resolution (see Chapter 17). Figure 22.13 illustrates a washing experiment performed by using ellipsometry. The starting surface is one with an adsorbed layer of oily soil. Initially, there is an increase in the amount of substance on the surface, corresponding roughly to adsorption of a surfactant monolayer. As emulsification and/or solubilization proceeds, the total amount of material on the surface gradually decreases. After rinsing, a totally clean surface is obtained.

Other Surfactant Types can also be Used

The discussion above has focused on non-ionic surfactants. However, the same principles can be applied to other surfactant types, although for ionic surfac-tants the phase behaviour is preferably adjusted by variation of the electrolyte concentration rather than by temperature. Most of the common anionics tend to be too hydrophilic, i.e. have too small a CPP value, to give efficient soil removal by the emulsification–solubilization mechanism. However, by combin-ing them with a hydrophobic alcohol ethoxylate, or even a fatty alcohol, or by washing at high salt concentrations, excellent detergent action can be obtained.

The observation that a mixture of two surfactants often gives better deter-gency than either of the two when used alone is sometimes referred to as a 'synergistic' effect. However, in most cases it is merely the result of the mixture

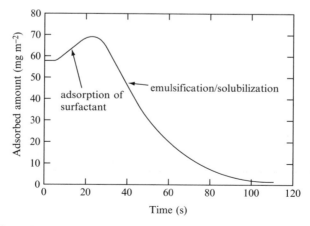

Figure 22.13 Monitoring the washing process by ellipsometry

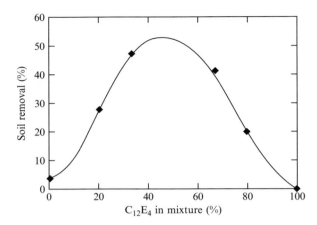

Figure 22.14 Oily soil removal as a function of the $C_{12}E_4$ content (%) of a mixture of the hydrophobic alcohol ethoxylate $C_{12}E_4$ and the hydrophilic anionic surfactant sodium octylbenzene sulfonate. Reprinted with permission from M. Malmsten and B. Lindman, *Langmuir*, **5** (1989) 1105. Copyright (1989) American Chemical Society

having a CPP value close to one and the individual components either being too hydrophilic, i.e. having too low a CPP, or too hydrophobic, i.e. having too high a CPP. The effect is illustrated in Figure 22.14 for the combination of a hydrophobic non-ionic surfactant, $C_{12}E_4$, and the hydrophilic anionic surfactant, sodium octylbenzene sulfonate. The same trend would have been obtained for a mixture of a hydrophilic non-ionic surfactant, such as $C_{12}E_7$, and a hydrophobic anionic surfactant, e.g. sodium dodecylbenzene sulfonate, at a relatively high electrolyte concentration.

Microemulsion-Based Cleaning Formulations are Efficient

Microemulsions, being microheterogeneous mixtures of oil, water and surfactant, are excellent solvents for non-polar organic compounds as well as for inorganic salts. The capability of microemulsions to solubilize a broad spectrum of substances in a one-phase formulation has been found useful for the cleaning of hard surfaces—dirt is often a complex mixture of hydrophilic and hydrophobic components. Of particular interest from a practical point of view is the possiblity of replacing formulations based on halogenated or aromatic hydrocarbons with microemulsions containing non-toxic aliphatic hydrocarbons. A typical example is given in Figure 22.15.

Microemulsions, mainly based on non-ionic surfactants, have an established position in the industrial cleaning of hard surfaces. These are usually sold as concentrated mixtures which need to be diluted with water before use. Hence,

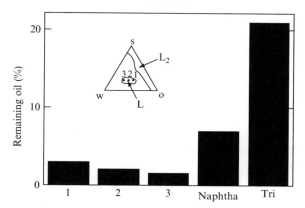

Figure 22.15 Removal of lubricating oil by three microemulsions all lying within a bicontinuous microemulsion phase (indicated in the phase diagram). The surfactant was a mixture of two ethoxylates with a mean HLB value of 10.7, while aliphatic hydrocarbon had a b.p. of 190–240°C. Oil removal by hydrocarbon only and by using trichloroethane are shown as references. The remaining oil is determined by (remaining) fluorescence measurements

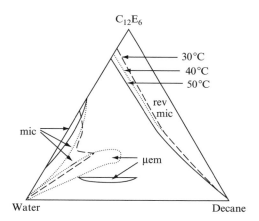

Figure 22.16 Phase diagrams for the $C_{12}E_6$–decane–water system at 30, 40 and 50°C. At 50°C, two small isotropic domains exist in the water-rich region, while at 40°C these have grown bigger but are still disconnected. At 30°C, one larger isotropic domain appears in the water corner

the isotropic domain should preferably extend to the water corner. A typical example of a suitable model system is that of $C_{12}E_6$, decane and water, employed at 30°C, as shown in Figure 22.16. Non-ionics are suitable surfactants in these formulations since they can be formulated with ionic 'builders' such as phosphate or citrate. Such systems suffer from the drawback of considerable temperature sensitivity, however, as is also illustrated in Figure 22.16. One way to increase the temperature interval is to use a mixture of non-ionics, e.g. a blend of alcohol ethoxylates with polyoxyethylene chains length below and above that of the optimum compound. Commercial ethoxylates have a very broad homologue distribution by themselves and give microemulsions with broader temperature regions than homologue-pure model compounds. By mixing commercial non-ionics, the temperature intervals can be further extended.

Microemulsions were once Believed to be the Solution to Enhanced Oil Recovery

Oil fields consist of porous rock, usually limestone or sandstone, in which the pores are filled with petroleum and brine. The porous rock formation is surrounded by impermeable rock. The permeability depends on the pore size, which is typically 50–1000 nm. In a normal oil field, 10–25% of the pore volume is occupied by brine, 55–80% by oil and the rest is void volume. The pressure in the reservoir is normally high and the temperature typically in the range 70–100°C.

When the first wells are dug, oil comes out under its own pressure. This spontaneous production is later supplemented by pumping action. Together, these two processes are referred to as the primary recovery. On average, this stage leads to a recovery of 15–20% of the oil in place. In the next stage, the secondary recovery, water is used to sweep out additional oil. In this process water is pumped down the injection well and moves outwards in a piston-like fashion, thus displacing the oil. The immobilized oil is recovered via production holes, as shown in Figure 22.17. The so-called sweep efficiency is sometimes not very good, however, particularly when the oil is of higher viscosity than the displacing water. The primary and secondary recovery together often manage to recover considerably less than half of the total amount of petroleum in the reservoir.

Any oil recovery process following water flooding is referred to as enhanced oil recovery (EOR) or tertiary oil recovery. Surfactant flooding, sometimes called microemulsion flooding, is the technique of relevance to us here. The interest in microemulsions for this purpose derives from their ability to reduce the oil–water interfacial tension to ultra-low values. Surfactant flooding was once believed to have a very large economic potential. Today, however, with improved drilling techniques and with a considerably slower rise in oil prices than was once predicted, the interest in using microemulsions for oil recovery has declined. Since very large efforts were once put into the development of surfactant flooding and since that research contributed very much to the general understanding of phase behaviour of oil–water–surfactant systems, a short treatise of the topic will be given in the following.

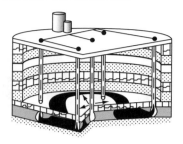

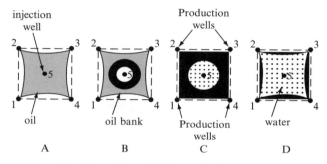

Figure 22.17 Schematic illustration of oil production. From *Surface Phenomena in Enhanced Oil Recovery*, D. O. Shah (Ed.), 1981, p. 3, 'Fundamental aspects of sufactant–polymer flooding process', D. O. Shah, Figure 1, with kind permission of Kluwer Academic Publishers

One main reason for the inefficiency of water flooding through the reservoir rock is that oil is trapped by capillary action in the form of disconnected 'ganglia'. Figure 22.18 shows two different mechanisms for the capillary trapping of oil, i.e. a snap-off process that traps oil in the wider sections of a pore, and a bypass process, caused by competition of flow between pores. Snap-off occurs in pores with a large ratio between the pore body and pore throat. The wetting phase (water in Figure 22.18) forms a collar around the non-wetting phase which eventually breaks at the narrow throats. Bypassing, on the other hand, is caused by differences in pore size. Viscous forces make the fluid flow faster in the larger channels, whereas capillary forces draw the displacing phase into the smaller pores. Thus, under conditions when the flow is an imbibition process, i.e. at low injection rates and with a low viscosity of the displacing phase, oil in the large pores will be trapped.

The amount of oil retained in the reservoir after water flooding will depend on the ratio between viscous forces trying to displace the oil and capillary forces trapping the oil in the pores. A dimensionless number, the capillary number, N_c, is often used to describe the relationship between viscous and capillary forces, according to the following expression:

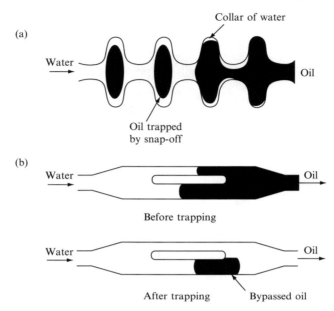

Figure 22.18 Oil trapping by (a) a snap-off mechanism, and (b) bypassing. From K. Taugböl, *Chemical Flooding of Oil Reservoirs*, Ph.D. Thesis, University of Bergen, Norway, 1995

$$N_c = \mu v / \gamma_{o/w} \qquad (22.1)$$

where μ is the viscosity and v the velocity of the displacing fluid.

It has been experimentally shown that the residual oil saturation in the reservoir is constant below a certain value of N_c, typically in the range $10^{-4} - 10^{-5}$. Normal water flooding yields N_c values below this level. Above a critical N_c, the residual saturation after flooding decreases almost linearly with $\log N_c$.

Thus, the key to achieving good oil recovery is to attain a high enough N_c. In principle, an increase in N_c can be attained by (a) increasing the viscous forces (viscosity and/or velocity of the water flood), (b) decreasing the capillary forces, or (c) a combination of (a) and (b). In practice, there is a limit to the increase of the viscous forces that can be employed: a high water pressure will fracture the reservoir rock and large fractures cause a considerable decrease in sweep efficiency. Thus, the remaining variable is the oil–water interfacial tension, $\gamma_{o/w}$, which needs to be reduced to very low values. It can be shown that, at least in water-wet rocks, reduction of the interfacial tension to values of the order of 10^{-3} mN/m is needed in order to obtain substantial mobilization and recovery of oil.

Ultra-low Oil–Water Interfacial Tension is Needed

The phase behaviour of oil–water–surfactant systems was discussed earlier in Chapter 6. It was demonstrated that on increasing the surfactant hydrophobicity a transition occurs from a two-phase system consisting of an o/w microemulsion in equilibrium with excess oil (Winsor I), via a system comprising a microemulsion coexisting with both oil and water (Winsor III), to a two-phase system consisting of a w/o microemulsion in equilibrium with excess water (Winsor III) (Figures 6.2 and 6.3). It has been found that the oil–water interfacial tension has a deep minimum in the three-phase region, i.e. $\gamma_{o/w}$ decreases as a system goes from Winsor I to Winsor III, has a minimum in the middle of the Winsor III region and increases as it moves on into the Winsor II regime (Figure 22.19). (The Winsor I → III → II transition can be achieved by raising the temperature for a system based on a non-ionic surfactant and by increasing the salinity for a system based on an ionic surfactant.) The condition at which the hydrophilic and lipophilic properties of a surfactant are balanced is called the phase inversion temperature (PIT) (see earlier) for non-ionics, with temperature usually being the most important variable (Chapter 21). For ionic surfactants, this state is often referred to as optimal conditions, e.g. optimal salinity. At this point, there is equal volumetric solubilization of oil and water in the middle-phase microemulsion and the interfacial tensions are at their minimum.

It has been demonstrated in laboratory flooding experiments that surfactant formulations capable of forming Winsor III microemulsions with the specific

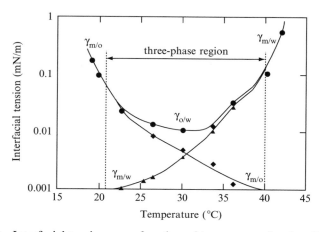

Figure 22.19 Interfacial tensions as a function of temperature for the oil–water–non-ionic surfactant system. The symbols $\gamma_{o/w}$, $\gamma_{m/o}$ and $\gamma_{m/w}$ indicate the interfacial tensions for oil–water, middle phase–oil and middle phase–water, respectively. The system is balanced at 30°C

oil and brine present in the rock at the specific temperature of the reservoir could give a remarkable yield of recovered oil.

Cosurfactant-Free Formulations are Required

Since particularly in offshore reservoirs the distances between injection and production holes are long, the formulations used should preferably contain as few surface active components as possible. Mixtures of different surfactants or surfactant–cosurfactant combinations are less suitable since they are likely to separate during their way through the porous rock. In this respect, the rock can be expected to act like a long chromatography column. Therefore, surfactants have been sought that display the following properties:

(1) Form Winsor III systems with the specific reservoir oil and brine at the reservoir temperature.

(2) Are hydrolytically stable for an extended period of time under reservoir conditions.

(3) Do not precipitate in hard water.

(4) Do not adsorb extensively at the mineral surfaces of the reservoir.

In addition to the above requirements, 'trivial' aspects such as cost, toxicity and biodegradability need to be taken into account. Several laboratories have come up with branched ether sulfonates (or possibly ether sulfates) as a suitable choice of surfactant type. Two representative compounds are shown schematically in Figure 22.20.

Figure 22.21 shows the relative phase volumes as a function of surfactant concentration for a system containing a branched-tail ether sulfonate of the general structure shown in Figure 22.20, designed for specific reservoir conditions. Note that there is a Winsor I → Winsor III → Winsor II transition on dilution of this sulfonate, which must be taken into account in the optimization of the surfactant. The system shown in Figure 22.21 goes from Winsor I to

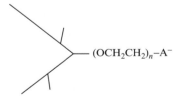

Figure 22.20 Branched-chain ether sulfonate (A = SO₃) or ether sulfate (A = OSO₃) surfactants suitable for use in enhanced oil recovery

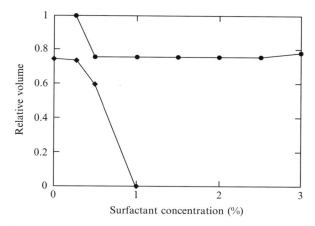

Figure 22.21 Relative phase volume as a function of surfactant concentration for a branched-tail ether sulfonate at a water-to-oil ratio of 4. Above 1% and below 0.25% surfactant, two phases form, i.e. an o/w microemulsion in equilibrium with oil and a w/o microemulsion in equilibrium with water, respectively. At intermediate surfactant concentrations, a three-phase system appears, consisting of a middle-phase microemulsion in equilibrium with excess oil and water

Winsor III at 1.0% surfactant and from Winsor III to Winsor II at around 0.25% surfactant. A very high displacement efficiency was obtained in core flooding experiments with reservoir oil as the oil phase. A 0.5 pore volume slug containing 2 wt% surfactant in sea water produced 89% of the oil left in the core after water flooding.

A more recent extension of the surfactant flooding technology is to use a combination of an anionic surfactant and a similarly charged water-soluble polymer in the displacing fluid. This concept, known as 'low-tension polymer flooding' or, better, 'polymer-assisted surfactant flooding', enables an ultra-low interfacial tension to occur with a considerably reduced amount of surfactant. The polymer, which is used in minute concentrations, displaces the surfactant from the bulk aqueous phase, thus promoting its accumulation at interfaces. Another effect of the added polymer is a reduction of the surfactant CMC. Anionic polysaccharides such as xanthan and scleroglucane and synthetic polymers such as sulfonated polyacrylamide have been found useful for the purpose. The demands on the surfactant are the same as in normal surfactant flooding, i.e. it should form a microemulsion without any cosurfactant, etc. Similar to normal flooding, the displacing fluid is pushed forward by a viscous plug containing a polymer which may, or may not, be the same as the one used together with the surfactant.

Bibliography

Azemar, N., The role of microemulsions in detergency processes, in *Industrial Applications of Microemulsions*, C. Solans and H. Kunieda (Eds), Surfactant Science Series, 66, Marcel Dekker, New York, 1997.

Bourrel, M. and R. S. Schechter, *Microemulsions and Related Systems*, Surfactant Science Series, 34, Marcel Dekker, New York, 1988.

Miller, C. A. and K. H. Raney, *Colloid Surf., A*, **74** (1993) 169.

Shah, D. O. and R. S. Schechter (Eds), *Improved Oil Recovery by Surfactant and Polymer Flooding*, Academic Press, New York, 1977.

Solans, C. and N. Azemar, Detergency and the HLB temperature, in *Organized Solutions*, S. E. Friberg and B. Lindman (Eds), Surfactant Science Series, 44, Marcel Dekker, New York, 1992.

von Rybinski, W., Surface chemistry in detergency, in *Handbook of Applied Surface and Colloid Chemistry*, K. Holmberg (Ed.), Wiley, Chichester, UK, 2001.

23 CHEMICAL REACTIONS IN MICROHETEROGENEOUS SYSTEMS

Microemulsions can be used as Minireactors for Chemical Reactions

Microemulsions, being microheterogeneous mixtures of oil, water and surfactant, are excellent solvents for non-polar organic compounds as well as inorganic salts. The capability of microemulsions to solubilize a broad spectrum of substances in a one-phase formulation has been found useful for many technical applications. As mentioned in Chapter 22, microemulsions are, for instance, unsurpassed at cleaning of hard surfaces—dirt is often a complex mixture of hydrophilic and hydrophobic components.

In recent years, microemulsions have been evaluated as reaction media for a variety of chemical reactions. In preparative organic chemistry, microemulsions may be used to overcome reactant solubility problems and can also increase the reaction rate, as will be discussed below. Microemulsions are also of interest as media for inorganic reactions. Water-in-oil type microemulsions have been found useful in preparing small particles of metals and inorganic salts.

Another use of microemulsions for particle formation is in the preparation of monodisperse lattices of very small droplet size. By polymerization in w/o microemulsions, ultrafine particles containing very high molecular weight polymers are obtained. In bioorganic synthesis, w/o microemulsions can be employed as 'minireactors', both for condensation and for hydrolysis reactions. Most publications deal with lipase-catalysed reactions using hydrophobic substrates. The enzyme is located in the water pools and reaction is believed to take place at the oil–water interface.

The four types of reactions in microemulsions mentioned above, i.e. organic synthesis, formation of inorganic particles, polymerization and bioorganic synthesis, can all be seen as emerging technologies of considerable current interest. In slightly different ways, they all take advantage of the large interior interfacial area of microemulsions. The four reaction types will be treated individually below.

In the literature related to reactions in microheterogeneous media, there is often no clear distinction between microemulsions and micellar systems. Systems with a large water-to-oil ratio are sometimes referred to as micellar systems and those with a large oil-to-water ratio as reversed micellar systems. The term 'microemulsion' is used for those systems that contain a considerable amount of both oil and water. In this present chapter, no such distinction is made. All systems containing oil and water, together with surfactant, are termed microemulsions, regardless of the relative component proportions. The term micellar system is restricted to systems containing water and surfactant only.

Surface Active Reagents may be Subject to Micellar Catalysis

The catalytic effect on organic reactions exerted by micelles has been widely studied, with a typical reaction being the base catalysed hydrolysis of lipophilic esters. This type of rate enhancement is normally referred to as micellar catalysis. The analogous effect occurring in microemulsions may be called microemulsion catalysis.

The rate enhancement due to the microstructure of the reaction medium may be substantial. The effect of micelle formation on reaction rates is primarily a consequence of reactant compartmentalization. Inclusion or exclusion of reactants from the micelle–bulk interfacial region has either a catalytic or an inhibitory effect on the reaction rate, depending on the reaction type and nature of the micelle. The most widely used model to describe these phenomena is the pseudo-phase kinetic model. This approach assumes that micelles act as a phase (pseudo-phase) apart from water. Furthermore, it is assumed that the effects on reaction rates are due primarily to the distribution of reactants between the micellar and aqueous pseudo-phases. The rate of a chemical reaction is then the sum of the adjusted rates in the aqueous and micellar pseudo-phases. There are numerous examples of marked rate enhancements in reactions between a positively charged surface active compound and a negatively charged water-soluble species, or vice versa. A thoroughly studied example of micellar catalysis is the alkaline hydrolysis of long-chain betaine esters. This reaction is presented below.

Betaine esters of straight-chain alcohols in the C10 to C14 range are of interest as bactericides with limited half-lives. The cationic charge in close proximity to the ester bond gives a partial positive charge to the carbonyl carbon atom. The increased electrophilicity of this group makes such an ester extremely susceptible to alkaline hydrolysis and correspondingly stable to acid. (This type of product was described in Chapter 11 as an example of a surfactant with a weak bond deliberately built in.) The reaction involved is shown in Figure 23.1.

Figure 23.2 illustrates the reaction kinetics for a series of betaine esters of alcohols ranging from C3 to C14. As can be seen, hydrolysis of the propyl ester (C3) is concentration-independent. The surface active esters, on the other hand,

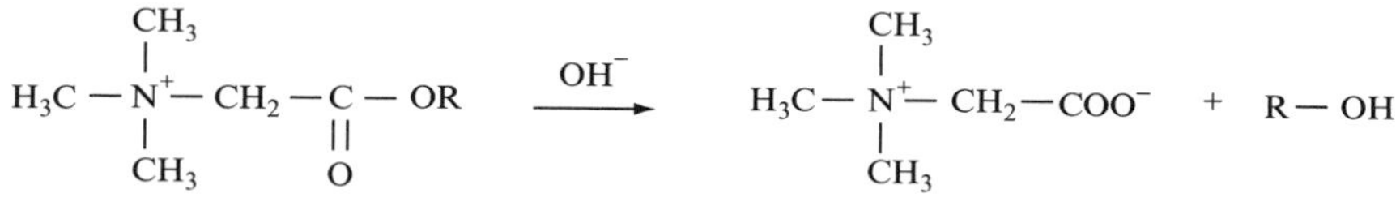

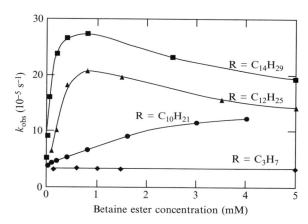

Figure 23.1   Alkaline hydrolysis of betaine esters

Figure 23.2   Concentration dependence of the rate constant, k_{obs}, for the hydrolysis of betaine esters of four different alcohols; R refers to the alkyl group of the alkylbetainates of Figure 23.1. Reproduced by permission of Academic Press from R. A. Thompson and S. Allenmark, *J. Colloid Interface Sci.*, **148** (1992) 241

show considerable concentration-dependence. The curves indicate micellar catalysis. The rapidly ascending branches of the curves can be accounted for by the formation of micelles. The ester bonds and the hydroxide ions are concentrated in the small pseudo-phase formed around the micelle. The descending branches of the curves above the CMCs are caused by displacement of hydroxide ions as a consequence of the increasing concentrations of bromide counterions.

The influence of electrolyte on micellar catalysis is of practical importance. Figure 23.3 shows the effect of the addition of common salts on the rate of hydrolysis of the tetradecyl ester. It is evident from this figure that salts drastically reduce the hydrolysis rate. Sodium bromide even brings down the rate constant to values below that of the non-surface active propyl ester. The effect can be explained by the pseudo-phase model in which the distribution of ions between the aqueous and micellar phases can be described by an ion-exchange mechanism. The small, high-charge-density, hydroxide ions compete more poorly for the micellar phase than do the larger ions of lower charge density such as chloride or bromide. Such ions are polarizable and interact

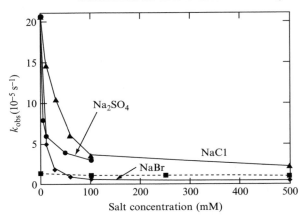

Figure 23.3 Effect of inorganic salts on the rate of hydrolysis of tetradecylbetainate; the dotted line represents the hydrolysis rate of propylbetainate. Reproduced by permission of Academic Press from R. A. Thompson and S. Allenmark, *J. Colloid Interface Sci.*, **148** (1992) 241

strongly with the micelle surface. As can be seen from this figure, bromide ions are very efficient in displacing hydroxide ions from the micellar pseudo-phase. In other words, in the presence of a bromide salt, the local hydroxyl ion concentration in the vicinity of the ester bonds is low, thus leading to slow hydrolysis. This is shown schematically in Figure 23.4.

The catalytic effect exerted by micelles also disappears in the presence of hydrophobic surface active species of opposite charge. As can be seen from Figure 23.5, the medium chain acids, hexanoic acid and *p*-toluenesulfonic acid, decrease hydrolysis to approximately the same value as that of propylbetainate. Tetradecanoic acid almost completely stops the hydrolysis at a concentration as low as 5 mM. This is probably due to the formation of mixed micelles of tetradecylbetainate and tetradecanoate. Above a certain concentration of the latter species, the net charge of the micelles becomes negative, hence leading to repulsion of hydroxide ions from the micellar pseudophase.

Microemulsions are Good Solvents for Organic Synthesis

The great utility of microemulsions as solvents for organic reactions lies in their ability to solubilize polar and non-polar substances and to compartmentalize and concentrate reagents. The reactions discussed below are organized into three categories: (a) the use of microemulsions to overcome reagent incompatibility problems, (b) reactions where a specific catalytic effect is obtained, and (c) the use of microemulsions to influence reaction regioselectivity.

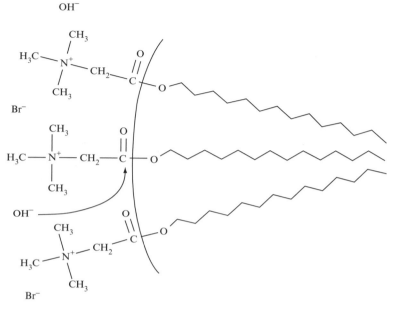

Figure 23.4 Micellar catalysis is caused by a high concentration of reactive species, e.g. hydroxide ions, at the micelle surface. The reactive species can be displaced by unreactive ones, e.g. bromide ions, in which case the catalytic effect disappears

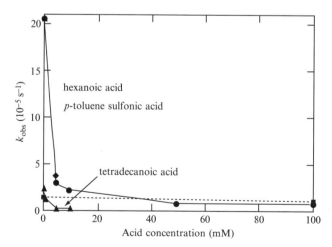

Figure 23.5 Effect of hydrophobic surface active species on the rate of hydrolysis of tetradecylbetainate; the dotted line represents the hydrolysis rate of propylbetainate. Reproduced by permission of Academic Press from R. A. Thompson and S. Allenmark, *J. Colloid Interface Sci.*, **148** (1992) 241

Overcoming Reagent Incompatibility Problems

A common practical problem in synthetic organic chemistry is to attain proper phase contact between non-polar organic compounds and inorganic salts. There are many examples of important reactions where this is a potential problem: hydrolysis of esters with alkali, oxidative cleavage of olefins with permanganate–periodate, addition of hydrogen sulfite to aldehydes and to terminal olefins, preparation of alkyl sulfonates by treatment of alkyl chlorides by sulfite or by addition of hydrogen sulfite to α-olefin oxides. The list can be extended further. In all examples given, there is a compatibility problem to be solved if the organic component is a large non-polar molecule.

There are various ways to solve the problem of poor phase contact in organic synthesis. One way is to use a solvent or a solvent combination capable of dissolving both the organic compound and the inorganic salt. Polar, aprotic solvents are sometimes useful for this purpose but many of these are unsuitable for large-scale work due to toxicity and/or difficulties in removing them by low-vacuum evaporation.

Alternatively, the reaction may be carried out in a mixture of two immiscible solvents. The contact area between the phases may be increased by agitation. Phase-transfer reagents, in particular quaternary ammonium compounds, are useful aids in many two-phase reactions. In addition, crown ethers are very effective in overcoming phase-contact problems; however, their usefulness is limited by high price. (Open-chain polyoxyethylene compounds often give a 'crown ether effect' and may constitute practically interesting alternative phase-transfer reagents.)

Microemulsions are excellent solvents both for hydrophobic organic compounds and for inorganic salts. Being macroscopically homogeneous, yet microscopically dispersed, they can be regarded as something between a solvent-based one-phase system and a true two-phase system. In this context, microemulsions should be seen as an alternative to two-phase systems with phase-transfer reagents. This is illustrated below by the use of microemulsions for the detoxification of 'mustard gas', $ClCH_2CH_2SCH_2CH_2Cl$.

Mustard gas is a well-known chemical warfare agent. Although it is susceptible to rapid hydrolytic deactivation in laboratory experiments, where rates are measured at low substrate concentrations, its deactivation in practice is not easy. Due to its extremely low solubility in water, it remains for months on the water surface. Addition of strong 'caustic' does not increase the rate of reaction very much. Microemulsions have been explored as media for both the hydrolysis and oxidation of 'half-mustard', $CH_3CH_2SCH_2CH_2Cl$, a mustard-gas model (Figure 23.6). Oxidation with hypochlorite turned out to be extremely rapid in both o/w and w/o microemulsions. In formulations based on either anionic, non-ionic or cationic surfactants, oxidation of the half-mustard sulfide to sulfoxide was complete in less than 15 s. The same reaction takes

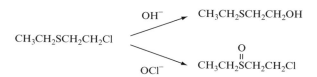

Figure 23.6 Transformation of 2-chloroethylethyl sulfide (half-mustard) into 2-hydroxyethylethyl sulfide (by alkali) or into 2-chloroethylethyl sulfoxide (by hypochlorite). Reprinted with permission from F.M. Menger and A.R. Elrington, *J. Am. Chem. Soc.*, **113** (1991) 9621. Copyright (1991) American Chemical Society

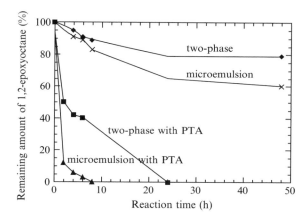

Figure 23.7 Reaction rate of ring-opening of a lipophilic epoxide with hydrogen sulfite. This reaction is performed in a two-phase system with and without added phase-transfer agent (PTA), in a microemulsion, and in a microemulsion with added phase transfer agent. Reprinted from *Tetrahedron Lett.*, **41**, M. Häger and K. Holmberg, 'Microemulsions as reaction medium for surfactant synthesis', 1245–1248, Copyright (2000), with permission from Elsevier Science

20 min when a two-phase system, together with a phase-transfer reagent, is employed.

The use of microemulsions as reaction media may be seen as an alternative to phase-transfer catalysis. The two approaches may also be combined. Figure 23.7 shows that the rate of the following reaction:

is very high when carried out in a microemulsion with a phase-transfer catalyst added to the reaction mixture.

Organic reactions in microemulsions need not be performed in one-phase systems. It has been found that most reactions work well in two-phase Winsor I or Winsor III systems (see Chapter 6). The transport of reactants between the microemulsion phase, where the reaction takes place, and the excess oil or water phase is evidently fast when compared to the rate of the reaction. This is a important practical aspect of the use of microemulsions as media for chemical reactions, because it simplifies the formulation work. Formulating a Winsor I or Winsor III system is usually much easier than formulating a one-phase microemulsion of the whole reaction mixture.

Specific Rate Enhancement

By a proper choice of surfactants, rate enhancement analogous to micellar catalysis can be achieved. In microemulsion systems, the effect may be referred to as microemulsion catalysis.

The importance of the choice of surfactant on reaction yield can be illustrated by the reaction between decyl bromide and sodium sulfite to give decyl sulfonate, a surface active agent:

$$C_{10}H_{21}Br + Na_2SO_3 \longrightarrow C_{10}H_{21}SO_3Na + NaBr$$

Reactions were carried out in microemulsions based on decyl bromide dissolved in dodecane as the oil component, aqueous Na_2SO_3 as the water component and either a non-ionic surfactant or a non-ionic surfactant plus a small amount of ionic surfactant as the amphiphile.

The reaction was extremely slow in a surfactant-free mixture of the oil and water components, fairly sluggish in the liquid crystalline phase and fast in the two microemulsion systems. Figure 23.8 shows the influence of the addition of a small amount of ionic surfactant, sodium dodecyl sulfate (SDS), which gives a negative charge to the droplet surface, to bring about a reduction in the reaction rate. This is the expected result since electrostatic double-layer forces will render approach of the sulfite ion into the interfacial region difficult.

It may seem surprising that the cationic surfactant $C_{14}TAB$ (see Figure 23.8) also decreases the reaction rate. The other cationic surfactant, $C_{14}TAAc$, on the other hand, gives a considerable increase in reactivity. Evidently, the choice of counterion is decisive for the reaction rate. A large, polarizable counterion, such as the bromide ion, interacts so strongly with the surfactant palisade layer that approach of the anionic reactant, the sulfite ion, is prevented. Interaction with the acetate ion is much weaker and sulfite ions are allowed to diffuse into the interfacial region where reaction occurs. The electrolyte effect is analogous to that discussed on p. 495 for the alkaline hydrolysis of surface active betaine esters.

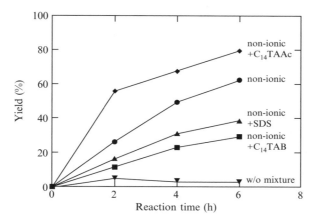

Figure 23.8 Effect of ionic surfactants on the rate of formation of decyl sulfonate from decyl bromide and sodium sulfite in microemulsions based on the non-ionic surfactant $C_{12}E_5$ with the addition of 2% ionic surfactant. SDS, C_{14}TAAc and C_{14}TAB stand for sodium dodecyl sulfate, cetyltrimethylammonium acetate and cetyltrimethylammonium bromide, respectively. From K. Holmberg, S.-G. Oh and J. Kizling, *Prog. Colloid Polym. Sci.*, **100** (1996) 281

Effect on Regioselectivity

The presence of an oil–water interface may induce orientation of reactants in microemulsion systems, which in turn may affect the regioselectivity of organic reactions. The principle is shown in Figure 23.9. In reactions with difunctional

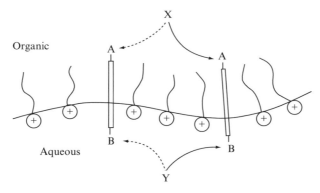

Figure 23.9 Orientation of a reactant at the oil–water interface of a microemulsion may induce regioselectivity. A water-soluble reactant, Y, will attack the compound with two reactive groups from the aqueous side and an oil-soluble species, X, will attack from the oil side

substances which align at the interface, a water-soluble species will attack from the water side and an oil-soluble reactant will approach from the oil side. Regioselectivity is an important issue in organic synthesis and the use of an oil–water interface as a template for the reaction is an alternative to other approaches used to obtain regiochemical control.

Microemulsions are Useful as Media for Enzymatic Reactions

The major potential advantages of employing enzymes in media of low water content are as follows:

(1) Increased solubility of non-polar reactants.

(2) Possibility of shifting thermodynamic equilibria in favour of condensation.

(3) Improvement of thermal stability of the enzymes, thus enabling reactions to be carried out at higher temperatures.

In fact, the use of enzymes in 'water-poor media' is not unnatural. Many enzymes, including lipases, esterases, dehydrogenases and oxidoreductive enzymes, often function in the cells in microenvironments that are hydrophobic in nature. In addition, the use of enzymes in microemulsions is not an artifical approach *per se*. In biological systems, many enzymes operate at the interfaces between the hydrophobic and hydrophilic domains, and such interfaces are often stabilized by polar lipids and other natural amphiphiles.

Enzymatic catalysis in microemulsions has been used for a variety of reactions such as the synthesis of esters, peptides and sugar acetals, transesterifications, various hydrolysis reactions and steroid transformations. The enzymes employed include lipases, phospholipases, alkaline phosphatase, pyrophosphatase, trypsin, lysozyme, α-chymotrypsin, peptidases, glucosidases and oxidases.

By far the most widely used class of enzymes in microemulsion-based reactions are the lipases—of microbial as well as of animal origin. Some examples of lipase-catalysed ester synthesis and transesterification will be given below. Before discussing specific reactions, some general aspects of enzymatic catalysis in microemulsions will be illuminated.

Enzymatic Activity in W/O Microemulsions

The role of the enzyme-bound water for biocatalysis has still not yet been fully clarified. However, the water dependency clearly differs from one enzyme to another. For instance, some lipases exhibit high activity and good stability in organic solvents containing only traces of water, while other lipases reach optimal activity at a relatively high water-to-organic solvent ratio. For a wide variety of enzymes, it has been found that there is a 'bell-shaped' dependence of

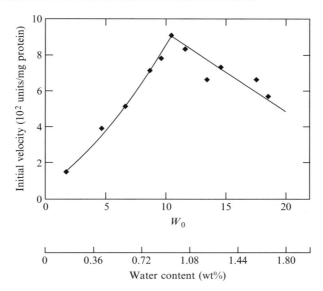

Figure 23.10 Initial velocity of lipase in w/o microemulsions (50 mM AOT in isooctane) of varying molar ratio of water to surfactant, W_0. Reproduced by permission of John Wiley & Sons, Inc. from D. Han and J. S. Rhee, *Biotechnol. Bioeng.*, **28** (1986) 1250

activity on the ratio of water to surfactant in the reaction medium. (The molar ratio of water to surfactant is often referred to as W_0). Figure 23.10 shows a representative example.

In general, it seems that maximum activity occurs around a value of W_0, at which the size of the droplet is somewhat larger than that of the entrapped enzyme. The latter sometimes exhibits enhanced activity (superactivity) when compared to that in the bulk water.

The anomalous activity characteristics have been attributed to conformational changes of the solubilized enzymes, but more recent spectrosopic studies seem to indicate that this is not the main cause. Solubilization of an enzyme into reverse micelles does not normally lead to major conformational alterations, as indicated, for example, by fluorescence and phosphorescence spectral investigations. The situation is complex, however, and circular dichroism measurements suggest that the influence of the oil–water interface on enzyme conformation may vary even between enzymes belonging to the same class.

Choice of Solvent

The organic solvent used in the microemulsion formulation should be non-polar. The hydrophobicity of the solvent seems to be a key factor for the catalytic activity of the enzyme. A good correlation between hydrophobicity of the organic solvent and biocatalytic activity is obtained by the use of log P values of solvents. (P is the partition coefficient of the solvent in the octanol–water system.) In general, the enzyme stability and activity in microemulsions are poor with relatively hydrophilic solvents in which log P is < 2, they are moderate in solvents where log P is between 2 and 4, and high in hydrophobic solvents with log $P > 4$. Even relatively small changes in the choice of solvent, such as going from cyclohexane to nonane, can give rise to large improvements in enzyme stability. The rationale behind this division is that with very hydrophobic solvents an effective segregation is obtained between the water and oil domains. Thus, such solvents do not distort the essential water layer around the enzyme, thereby leaving the catalyst in an active state. This limits the choice of organic solvent to aliphatic hydrocarbons with seven or more carbon atoms. On the other hand, lower hydrocarbons are preferred from a work-up point of view since they can be readily removed by evaporation after reaction. As a compromise, heptane, octane and nonane are the solvents of choice and these hydrocarbons have been used in almost all enzymatic reactions in microemulsion media.

Structure of Protein-Containing Microemulsion Droplets

Most enzymatic reactions in microemulsions have been carried out in w/o-type systems with the enzyme confined in the water pools. A common way to picture the situation is as a system of water droplets in oil, some of which contain an enzyme, and some of which are empty. Not more than one enzyme resides in one droplet. The enzyme is normally surrounded by water on all sides. There are indications that the enzyme-filled droplets are larger than the empty ones.

This 'water shell model', picturing the enzyme molecule residing in the interior of the water droplet, will only apply to hydrophilic proteins that do not contain large hydrophobic domains. Many enzymes, including lipases, are surface active and interact strongly with oil–water interfaces. Such proteins are, of course, not confined to the interior of the water pools. In fact, lipases seem to need a hydrophobic surface in order to 'open the lid' covering the active site. No water layer is therefore likely to separate the lipase active site from the continuous hydrocarbon domain.

Lipase-Catalysed Ester Synthesis

Lipase catalysed synthesis of esters of monofunctional alcohols proceeds in good yield in microemulsions. Different lipases exhibit different preferences with regard to the chain length of the acid and the type of alcohol. This selectivity is probably related to the localization of the enzyme molecule at the oil–water interface. Hence, the hydrophilic–lipophilic character of the protein, and not only its specificity as expressed in aqueous solution, is responsible for the selectivity. This illustrates the important point that the regioselectivity of bioorganic (and organic) reactions may differ between homogeneous and microheterogeneous media. Esterification of fatty acids with simple sugars, such as glucose and mannitol, in w/o microemulsions gives a very low yield of sugar ester. This is probably due to poor phase contact between the very hydrophilic sugar molecule in the water pool and the fatty acid that resides in the hydrocarbon domain. Sugar monoesters can be produced in high yields by lipase-catalysed esterification in a water-free medium.

The main practical problem in the large-scale use of biocatalysis for the synthesis of hydrophobic esters is that of work-up. Separating the surfactant from the product is not a trivial issue since normal purification procedures, such as extraction and distillation, tend to be troublesome due to the well-known problems of emulsion forming and foaming caused by the surfactant.

An interesting extension of the use of microemulsions for synthesis is to use microemulsion-based gels (MBGs) as reaction media. The MBGs are made by mixing normal w/o microemulsions with aqueous gelatin solutions above the gelling temperature of the gelatin and then allowing the mixture to cool, for instance, in a column, so that a stiff gel is formed (Figure 23.11).

Various spectroscopic techniques have revealed that the microemulsion structure is retained in the gel. The latter can be seen as an immobilized enzyme-containing microemulsion. This gel is resistant to hydrocarbon solvents. By charging a hydrocarbon solution of an acid and an alcohol at the top of the column, the corresponding ester can be recovered from the eluent. This is

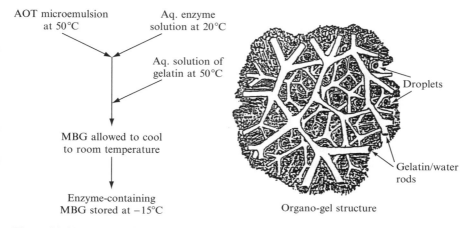

AOT microemulsion
at 50 °C

Aq. enzyme
solution at 20 °C

Aq. solution of
gelatin at 50 °C

MBG allowed to cool
to room temperature

Enzyme-containing
MBG stored at −15°C

Droplets

Gelatin/water
rods

Organo-gel structure

Figure 23.11 Preparation of an enzyme-containing microemulsion-based gel (MBG). Reprinted from *Biochim. Biophys. Acta*, **1073**, G. D. Rees, M. G. Nascimento, T. R. J. Jenta and B. H. Robinson, 'Reverse enzyme synthesis in microemulsion-based organogels', 493–501, Copyright (1991) with permission from Elsevier Science

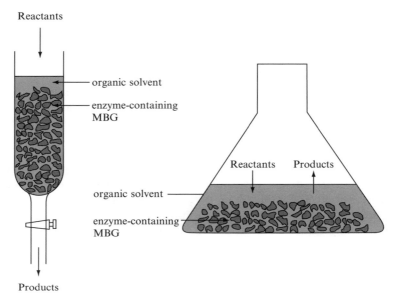

Figure 23.12   The microemulsion-based gel (MBG) can be formed and used in a column or placed in pieces in a beaker

an elegant way of avoiding the problem of separation of the product and surfactant from the reaction mixture (Figure 23.12).

Lipase-containing MBGs have been used to synthesize, on a preparative scale, a variety of different esters under mild conditions, and both regio- and stereoselectivity have been demonstrated.

Lipase-Catalysed Transesterification

Lipase-catalysed transesterification, i.e. replacement of one acyl group in a triglyceride by another acyl group, can be performed in microemulsions of low water content. Special attention has been directed towards production from inexpensive starting material, such as palm oil, of a triglyceride mixture that corresponds to natural cocoa butter. This reaction requires a partial replacement of palmitoyl groups by stearoyl groups in the 1(3)-position, while leaving the 2-position essentially unaffected. A high degree of conversion can be obtained in lipase-containing microemulsions.

In a study aimed at incorporating γ-linolenic acid (GLA) into a saturated triglyceride, tristearin was transesterified with GLA using lipase as the catalyst. Reactions were carried out at four different compositions, all situated in the hydrocarbon-rich corner of the ternary phase diagram. As seen in Figure 23.13,

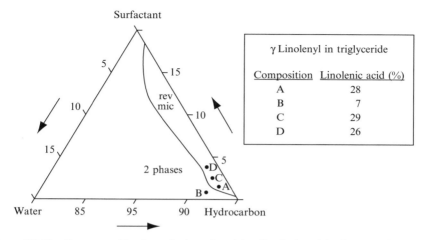

Figure 23.13 Transesterification of tristearin with γ-linolenic acid using reaction media of slightly different composition. The percentage of linolenic acid given refers to the percentage incorporated in triglyceride after 24 h reaction time. Reprinted from *Adv. Colloid Interface Sci.*, **51**, K. Holmberg, 'Organic and bioorganic reaction in Microemulsions', 137–174, Copyright (1994) with permission from Elsevier Science

three samples lie in the isotropic L2 region, whereas one sample falls within the two-phase region. The reactions were carried out under stirring. As also shown in Figure 23.13, the reaction in the two-phase region (emulsion) is incomplete compared to the reactions in microemulsions. This set of experiments is a good illustration of the benefits of the much larger oil–water interfacial area of microemulsions when compared to emulsions.

Microemulsions can be Used to Prepare Nanosized Lattices

Normal emulsion polymerization typically gives lattices with a particle size in the range between 0.1 and 0.5 μm. Compared with free-radical polymerization performed in a homogeneous solution, this type of polymerization yields high molecular weight polymers at high reaction rates since the free radicals grow in relative isolation. Due to the low viscosity of the medium, water being the continuous phase, good control of heat transfer is made possible. These lattices are used in large-scale applications in areas such as paints and paper coatings, carpet backings, glues and adhesives, etc.

By carrying out polymerizations in microemulsions, thermodynamically stable lattices in the nanosize range (20–50 nm) can be obtained. Such lattices are of interest for specific applications where small particle size and extreme dispersion stability are essential.

In principle, polymerization can take place in all types of microemulsions, w/o, o/w, bicontinuous and also Winsor I, II and III systems. The main

practical interest relates to polymerizations in w/o and bicontinuous microemulsions and pertinent examples of such systems are given below.

Polymerization in W/O Microemulsions

The most widely used monomer is acrylamide (AM) and the surfactant
employed has most often been sodium bis(2-ethylhexyl)sulfosuccinate (AOT).
Aliphatic or aromatic hydrocarbons, such as octane, decane or toluene, have
been used as the oil component. By using AOT–hydrocarbon mixtures, relatively large amounts of water can be solubilized without the need for a cosurfactant.

Addition of AM to the ternary system toluene–water–AOT leads to an
extension of the L2 microemulsion (w/o) domain. AM accumulates at the oil–
water interface, acting like a cosolvent, and enhances the solubilization capacity
of the system. The water droplets obtained have diameters of the order of
5–10 nm.

During polymerization in w/o microemulsions, there is a continuous growth
of the droplets. The diameter of the final latex particle is somewhat dependent
on the initial composition of the system but is of the order of 25–50 nm. This
indicates that nucleated particles grow by addition of monomer from other
micelles, either by diffusion of monomer through the continuous phase or by
coalescence with neighbouring micelles. Both mechanisms are feasible in oil-
continuous microemulsions where droplet stability is attained by steric stabilization. The two routes are illustrated in Figure 23.14.

An important characteristic of microemulsion polymerization, regardless of
microemulsion structure, is that after the completed reaction each droplet
contains a very low number of individual polymer molecules with very high

 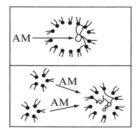 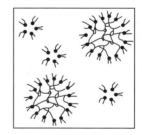

Figure 23.14   Polymerization of acrylamide (AM) in a w/o microemulsion. Polymerization occurs in some of the micelles which grow through incorporation of new monomer.
Supply of new AM may occur either by monomer diffusion through the continuous
medium (upper picture) or by transport in micelles which coalesce with the nucleated particles (lower picture). From *Scientific Methods for the Study of Polymer Colloids and their
Applications*, F. Candau and R. H. Ottewill (Eds), 1990, p. 73, 'An introduction to polymer
colloids', F. Candau, Figure 9, with kind permission of Kluwer Academic Publishers

molecular weights. The number of molecules is often as low as 2–5 and in the case of AM polymerized in AOT-based microemulsion with the use of a proper initiator, each final latex particle consists of a single macromolecule of very high molecular weight. It has been demonstrated by electron microscopy that the particles grow rapidly until a certain diameter is reached. The diameter of each particle then remains constant while the number of particles grows steadily during the course of the reaction. Evidently, the system undergoes new particle nucleation during the entire polymerization process. This is contrary to conventional emulsion polymerization where the first nucleation stage is followed by a particle growth at constant particle number. The difference is shown schematically in Figure 23.15.

At the start of the microemulsion polymerization, the number of monomer-swollen micelles is very large, typically of the order of 10^{21} per litre. After the reaction is completed, the number of microlatex particles is about three orders of magnitude smaller. Consequently, only a small fraction of the surfactant molecules is needed to stabilize the latex. This means that after reaction there is an excess of surfactant which forms small micelles with a diameter of 2–4 nm which coexist with the polymer latex particles with a diameter of 25–50 nm, as illustrated in Figure 23.15.

It is interesting that a very large polymer molecule, such as a polyacrylamide of molecular weight 4×10^6, can be confined in a water droplet with a radius of 15 nm. The radius of gyration of this macromolecule dissolved in water is 125 nm. This indicates that the molecule has undergone a conformational collapse and that the water in the particle core acts more as a plasticizer than as a solvent for the polymer.

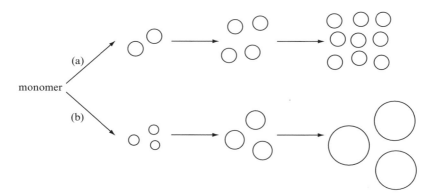

Figure 23.15 Illustration of the growth of polymer particles in (a) microemulsion polymerization and (b) conventional latex polymerization. Reproduced by permission of Academic Press from F. Candau, Y. S. Leong, G. Pouyet and S. Candau, *J. Colloid Interface Sci.*, **101** (1984) 167

Polymerization in Bicontinuous Microemulsions

Acrylamide has been solubilized in high concentrations (up to 25%) in bicontinuous microemulsions based on non-ionic surfactants. Immediately after the onset of polymerization, the transparent solution becomes turbid and the viscosity increases. The rise in viscosity is due to the formation of very large macromolecules in the aqueous domains. Towards the end of the reaction, the system again becomes clear with low viscosity. As with polymerization in w/o systems, the reaction is extemely fast and 100% conversion is reached within less than 20 min.

The bicontinuous structure breaks immediately at the onset of polymerization, probably as a result of consumption of the monomer, which to a large extent resides at the oil–water interface and acts as a cosolvent. A two- or multiphase system is formed which is gradually transformed into the final structure, i.e. microlatex droplets dispersed in an aqueous phase. The particle size is usually somewhat larger than that obtained in w/o microemulsions, but this is mainly due to the higher monomer concentration which is employed. (The particle size is essentially dependent on the monomer-to-surfactant ratio: the higher the ratio, then the bigger the size.) The possibility of using a high monomer concentration is a major advantage of bicontinuous when compared to droplet-type microemulsions.

The size distribution of the particles, expressed as d_w/d_n, is more narrow than is obtained in conventional latex polymerization. A typical value for acrylamide polymerized in bicontinuous microemulsions is 1.10–1.15 whereas a value of around 2 is common in emulsion polymerizations.

Properties and Uses

As mentioned above, a characteristic feature of microlattices is that each particle contains only very few molecules of very high molecular weight, typically in the range of 10^6-10^7. The microlattices can be used as such, and they are of current interest for various biomedical applications. Targeted drug delivery is an example of such an application. The small lattices may also be used for microencapsulation of cells or polymerization may be performed in the presence of a biologically active substance which becomes embedded in the microlatex droplets. The polymerization process can also be seen as a way to produce high molecular weight macromolecules. The polymers can be precipitated by the addition of a non-solvent. After washing with a suitable solvent to get rid of the surfactant (which may be re-used), the polymer is obtained in a fairly pure form. Examples of potential uses of such polymers are as follows:

● Flocculating agents, e.g. for water purification

● Retention aids for paper making

- Binders for glues and adhesives
- Thickeners and rheological control aids

High molecular weight polymers are today being prepared commercially by polymerization in bicontinuous microemulsions, followed by separation through precipitation.

Some of the advantages and disadvantages of microemulsion polymerization when compared to conventional emulsion polymerization are summarized in the following:

(1) The polymerization rate is very high, leading to short reaction times.

(2) No coagulum is formed in the polymerization.

(3) The microlatex formed exhibits high stability.

(4) The particles are uniform in size.

(5) In copolymerization of one hydrophilic and one hydrophobic monomer, a product more homogeneous in composition is obtained.

(6) The polymer formed is of a very high molecular weight.

(7) The amount of surfactant needed is very high. When the process is used for preparing high molecular weight polymers, the surfactant normally needs to be separated from the product.

(8) The maximum amount of monomer in the formulation is low for o/w and w/o microemulsions (but reasonably high for bicontinuous systems).

Nanosized Inorganic Particles can be Prepared in Microemulsions

Microemulsions, particularly of the w/o type, have become of interest as a medium for synthesis of inorganic particles, ranging from metals to complex crystallites, of very small size. The interest derives from the well-known fact that the properties of advanced materials are critically dependent on the microstructure of the sample. Control of size, size distribution and morphology of the individual grains or crystallites is of the utmost importance in order to obtain the material characteristics desired. Chemical reactions in microemulsions have been explored as one route to achieve fine particles. Table 23.1 gives examples of areas where this technique has been used.

The principle employed is straightforward. In the simplest case, involving two reagents that are soluble in water but insoluble in oil, one reagent is dissolved in the water pools of w/o microemulsion A, while the other reagent is dissolved in the water pools of w/o microemulsion B. The two microemulsions are subsequently mixed. Due to their small size, the droplets are subject to

Table 23.1   Applications of the microemulsion technique to prepare small inorganic particles

Application	Examples
Semiconductors	CdS, CdSe
Supercoductors	Y–Ba–Cu–O, Bi–Pb–Sr–Ca–Cu–O
Catalysts	Pt, Pd, Rh
Magnetic particles	Fe or Fe alloys[a], $BaFe_{12}O_{19}$[b]

[a]Prepared, for example, from microemulsions of $FeCl_2$ and $NaBH_4$.
[b]Prepared, for example, by mixing a microemulsion containing $Ba(NO_3)_2$ and $Fe(NO_3)_3$ with one containing NH_4NO_3.

Brownian motion. They collide continuously and in doing so dimers and other aggregates will form. These aggregates are short-lived and rapidly disintegrate into droplets of the original size. As a result of the continuous coalescence and decoalescence process, the content of the water pools of microemulsions A and B will be distributed evenly over the entire droplet population and reaction will occur in the droplets. The product will eventually precipitate out.

The majority of work has been carried out with sodium bis(2-ethylhexyl)sulfosuccinate (AOT) as the surfactant since microemulsions formed using this surfactant have a large L2 (water-in-oil) region and consist of well-defined droplets. The size of the droplets is expected, on the basis of simple geometrical arguments, to be proportional to the molecular ratio, W_0, of water to surfactant. This has also been confirmed experimentally by scattering techniques. In fact, the droplet size may be calculated to a first approximation by using the following equation:

$$r = 0.18W_0 + 1.5 \tag{23.1}$$

where r is the hydrodynamic radius in nm.

Consequently, the droplet size can be varied simply by adjusting W_0 and the compartmentalized water droplets can then be used to control the particle growth. The situation is, however, more complex because after initial particle growth there are secondary processes that affect the final particle size, such as Ostwald ripening and flocculation.

In the following, two examples of inorganic particle formation in w/o microemulsions will be given, namely cadmium sulfide and platinum.

Formation of CdS Particles

Figure 23.16 illustrates the formation of CdS particles according to the following reaction:

$$Cd(NO_3)_2 + Na_2S \longrightarrow CdS + 2NaNO_3$$

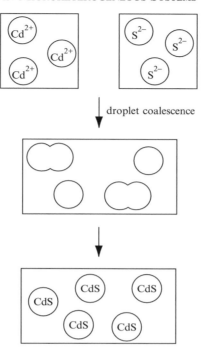

Figure 23.16 Preparation of cadmium sulfide crystals from water-soluble cadmium nitrate and sodium sulfide

Nanosized particles of cadmium sulfide are of interest for semiconductors. There is a particular need for extremely small particles for this application since the material properties are strongly dependent on the sizes of the crystallites.

As mentioned above, the main feature of the process is the contact between the two water-soluble salts as a result of the transient fusion of droplets. This is a very fast process which in this system has a second-order rate constant of $10^6 - 10^7$ l/mol s with some dependence on droplet size. Several different processes, listed in Table 23.2, are involved in the formation of the final stable CdS microparticles.

In the CdS process, it has been established that the rate of the nucleation/growth reaction in the microemulsion medium is controlled by the rate of interdroplet exchange. For an AOT-based microemulsion with $W_0 = 5$, the time to form a nucleus is around 1 ms.

Figure 23.17 shows the relationship between the particle size and initial droplet size for CdS. The x-axis gives a molar ratio of water to surfactant, W_0, which can be transformed into the droplet radius, r, via equation (23.1). It can be seen that there is a certain, but far from linear, correlation between the particle size and droplet radius.

Table 23.2 Events in the formation of CdS particles. From T. F. Towey, A. Khan-Lodhi and B. H. Robinson, *J. Chem. Soc., Faraday Trans.*, **86** (1990) 3757. Reproduced by permission of The Royal Society of Chemistry

---

Formation of primary CdS particles.
Fusion of small particles into larger aggregates:

$$(CdS)_m + (CdS)_n \longrightarrow (CdS)_{n+m}$$

Addition of S^{2-} and Cd^{2+} to growing particles:

$$(CdS)_n + S^{2-} \longrightarrow (Cd_nS_{n+1})^{2-}$$
$$(CdS)_n + Cd^{2+} \longrightarrow (Cd_{n+1}S_n)^{2+}$$

Ostwald ripening.

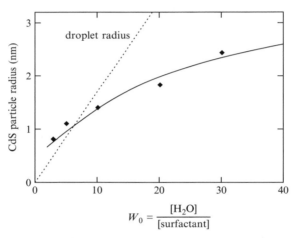

Figure 23.17 Cadmium sulfide particle size as a function of molar ratio of water to surfactant, W_0. From *The Structure, Dynamics and Equilibrium Properties of Colloidal Systems*, D. M. Bloor and E. Wyn-Jones (Eds), 1990, p. 373, 'Microparticle formation in reverse micelles', A. Khan-Lodhi, B. H. Robinson, T. Towey, C. Hermann, W. Knoche and U. Thesing, Figure 4, with kind permission of Kluwer Academic Publishers

Formation of Pt Particles

Small platinum particles, of interest for catalyst applications, can be prepared by reduction of a water-soluble platinum complex in a w/o microemulsion. A microemulsion is prepared with hexachloroplatinic acid in the water pools. Reduction to metallic platinum can be achieved by mixing this microemulsion with a microemulsion containing a reducing agent such as sodium borohydride or hydrazine in the water pools. The reaction is shown below for borohydride reduction:

$$[Pt^{+IV}Cl_6]^{2-} + BH_4^- + 3H_2O \longrightarrow Pt^0 + H_2BO_3^- + 6Cl^- + 4H^+ + 2H_2$$

A fine suspension of platinum is obtained. The size of the platinum particles is roughly that of the droplets of the starting microemulsion. For catalyst applications, the particles need to be deposited on a solid support such as alumina. This can be carried out by adding a solvent that dissolves the surfactant and is miscible with both oil and water. The surfactant will then be removed from the surface of the small particles and precipitation will occur due to gravitational forces.

The kinetics of platinum particle formation have been found to depend on the type of surfactant used to create the microemulsion. As can be seen from Figure 23.18, the reaction is fast when a non-ionic surfactant, $C_{12}E_5$, is used and more sluggish when an anionic surfactant, such as AOT, is employed. The difference in reactivity probably reflects the difference in solubility of the two surfactants in the continuous oil domain. Whereas AOT, being an ionic species, will remain at the interface, the alcohol ethoxylate, which has a high oil solubility, will partition between the continuous oil domain and the interface. The latter will give a much less rigid interfacial layer which, in turn, leads to less resistance to droplet coalescence.

Nanoparticles of other noble metals, such as palladium, and also alloys of noble metals, e.g. Pd–Pt, can be prepared analogously. The possibility of making small alloy particles by starting with an aqueous mixture of reducable salts or complexes of two metals is of considerable practical interest.

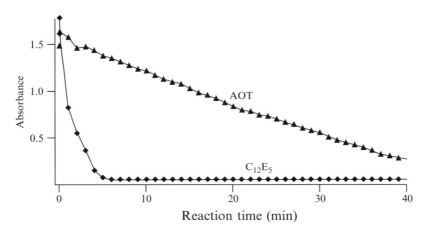

Figure 23.18 Formation of platinum particles as a function of reaction time. Two different surfactants were used to create the microemulsion, i.e. the anionic surfactant AOT and the non-ionic surfactant $C_{12}E_5$. The figure shows the disappearance of UV absorption at 265 nm, characteristic of the starting platinum complex. From H. Häre-lind Ingelsten *et al., J. Colloid Interface Sci.*, **241** (2001) 104

Mesoporous Materials can be Prepared from Surfactant Liquid Crystals

Mesoporous materials are inorganic solids with well-defined pore sizes of 2 to 50 nm. (Below 2 nm pore size, the material is referred to as microporous, with zeolites being the prime example.) Mesoporous materials have very large surface areas and long-range ordering of the packing of pores. Aerogels, pillared layered clays and a class of molecular sieves often referred to as MCM materials are well-known examples. The latter class of mesoporous materials can be produced from surfactant liquid crystals which template the inorganic component during synthesis.

The original MCM material is an aluminosilicate with hexagonally packed cylindrical mesopores. It can be made from an aqueous surfactant system in which the surfactants are in the form of hexagonal liquid crystals. The voids between the cylindrical surfactant aggregates are filled with soluble aluminosilicate. The aqueous phase is allowed to gel by adjustment of pH, so leading to condensation into a solid, continuous framework. Removal of the surfactant by washing or by heat treatment produces the mesoporous structure. The principle is shown in Figure 23.19. The same type of mesoporous structure can be obtained even if the concentration of the starting surfactant solution is below the concentration needed to form liquid crystals. It has been proposed that monolayers of inorganic precursor will then deposit onto isolated micellar rods and that these rods eventually pack into hexagonal mesostructures, as also indicated in Figure 23.19. Regardless of the mechanism, the structures obtained can be highly ordered with a narrow pore size distribution. Mesoporous materials with a cubic geometry can be prepared starting from bicontinuous cubic liquid crystals.

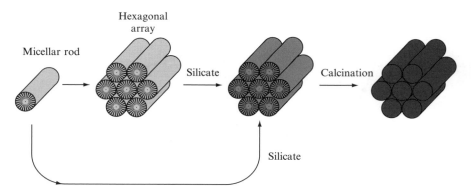

Figure 23.19 Pathways for formation of a mesoporous material with a hexagonal structure. Reprinted with permission from J. S. Beck *et al.*, *J. Am. Chem. Soc.*, **114** (1992) 10834. Copyright (1992) American Chemical Society

The pore diameter of mesoporous materials of both hexagonal and cubic geometry should theoretically be twice the length of the surfactant molecule. This has also been found to be the case. This means that the pore size is governed by the choice of surfactant. Mesoporous, inorganic materials are of considerable interest for catalysis applications.

Bibliography

Candau, F., Polymerization in microemulsion, in *Polymerization in Organized Media*, C. M. Paleos (Ed.), Gordon and Breach, Philadelphia, PA, 1992, Ch. 4.

Holmberg, K., *Adv. Colloid Interface Sci.*, **51** (1994) 137.

Holmberg, K., Microemulsions in biotechnology, in *Industrial Applications of Microemulsions*, C. Solans and H. Kunieda (Eds), Surfactant Science Series, 66, Marcel Dekker, New York, 1997.

Pileni, M. P., *J. Phys. Chem.*, **97** (1993) 6961.

Pillai, V., P. Kumar, M. J. Hou, P. Ayyub and D. O. Shah, *Adv. Colloid Interface Sci.*, **55** (1995) 241.

Sjöblom, J., R. Lindberg and S. E. Friberg, *Adv. Colloid Interface Sci.*, **95** (1996) 125.

Ying, J. Y, C. P. Mehnert and M. S. Wong, *Angew. Chem. Int. Ed. Engl.*, **38** (1999) 56.

APPENDIX 1. List of Surfactant Trade Names

More extensive information about surfactants, along with the addresses of surfactant manufacturing companies, can be found in McCutcheon's, Volume 1, *Emulsifiers and Detergents*, MC Publishing, Glen Rock, NJ, 1999, and in *Surfactants Europa*, 3rd Edn, G. L. Hollis (Ed.), Royal Society of Chemistry, Cambridge, UK, 1995.

Name	Type	Producer
Abil	Silicone-based surfactants	Th. Goldschmidt
Aerosol	Mostly sulfosuccinates	Cytec Industries
Aerosol OT (AOT)	Sodium dioctylsulfosuccinate	Cytec Industries
Akypo	Mostly alkyl ether carboxylates	Condea
Alkamide	Alkanolamides	Rhodia
Amiet	Ethoxylated amines and amides	Kao
Ammonyx	Amine oxides	Stephan
Ampholak	Amphoterics	Akzo Nobel Surface Chemistry
Arlacel	Fatty acid esters and ethoxylated fatty acid esters	ICI Surfactants
Arlatone	Ethoxylated fatty acid esters	ICI Surfactants
Armeen	Fatty amines	Akzo Nobel Surface Chemistry
Atlas	Mostly ethoxylated products	ICI Surfactants
Atlox	Surfactants and surfactant mixtures for pesticide formulations	ICI Surfactants

Berol	Various, mostly ethoxylated products	Akzo Nobel Surface Chemistry
Biodac	Ethoxylated C10 alcohol	Condea
Brij	Ethoxylated fatty alcohols	ICI Surfactants
Britex	Ethoxylated fatty alcohols	Cesalpinia Chemicals
Calgene	Various ester surfactants	Calgene Chemicals
Chemal	Ethoxylated fatty alcohols	Chemax
Chemax	Various ethoxylated products	Chemax
Chimipal	Alkanolamides and various ethoxylated products	Cesalpinia Chemicals
Cithrol A, ML, MO and MS	Ethoxylated fatty acids	Croda
Cithrol, others	Glycol and glycerol ethers	Croda
Crodamine	Amine oxides	Croda
Crodet	Ethoxylated fatty acids	Croda
Dehydol	Ethoxylated fatty alcohols	Cognis
Dehydrophen	Ethoxylated alkylphenols	Cognis
Dehypon	Ethoxylated fatty alcohols, specialities, e.g. end-blocked	Cognis
Dobanol	Ethoxylated fatty alcohols	Shell
Dowfax	Diphenyloxide-based sulfonates	Dow Chemicals
Elfan	Various sulfates and sulfonates	Akzo Nobel Surface Chemistry
Emal	Sulfates of alcohols and ethoxylated alcohols	Kao
Emcol	Various	Witco
Empicol	Sulfates of alcohols and ethoxylated alcohols, alkyl ether carboxylates	Rhodia
Empilan	Alkanolamides and various ethoxylated products	Rhodia
Empimin	Sulfates and sulfosuccinates	Rhodia
Emulan	Various ethoxylated products	BASF
Emulgante	Ethoxylated C16–C18 alcohols	Condea
Emulson	Various	Cisalpinia Chemicals

Ethylan	Mostly ethoxylated products	Akzo Nobel Surface Chemistry
Eumulgin	Various ethoxylated products	Cognis
Findet	Various ethoxylated products	Kao
Fluorad	Fluorocarbon-based surfactants	3M
Genapol	Ethoxylated fatty alcohols	Clariant
Geropon	Mostly sulfosuccinates and taurates	Rhodia
Glucopon	Sugar-based surfactants	Cognis
Hamposyl	N-Acylsarcosinates	Dow Chemicals
Hostapur	Alpha olefin sulfonates and petroleum sulfonates	Clariant
Iconol	Various ethoxylated products	BASF
Igepal	Various ethoxylated products	Rhodia
Imbentin	Ethoxylated alkylphenols	Dr W. Kolb
Lialet	Ethoxylated fatty alcohols	Condea
Lipolan	Alpha olefin sulfonates	Lipo Chemicals
Lorodac	Ethoxylated fatty alcohols	Condea
Lutensol	Various ethoxylated products	BASF
Mackam	Amphoterics	The McIntyre Group
Macol	Various ethoxylated products	BASF
Manro	Sulfates, sulfonates and alkanolamides	Hickson Manro
Marlipal	Ethoxylated fatty alcohols	Condea
Marlophen NP	Ethoxylated nonylphenol	Condea
Miranol	Imidazolines	Rhodia
Mirataine	Betaines	Rhodia
Monamid	Alkanolamides	Uniqema
Montane	Sorbitan derivatives	Seppic
Myverol	Monoglycerides	Eastman
Neodol	Ethoxylated fatty alcohols	Shell
Newcol	Various	Nippon Nyukazai
Nikkol	Various	Nikko Chemicals
Ninol	Alkanolamides	Stepan
Nissan Nonion	Various ethoxylated products	NOF

Nissan Plonon	Block copolymers	NOF
Nissan Soft Osen	Linear alkylbenzene sulfonates	NOF
Petronate	Alkylaryl sulfonates	Witco
Phospholan	Phosphate esters	Akzo Nobel Surface Chemistry
Plurafac	Ethoxylated fatty alcohols	BASF
Pluronic	Block copolymers	BASF
Polystep A	Various sulfonates	Stepan
Polystep B	Various sulfates and sulfonates	Stepan
Polystep F	Ethoxylated alkylphenols	Stepan
Poly-Tergent	Various ethoxylated products	Olin
Quimipol EA	Ethoxylated fatty alcohols	CPB
Quimipol ENF	Ethoxylated nonylphenol	CPB
Radiamuls	Various fatty acid esters	Fina Oleochemicals
Radiaquat	Quaternary ammonium surfactants	Fina Oleochemicals
Radiasurf	Polyol esters, including PEG esters	Fina Oleochemicals
Remcopal	Various ethoxylated products	Elf Atochem
Rewoteric	Various zwitterionics	Witco
Rhodafac	Phosphate esters	Rhodia
Rhodapex	Sulfates of ethoxylated fatty alcohols	Rhodia
Rhodapon	Sulfates of fatty alcohols	Rhodia
Rhodasurf	Ethoxylated fatty alcohols	Rhodia
Rolfor	Ethoxylated fatty alcohols and esters	Cesalpinia Chemicals
Schercomid	Alkanolamides	Scher Chemicals
Schercotaine	Betaines	Scher Chemicals
Sermul	Ethoxylated nonylphenol and various sulfates and sulfonates	Condea
Servoxyl	Phosphate esters	Condea
Simulsol	Various ethoxylated products	Seppic
Sinopol	Fatty acid esters and various ethoxylated products	Sino-Japan Chemicals

Span	Sorbitan esters of fatty acids	ICI Surfactants
Standapol	Sulfates of fatty alcohols and ethoxylated fatty alcohols	Cognis
Steol	Sulfates of ethoxylated fatty alcohols	Stepan
Stepanol	Sulfates of fatty alcohols	Stepan
Sulfopon	Sulfates of fatty alcohols	Cognis
Sulfotex	Sulfates of fatty alcohols and ethoxylated fatty alcohols	Cognis
Surfax	Mostly sulfated and sulfonated products	Houghton International
Surfonic	Various ethoxylated products	Huntsman
Surfynol	Non-ionics	Air Products
Surfynol 104	Tetramethyldecynediol	Air Products
Synperonic	Various alkoxylated products	ICI Surfactants
T-Det	Various ethoxylated products	Harcros Chemicals
Tauranol	Taurates and isethionates	Finetex
Tegin	Glycerol esters	Th. Goldschmidt
Tego Betain	Betaines	Th. Goldschmidt
Tegomuls	Glycerol esters	Th. Goldschmidt
Tergitol	Various ethoxylated products	Dow Chemicals
Teric	Mostly ethoxylated products	ICI Australia
Texapon	Sulfates of fatty alcohols and ethoxylated fatty alcohols	Cognis
Tridac	Ethoxylated C13 alcohol	Condea
Triton BG-10	Sugar esters	Dow Chemicals
Triton CF	Various ethoxylated products	Dow Chemicals
Triton DF	Ethoxylated fatty alcohols	Dow Chemicals
Triton X	Ethoxylated octylphenol and various sulfates and sulfonates	Dow Chemicals
Trycol	Ethoxylated fatty alcohols and nonylphenol	Cognis

Trydet	Ethoxylated fatty acids and esters	Cognis
Trylon	Ethoxylated fatty alcohols and oils	Cognis
Tween	Ethoxylated sorbitan esters	ICI Surfactants
Ufarol	Alkylbenzene sulfonates and sulfates of fatty alcohols and ethoxylated fatty alcohols	Unger Fabrikker
Ufaryl	Alkylbenzene sulfonates	Unger Fabrikker
Ufasan	Alkylbenzene sulfonates	Unger Fabrikker
Ungerol	Sulfates of ethoxylated fatty alcohols	Unger Fabrikker
Varamide	Alkanolamides	Witco
Variquat	Quaternary ammonium surfactants	Witco
Volpo	Ethoxylated fatty alcohols	Croda Chemicals
Witcamide	Alkanolamides	Witco
Witconate	Alkylaryl sulfonates	Witco
Witconol	Glycol and glycerol esters	Witco

APPENDIX 2. Common Acronyms of Surfactants, Surfactant Classes and Surfactant Raw Materials

Acronym	Full name
AE	Alcohol ethoxylate
AES	Alcohol ether sulfate
AOS	Alpha olefin sulfonate
APE	Alkylphenol ethoxylate
APG	Alkyl polyglucoside
AS	Alcohol sulfate
CTAB	Cetyltrimethylammonium bromide (cetyl = hexadecyl)
DEA	Diethanolamine
EO	Ethylene oxide
LABS	Linear alkylbenzene sulfonate
LAS	Linear alkylbenzene sulfonate
MEA	Monoethanolamine
NP	Nonylphenol
NPE	Nonylphenol ethoxylate
PEG	Poly(ethylene glycol)
PEO	Poly(ethylene oxide)
PO	Propylene oxide
PPG	Poly(propylene glycol)
PPO	Poly(propylene oxide)
SAS	Secondary alcohol sulfonate (paraffin sulfonate, secondary alkane sulfonate)
SDS	Sodium dodecyl sulfate
TEA	Triethanolamine

INDEX